Astronomical Optics

Astronomical Optics

DANIEL J. SCHROEDER

Department of Physics and Astronomy
Beloit College
Beloit, Wisconsin

ACADEMIC PRESS, INC.
Harcourt Brace Jovanovich, Publishers
San Diego New York Berkeley Boston
London Sydney Tokyo Toronto

Photograph on cover and Figure 17.2 courtesy of National Optical Astronomy Observatories. By accepting this product of the National Optical Astronomy Observatories (NOAO), the user agrees that the material may not be used to state or imply the endorsement by NOAO, or by any NOAO employee, of any individual, philosophy, commercial product, process, or service.

COPYRIGHT © 1987 BY ACADEMIC PRESS, INC.
ALL RIGHTS RESERVED.
NO PART OF THIS PUBLICATION MAY BE REPRODUCED OR
TRANSMITTED IN ANY FORM OR BY ANY MEANS, ELECTRONIC
OR MECHANICAL, INCLUDING PHOTOCOPY, RECORDING, OR
ANY INFORMATION STORAGE AND RETRIEVAL SYSTEM, WITHOUT
PERMISSION IN WRITING FROM THE PUBLISHER.

ACADEMIC PRESS, INC.
1250 Sixth Avenue, San Diego, California 92101

United Kingdom Edition published by
ACADEMIC PRESS INC. (LONDON) LTD.
24–28 Oval Road, London NW1 7DX

Library of Congress Cataloging in Publication Data

Schroeder, D. J.
  Astronomical optics.

  Includes bibliographies and index.
  1. Astronomical instruments.   2. Optics.   I. Title.
QB86.S35   1987   681'.412   87–1213
ISBN 0–12–629805–X (alk. paper)

PRINTED IN THE UNITED STATES OF AMERICA

87 88 89 90   9 8 7 6 5 4 3 2 1

Contents

Foreword   ix
Preface   xi

Chapter 1   Introduction

I.	A Bit of History	1
II.	Approach to Subject	3
III.	Outline of Book	4

Chapter 2   Preliminaries: Definitions and Paraxial Optics

I.	Sign Conventions	7
II.	Paraxial Equation for Refraction	8
III.	Paraxial Equation for Reflection	11
IV.	Two-Surface Refracting Elements	12
V.	Two-Mirror Telescopes	15
	Bibliography	21

Chapter 3   Fermat's Principle: An Introduction

I.	Fermat's Principle in General	23
II.	Examples Using Fermat's Principle	28
III.	Physical Interpretation of Fermat's Principle	32
IV.	Fermat's Principle and Reflecting Surfaces	33
V.	Conic Sections	37
VI.	Concluding Comments	39
	References	40
	Bibliography	40

Chapter 4 Introduction to Aberrations

I.	Reflecting Conics and Focal Length	42
II.	Spherical Aberration	43
III.	Off-Axis Aberrations	49
IV.	Aberration Compensation	52
	References	60
	Bibliography	60

Chapter 5 Fermat's Principle and Aberrations

I.	Application to Surface of Revolution	61
II.	Evaluation of Aberration Coefficients	65
III.	Ray and Wavefront Aberrations	67
IV.	Summary of Aberration Results, Stop at Surface	70
V.	Aberrations for Displaced Stop	74
VI.	Aberrations for Multisurface Systems	79
VII.	Curvature of Field	82
VIII.	Aberrations for Decentered Pupil	88
IX.	Concluding Remarks	92
	References	93
	Bibliography	93

Chapter 6 Reflecting Telescopes

I.	Paraboloid	95
II.	Two-Mirror Telescopes	97
III.	Alignment Errors in Two-Mirror Telescopes	110
IV.	Three-Mirror Telescopes	114
V.	Testing of Large Mirrors	117
VI.	Concluding Remarks	120
	References	120
	Bibliography	121

Chapter 7 Schmidt Telescopes and Cameras

I.	General Schmidt Configuration	123
II.	Characteristics of Aspheric Plate	125
III.	Schmidt Telescope Example	132
IV.	Achromatic Schmidt Telescope	134
V.	Solid and Semisolid Schmidt Cameras	138
	References	141
	Bibliography	141

Chapter 8 Catadioptric Telescopes and Cameras

I.	Schmidt–Cassegrain Telescopes	142
II.	Cameras with Meniscus Correctors	152
III.	Semisolid Schmidt–Cassegrain Cameras	158
IV.	Concluding Remarks, All-Reflecting Systems	159
	References	160

Chapter 9 Auxiliary Optics for Telescopes

I.	Field Lenses, Flatteners	162
II.	Prime Focus Correctors	165
III.	Cassegrain Focus Correctors	170
IV.	Cassegrain Focal Reducers	173
	References	178
	Bibliography	178

Chapter 10 Diffraction Theory and Aberrations

I.	Perfect Image	180
II.	The Near-Perfect Image	189
	References	197
	Bibliography	198

Chapter 11 Transfer Functions; Hubble Space Telescope

I.	Transfer Functions and Image Characteristics	199
II.	The Hubble Space Telescope	213
III.	Concluding Remarks	219
	References	220
	Bibliography	220

Chapter 12 Spectrometry: Definitions and Basic Principles

I.	Spectrometer Types and Modes	222
II.	Parameters for Slit Spectrometers	224
III.	Parameters for Slitless Spectrometers	231
IV.	Parameters in Diffraction Limit	232
	References	235
	Bibliography	235

Chapter 13 Dispersing Elements and Systems

I.	Dispersing Prism	236
II.	Diffraction Grating; Basic Relations	238
III.	Grating Efficiency	243
IV.	Fabry–Perot Interferometer	251
V.	Fourier Transform Spectrometer	254
	References	256
	Bibliography	256

Chapter 14 Grating Aberrations; Concave Grating Spectrometers

I.	Application of Fermat's Principle to Grating Surface	258
II.	Grating Aberrations	262
III.	Concave Grating Mountings	267
	References	270
	Bibliography	271

Chapter 15 Plane Grating Spectrometers

I.	Plane Grating Mountings	272
II.	Echelle Spectrometers	284
III.	Nonobjective Slitless Spectrometers	294
IV.	Concluding Comments	304
	References	304
	Bibliography	305

Chapter 16 System Noise and Detection Limits

I.	Detector Characteristics	307
II.	Effects of Atmospheric Turbulence	313
III.	Signal-to-Noise Ratio	317
IV.	Detection Limits	319
V.	Concluding Comments	328
	References	328
	Bibliography	328

Chapter 17 Multiple-Aperture Telescopes

I.	Diffraction Images	330
II.	Slit Spectroscopy with Arrays	337
III.	Concluding Comments	341
	References	341
	Bibliography	341

Index 343

Foreword

This is a textbook on astronomical optics and as such is unique. There are many good optics texts available, but none that tells how astronomers "do it with mirrors." The largest lens-type telescope is the Yerkes Observatory 40-in. refractor. To provide greater light-collecting capability it was necessary to utilize back-supported reflecting optics. Thus all large modern astronomical telescopes are reflectors rather than refractors. Two other important advantages accrue from the use of reflecting optics: the absence of chromatic aberration and the extension to spectral regions where the choice of transmitting material is limited.

While the properties of the two-mirror telescope family are fully explored, the analytic tools that the author develops are equally capable of describing optical systems of all kinds. The first half of the book is devoted to geometrical optics, based on one powerful unifying concept, Fermat's principle. Within the framework of that principle, the design concepts and system aberrations of a variety of instruments, reflecting, refracting, and dispersing, are discussed in detail. The second part of the book expands the analysis to wave optics and the relationship between the point spread function and the modulation transfer function. The author has done an excellent job of making understandable the tools that he has used in the successful design of a number of instruments now in active use.

Two traditional approaches to the design of optical systems have been the analytic methods described here and the numerical technique of ray tracing. The evolution of a modern complex optical system usually follows both paths. The conceptual design and determination of the range of permissible parameters are probably best understood within the flexibility of the analytic approach. Final optical design may then be carried out using one of the powerful computer programs that optimize the system for dozens of aberrations simultaneously. For the optical engineer, the numerical techniques available today are essential. For the scientists who use optical instruments I believe that an understanding of the analytic basis of optical design is a valuable addition to their education.

For the astronomer, in particular, I believe that an acquaintance with astronomical optics is essential. For this reason I have from time to time taught such a course. This book partially owes its origin to such a course, which I enjoyed teaching in collaboration with Dan Schroeder. I believe that it is essential to an astronomer's training because an astronomer should understand the instruments that provide the observational basis on which discovery and insight are established. Often, too, new discoveries or insights suggest new measurements. Knowledge of optical design illuminates the possibilities for these new explorations. The astronomer should be able to develop conceptual designs for an instrument that he or she would like to have.

The next decade should see the emergence of a new breed of optical telescope, telescopes that depart radically from the kind that have served astronomers for the past half a century. The pressure to construct more efficient telescopes with much greater light-collecting power comes about from the many exciting discoveries about the origin and evolution of the universe. The ability to construct these supertelescopes comes about because of the remarkable advances in technology during the last decade. These include lightweight materials, computer-aided design and control, and ingenious optical configurations. The first of these new telescopes is to go into space. The author, who is one of the two telescope scientists for the Hubble Space Telescope, has called upon this experience in the preparation of this book. Some of the subtleties of telescope design only manifest themselves at the level of the precision required for diffraction-limited performance achievable in space. The final chapter describes the use of multiple-aperture telescopes to enhance the light-collecting power of telescopes of the future. This is an area currently being pursued by several groups.

While I expect that the nature of both astronomical optics and the research carried out will continue to change at an increasing rate, the basic principles enunciated here will remain valuable for many years. This volume should serve well as both a textbook and a reference book for future practitioners of the magic art of "doing it with mirrors."

ARTHUR D. CODE

Preface

The growth of optical astronomy in recent years is in large part a direct consequence of additions, improvements, and breakthroughs in instrumentation used by astronomers, ranging from 4-m telescopes and orbiting observatories to sophisticated data gathering and image analysis. The aim of this book is to give a reasonably complete treatment of the variety of telescopes and instruments that have made this growth possible and the optical foundations on which the design of such systems is based. In spite of the many excellent textbooks covering various areas of optics and review articles on selected types of astronomical instrumentation, there is no single source specifically covering the area of astronomical optics. It is my hope that this book will, at least in part, fill this gap.

It became apparent at the start of this project that it is impossible to cover all of astronomical optics in a single book of reasonable length, and I have chosen to limit the discussion to instruments used in the optical part of the electromagnetic spectrum. Thus, the treatment is one of telescopes and cameras that use near-normal incidence optics and spectrometers with dispersive elements or interferometers. There is no discussion of radio telescopes or grazing incidence instruments used by x-ray astronomers.

This is not a book on astronomical observing techniques, nor is the subject of detectors covered in detail. Rather, the emphasis is on the optical principles on which different systems are based and their application to specific instrument types. Numerous examples of instrument characteristics are given to illustrate the range of optical performance that can be expected. A detailed outline of the topics covered is given in Chapter 1.

The level of presentation and approach is appropriate for a graduate student in astronomy who is approaching the subject of astronomical optics for the first time. Thus, the presentation is not simply a compilation of telescope and spectrometer systems, but one that attempts to give the basics on which the design of such systems is based. Although the basic principles of optics are discussed as needed, it is assumed that the reader has the equivalent of an intermediate-level optics course at the undergraduate level.

This book should also serve as a useful reference for active researchers who want to learn more about specific systems they are using or intend to use. Numerous reference tables and examples are included for this purpose.

Each chapter has a list of references and a selected bibliography for further reading. Because this book is written as a textbook rather than a monograph, I have not tried to make the bibliography exhaustive. Many of the references cited, particularly review articles, have extensive lists of further references to the literature.

A number of persons have contributed directly or indirectly to this effort. First and foremost I want to thank Arthur Code, who gave me the opportunity to participate in the development of the Wisconsin Experiment Package of the first Orbiting Astronomical Observatory. Since that time I have been privileged to draw on his wealth of knowledge and to teach jointly with him on one occasion a course on astronomical optics. For his contributions to this book I am especially grateful.

I am indebted to Arthur Hoag for making possible an extended stay at Kitt Peak National Observatory, during which my study of the design principles of astronomical optics was started, and for his continued encouragement. My thanks also to Robert Bless and Donald Osterbrock for their help and support over the years, and to Robert O'Dell for his encouragement to take on the role of a telescope scientist on NASA's Hubble Space Telescope Project.

Finally, it is my pleasure to thank Lori Jones for her diligent work in preparing the figures, and the administration of Beloit College for providing the sabbatical leave during which the writing was done.

Astronomical Optics

Chapter 1 | Introduction

The increasing rate of growth in astronomical knowledge during the past few decades is a direct consequence of the increase in the number and size of telescopes and the efficiency with which they are used. Most celestial sources are intrinsically faint, and observations that required hours of observing time with small refracting telescopes and insensitive photographic plates are now done in minutes with large reflecting telescopes and efficient solid-state detectors. The increased efficiency with which photons are collected and recorded by modern instruments has indeed revolutionized the field of observational astronomy.

I. A BIT OF HISTORY

Early in the 1900s the desire for more light-gathering power led to the design and construction of the 100-in. Hooker telescope located on Mount Wilson. This reflecting telescope and its smaller predecessors were built following the recognition that refracting telescopes had reached a practical limit in size. With the 100-in. telescope, it was possible to begin systematic observations of nearby galaxies and begin to attack the problem of the structure of the universe.

Although the 100-in. telescope was a giant step forward for observational astronomy, it was recognized by Hale that still larger telescopes were necessary for observations of remote galaxies. Due largely to his efforts, work began on the design and construction of a 200-in. (5-m) telescope in the late 1920s. The Hale telescope was put into operation in the late 1940s and remained the world's largest until a 6-m telescope was built in Russia in the late 1970s.

The need for more large telescopes became an acute one in the 1960s as the boundaries of observational astronomy were pushed outward. Plans made during that decade and the following one resulted in the construction of a number of optical telescopes in the 4-m class during the 1970s and 1980s in both hemispheres. These telescopes, equipped with efficient detectors, have fueled an explosive growth in observational astronomy.

Although large reflectors are well suited for observations of small parts of the sky, typically a fraction of a degree in diameter, they are not suitable for surveys of the entire sky. A type of telescope suited for survey work was first devised by Schmidt in the early 1930s. The first large Schmidt telescope was a 1.2-m instrument covering a field about 6 degrees across, which was put into operation on Palomar Mountain in the early 1950s. Several telescopes of this type and size have since been built in both hemispheres. The principle of the Schmidt telescope has also been adapted to cameras used in many spectrometers.

The era of building large ground-based telescopes is still under way, although some of the systems now in the process of design and construction are somewhat different. It has long been recognized that single-mirror telescopes cannot be significantly larger than those now in operation, and attention has turned to the design of arrays of telescopes and segmented mirrors in the quest for more light-gathering power.

The array concept was first implemented with the completion of the multiple-mirror telescope (MMT) on Mount Hopkins; the MMT has six 1.8-m telescopes mounted in a common frame and an aperture equivalent to that of a single 4.5-m telescope. Beams of the separate telescopes are directed to a common focal plane and either combined in a single image or placed side by side on the slit of a spectrometer. The success of the MMT has led to the choice of this concept for the proposed national new technology telescope (NNTT), an array with four 7.5-m telescopes mounted in a common frame.

The segmented mirror approach is the choice for the Keck ten-meter telescope (TMT), with 36 hexagonal segments comprising the equivalent of a single filled aperture. This approach requires active control of the positions of the segments to maintain the mirror shape and image quality.

Instrumentation used on large telescopes has also changed dramatically since the time of the earliest reflectors. Noting only the development in spectrometers, small prism instruments were replaced by larger grating instruments at both Cassegrain and coudé focus positions to meet the demands for higher spectral resolution. In recent years many of these high-resolution coudé instruments have, in turn, been replaced by echelle spectrometers at the Cassegrain focus. On the largest telescopes, such as the TMT and NNTT, most large instrumentation will be placed at the Nasmyth focus position on a platform that rotates with the telescope, as it now is on the 6-m telescope in the U.S.S.R.

Although developments of ground-based optical telescopes and instruments during the past two decades have been dramatic, the same can also be said of Earth-orbiting telescopes in space. Since the first Orbiting Astronomical Observatory in the late 1960s, with its telescopes of 0.4-m aperture or smaller, the size and complexity of orbiting telescopes have increased markedly. The 2.4-m Hubble Space Telescope (HST), scheduled for launch soon after the Space Shuttle is again operational, is capable of observations to a depth in space not possible with ground-based telescopes. All indications are that observations with the HST will revolutionize astronomy.

This brief excursion into the development of telescopes and instruments up to the present and into the near future is by no means complete. It is intended only to illustrate the range of tools now available to the observational astronomer.

II. APPROACH TO SUBJECT

Most of the optical principles that serve as the starting point in the design and use of any optical instrument have been known for a long time. In intermediate-level optics texts these principles are usually divided into two categories: geometric optics and physical optics. Elements from both of these fields are required for full descriptions of the characteristics of optical systems.

The theory of geometric optics is concerned with the paths taken by light rays as they pass through a system of lenses and/or mirrors. Although the ray paths can be calculated by the simple application of the laws of refraction and reflection, a much more powerful approach is one starting with Fermat's principle. With the aid of this approach it is possible to determine both the first-order characteristics of an optical system and deviations from these characteristics. The latter lead to the theory of aberrations or image defects, a subject discussed in detail.

The theory of physical optics includes the effects of the finite wavelength of light and such topics as interference, diffraction, and polarization. Analysis of the characteristics of diffraction gratings, interferometers, and telescopes such as the Hubble Space Telescope requires an understanding of these topics. The basics of this theory are introduced prior to our discussions of these types of optical systems.

The approach, therefore, is to emphasize the basic principles of a variety of systems and to illustrate these principles with specific designs. Although the specifics of telescopes and instruments have changed, and will continue to change, the basic optical principles are the same.

III. OUTLINE OF BOOK

The 16 chapters that follow the Introduction can be grouped into five distinct categories. Chapters 2 through 5 cover the elements of geometrical optics needed for the discussion of optical systems. The first three chapters of this group are an introduction to this part of optics seen from the point of view of Fermat's principle, with Chapter 5 a detailed treatment of aberrations based on this principle.

Chapters 6 through 11 cover the characteristics of a variety of telescopes and cameras, including auxiliary optics used with them. The characteristics of diffraction-limited telescopes are covered in the last two chapters of this group, with application to the Hubble Space Telescope.

Chapters 12 through 15 are a discussion of the principles of spectrometry and their application to a variety of dispersing systems, with the emphasis on diffraction gratings. In this group Chapter 14 is the counterpart of Chapter 5, a treatment of grating aberrations from the point of view of Fermat's principle.

The remaining two chapters are distinct in themselves with each chapter drawing on results given in preceding chapters and applying these results to selected types of observations for both ground-based and space-based systems.

A closer look at the contents of each chapter is now in order. Chapter 2 is an introduction to the basic ideas of geometric optics, and the reader who is well versed in these ideas can cover it quickly. One topic covered in this chapter, not part of the usual course in optics, is the definition of normalized parameters for two-mirror telescopes.

Chapter 3 is an introduction to Fermat's principle with a number of examples illustrating its utility, including a discussion of atmospheric refraction and atmospheric turbulence. Chapter 4 is an introduction to aberrations,

III. Outline of Book

with emphasis on spherical aberration. The concept of aberration compensation is introduced and applied to two optical systems.

The discussions of the preceding three chapters set the stage for an in-depth discussion of the theory of third-order aberrations in Chapter 5. The results of the analysis are summarized in tables for easy reference.

In Chapter 6 we draw on the results from Chapter 5 to derive the characteristics of a number of types of reflecting telescopes. Comparisons of image quality are given for several of these types. Chapter 7 covers the characteristics of Schmidt systems, including a discussion of the achromatic Schmidt and solid and semisolid cameras.

Chapter 8 covers various types of catadioptric systems, including Schmidt–Cassegrain telescopes and cameras with meniscus correctors substituted for aspheric plates. The following chapter is a discussion of various types of auxiliary optics used with telescopes, including field lenses, field flatteners, prime and Cassegrain focus correctors, and focal reducers.

In Chapter 10 we discuss the basics of diffraction theory and aberrations and the characteristics of perfect and near-perfect images. Perfect and near-perfect images are discussed in terms of classical and orthogonal aberrations in Chapter 10, followed by a discussion in terms of transfer functions in Chapter 11. The results are illustrated with a discussion of the expected characteristics of the Hubble Space Telescope.

Chapter 12 covers the basic principles of spectrometry, followed by application of these principles to a variety of dispersing elements and systems in Chapter 13. The following two chapters are devoted entirely to the diffraction grating, with Chapter 14 an analysis of grating aberrations and concave grating mountings and Chapter 15 the application of these results to a variety of plane grating instruments.

Chapter 16 is a discussion of the detection limits that are reached at a given signal-to-noise level for several types of observations. Also included here is a discussion of detectors and the effects of atmospheric turbulence from the point of view of transfer functions. The final chapter is an introduction to multiple-aperture telescopes, with discussions of diffraction images given by telescope arrays and the unique features of slit spectroscopy with an array.

The reader approaching the topic of astronomical optics for the first time is encouraged to work through the basic theory. This exercise will facilitate the understanding of its application to a specific optical system and the bounds within which this system is usable. Other readers will be interested only in specific systems and their characteristics. We hope that their needs are met with the tables and equations that are given. Whatever the readers' motivation, a selected bibliography is given at the end of each chapter for additional reading.

Chapter 2 | Preliminaries: Definitions and Paraxial Optics

The analysis of any optical system generally proceeds along a well-defined route. First one arrives at a basic layout of optical elements—lenses, mirrors, prisms, gratings, and such—by using first-order or Gaussian optics. Such an analysis establishes such basic parameters as focal length, magnification, and locations of pupils, among others. The next step often involves using a ray-trace program on a digital computer to trace rays through the system and calculate aberrations of the image. Such an analysis might dictate changes in the basic layout in order to achieve image quality within certain specified limits. Ray-trace and optical analysis programs are now quite sophisticated and are particularly useful in systems with many optical elements. Tracing of rays is especially useful in optimizing system performance.

In order to use efficiently the results generated by a ray-trace program it is necessary to understand the theory of third-order aberrations. In subsequent chapters we go into considerable detail about the nature of these aberrations and how they can be eliminated or minimized in different kinds of optical systems. In many cases an analysis of aberrations is a useful intermediate step following the setup of the basic system and the analysis using a ray-trace program. Details of how such programs work are not discussed.

Each of the steps along this route requires a systematic approach to measurements of angles and distances. In this chapter we define the sign conventions used and determine the equations of first-order optics. We

I. SIGN CONVENTIONS

The coordinate system within which surface locations and ray directions are defined is the standard right-hand Cartesian frame shown in Fig. 2.1. For a single refracting or reflecting surface the z axis coincides with the optical axis, with the origin of the coordinate system at the vertex 0 of the surface. For an optical system in which the elements are centered, the optical axis is the line of symmetry along which the elements are located. In a system in which one (or more) of the elements is not centered, the optical axis for such an element will not coincide with that for a different element, a complication that is dealt with later. In the following discussion only centered systems are considered.

Figure 2.1 illustrates refraction at a spherical surface with an incident ray directed from left to right. Rays from an initial object are always assumed to travel in this direction. The indices of refraction are n and n' to the left and right of the surface, respectively, with points B, B', and C on the optical axis of the surface. The line PC is the normal to the interface between the two media at point P, and a ray directed toward B is refracted at P and directed toward B'.

The unprimed symbols in Fig. 2.1 refer to the ray before refraction, while the primed symbols refer to the ray after refraction. The slope angles are u and u', measured from the optical axis, and the angles of incidence and refraction, respectively, are i and i', measured from the normal to the

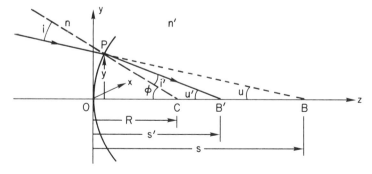

Fig. 2.1. Refraction at spherical interface. All angles and distances are positive in diagram; see text for discussion.

surface. The symbols s and s' denote the object and image distances, respectively, and R represents the radius of curvature of the surface, measured at the vertex.

The sign convention for distances is the same as for Cartesian geometry. Hence distances s, s', and R are positive when the points B, B', and C are to the right of the vertex, and distances from the optical axis are positive if measured upward. The sign convention for angles is chosen so that all of the angles shown in Fig. 2.1 are positive. Slope angles u and u' are positive when a counterclockwise rotation of the corresponding ray about B or B' brings the ray into coincidence with the z axis. The angles of incidence and refraction, i and i', are positive when a clockwise rotation of the corresponding ray about point P brings the ray into coincidence with the line PC. All rotations are made through acute angles.

The advantage of these conventions for distances and angles is that both refracting and reflecting surfaces can be treated with the same relations. As we show, formulas for reflecting surfaces are obtained directly by letting $n' = -n$ in the formulas for refracting surfaces. The meaning of a negative index of refraction is discussed in Section 2.III. The sign conventions for distances and angles are similar to those used by Born and Wolf and by Longhurst. Although the conventions for angles may at times seem awkward, they have the advantage of universal applicability and are especially appropriate in third-order analysis of complex systems.

II. PARAXIAL EQUATION FOR REFRACTION

In this section we develop some of the basics needed for a first-order analysis of an optical system. It is worth noting that our discussion is not intended as a comprehensive one, and should more details be needed the reader should refer to any of a number of excellent texts in optics. Examples of such texts are those by Longhurst, Hecht and Zajac, and Jenkins and White. You should be aware, however, that the sign conventions used in the latter two of these books are different from the one used here.

With the help of Fig. 2.1 we can easily determine the relation between s and s' when the distance y and all angles are small. By small we mean that point P is close enough to the optical axis that sines and tangents of angles can be replaced with the angles themselves. In this approximation any ray is close to the axis and nearly parallel to it, hence the term *paraxial approximation*.

The exact form of Snell's law of refraction is

$$n \sin i = n' \sin i',$$

II. Paraxial Equation for Refraction

which in the paraxial approximation becomes $ni = n'i'$. From Fig. 2.1 we find

$$i + u = \phi, \qquad i' + u' = \phi.$$

Solving these relations for i and i' and substituting into the paraxial form of Snell's law gives

$$n'u' - nu = (n' - n)\phi.$$

Applying the paraxial approximation to the distances, we get $\phi = y/R$, $u = y/s$, and $u' = y/s'$. Substituting, and canceling the common factor y, we get

$$n'/s' - n/s = (n' - n)/R. \tag{2.2.1}$$

The points at distances s and s' from the vertex are called *conjugate* points; that is, the image is conjugate to the object and vice versa. If either s or $s' = \infty$, then the conjugate distance is the *focal length*; that is, $s = f$ when $s' = \infty$ and $s' = f'$ when $s = \infty$.

A. POWER

In Eq. (2.2.1) we see that the right side of the equation contains factors related to the surface and surrounding media and not to the object and image. It is useful to denote this combination by P, where P is the *power* of the surface. The power is unchanged when the direction of light travel in Fig. 2.1 is reversed, provided n and n' are interchanged and each is made negative. This invariance of P to the direction of light travel makes it a useful parameter. Note also that s and s' change places when the light is reversed in Fig. 2.1, and Eq. (2.2.1) is unchanged.

Combining Eq. (2.2.1) with the defined focal lengths and power, we get

$$\frac{n'}{s'} - \frac{n}{s} = \frac{n' - n}{R} = P = \frac{n'}{f'} = -\frac{n}{f}. \tag{2.2.2}$$

This is the first-order or Gaussian equation for a single refracting surface and is the starting point for analyzing systems that have several surfaces. For multisurface systems the image formed by a given surface, say the ith one, serves as the object for the next surface, the $(i+1)$st in this case. A surface-by-surface application of Eq. (2.2.2), starting with the first surface, will be illustrated in examples to follow.

Equation (2.2.2) does not contain height y and hence applies to any ray passing through B before refraction, provided of course the paraxial approximation is valid. This equation also applies to object and image points that are not on the optical axis, provided these points are close to

B and B' and lie on a line passing through point C. This is illustrated in Fig. 2.2, where Q and Q' denote an object and image point, respectively, for a case where B and B' lie on opposite sides of the surface vertex. In Fig. 2.2 the line QCQ' can be thought of as a new axis of the spherical surface, where Q and Q' are conjugate points along the new axis just as B and B' are conjugate points on the original axis. If the angle ϕ in Fig. 2.2 is small, then the line segments BQ and $B'Q'$ can be taken perpendicular to the original axis. In general, of course, BQ and $B'Q'$ are short arcs of circles whose centers are at C.

B. MAGNIFICATION

The geometry in Fig. 2.2 can be used to determine the *transverse* or *lateral magnification* m, defined as the ratio of image height to object height. In symbols we have $m = h'/h$, where

$$h' = -\phi(s' - R), \qquad h = -\phi(s - R),$$

and the sign convention has been applied to each quantity. In Fig. 2.2 we have s' and $R > 0$ and s and $\phi < 0$, hence h and h' have opposite signs. Therefore

$$m = \frac{h'}{h} = \frac{s' - R}{s - R} = \frac{ns'}{n's}, \qquad (2.2.3)$$

where the final step follows by substitution of (2.2.1). Because h and h' have opposite signs in Fig. 2.2, the transverse magnification is negative for the case shown.

In Fig. 2.3 a ray joining conjugate points B and B' has slope angles u and u'. The *angular magnification* M is defined as $\tan u / \tan u'$, where from the geometry of Fig. 2.3 we see that $y = s \tan u = s' \tan u'$. Therefore

$$M = \frac{\tan u'}{\tan u} = \frac{s}{s'} = \frac{n}{n'm} = \frac{nh}{n'h'}. \qquad (2.2.4)$$

Equation (2.2.4) relates the transverse and angular magnifications for a pair

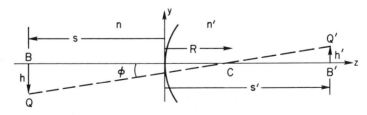

Fig. 2.2. Conjugate points in the paraxial region. B and B', Q and Q' are pairs of conjugate points. See Eq. (2.2.3) for definition of transverse magnification.

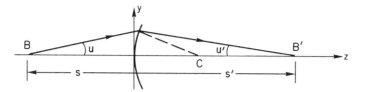

Fig. 2.3. Angular magnification. See Eq. (2.2.4) for definition.

of conjugate planes. Rewriting this relation we get

$$nh \tan u = n'h' \tan u', \quad (2.2.5)$$

which in the paraxial approximation becomes

$$nhu = n'h'u'. \quad (2.2.6)$$

If, as is customary, we let $H = nh \tan u$, then Eq. (2.2.5) states that H before refraction is the same as H after refraction. Thus in any optical system containing any number of refracting (or reflecting) surfaces, H is an invariant. This follows because the combination $n'h'u'$ for the first surface is nhu for the second surface, and so on. H, called the Lagrange invariant, is important in at least one other respect; the total flux collected by an optical system from a uniformly radiating source of light is proportional to H^2. Its invariance through an optical system is thus a consequence of conservation of energy.

III. PARAXIAL EQUATION FOR REFLECTION

With the aid of Fig. 2.4 we now find the Gaussian equation for a reflecting surface in the paraxial approximation. Applying the sign conventions to the symbols shown gives distances s, s', and R and angles i, ϕ, u, and u' as negative. The law of reflection is $i = -i'$, hence the angle of reflection i'

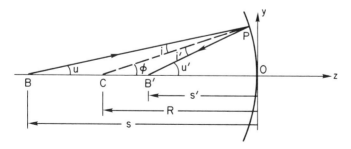

Fig. 2.4. Reflection at spherical surface. B and B' are conjugate axial points.

is positive in Fig. 2.4. From the geometry shown we get

$$i = \phi - u, \qquad i' = \phi - u', \qquad \phi = \frac{y}{R}, \qquad u = \frac{y}{s}, \qquad u' = \frac{y}{s'}.$$

Substituting into the law of reflection, $i = -i'$, gives

$$(1/s') + (1/s) = 2/R. \tag{2.3.1}$$

As in the case of Eq. (2.2.1), this relation applies generally to any object position provided we use the appropriate signs for the distances.

At this point it is important to point out that the law of reflection follows directly from Snell's law of refraction if we make the substitution $n' = -n$. Specifically, note that this substitution into Eq. (2.2.1) gives Eq. (2.3.1) directly. The fact that the relations for reflecting surfaces are thus directly obtained is very useful because we need only consider relations for refracting surfaces and simply put $n' = -n$ as needed. As an example we apply this substitution to Eqs. (2.2.2) and (2.2.3) and get

$$(1/s') + (1/s) = 2/R = -P/n = 1/f' = 1/f, \tag{2.3.2}$$

$$m = -s'/s. \tag{2.3.3}$$

Using Eq. (2.3.2) it is easy to verify that $P > 0$ for a concave mirror and $P < 0$ for a convex mirror, where a mirror is concave or convex as seen from the direction of the incident light. Note, however, that the focal length of a concave mirror changes sign when the direction of the incident light is reversed. This is expected because the reversal of Fig. 2.4, left for right, changes the signs of s and s'. But because n also changes sign in this reversal, P is invariant.

A negative index of refraction simply means that the light is traveling in the direction of the $-z$ axis, or from right to left. Consistent use of this convention, together with the other sign conventions in Eq. (2.2.2), allows one to work with any set of refracting and/or reflecting surfaces in combination.

In many situations it is convenient to take $f > 0$ for a concave mirror and $f < 0$ for a convex mirror, independent of the direction of the incident light. We will adopt this convention for convenience, keeping in mind that it violates the strict sign convention.

IV. TWO-SURFACE REFRACTING ELEMENTS

We now apply the results of Section 2.II to several systems with two refracting surfaces. We first consider a thick lens, a lens in which the second refracting surface is distance d to the right of the first surface.

A. THICK LENS

A schematic cross section of a thick lens is shown in Fig. 2.5. If we assume the lens has index n and is in air, then $n_1 = n_2' = 1$ and $n_1' = n_2 = n$. Applying Eq. (2.2.2) to each surface gives

$$\frac{n}{s_1'} - \frac{1}{s_1} = \frac{n-1}{R_1} = P_1, \qquad \frac{1}{s_2'} - \frac{n}{s_2} = \frac{1-n}{R_2} = P_2, \qquad (2.4.1)$$

where $s_2 = s_1' - d$.

With this system we are interested only in finding the net power P or, equivalently, the effective focal length f', where $P = 1/f'$. Figure 2.5 shows a ray with $s_1 = \infty$ intersecting the first surface at height y_1 and the second surface at height y_2. From similar triangles in Fig. 2.5 we get

$$y_2/y_1 = (s_1' - d)/s_1' = s_2'/f'. \qquad (2.4.2)$$

We can now find the effective focal length by setting $s_1 = \infty$ and $s_2 = s_1' - d$ in Eq. (2.4.1) and combining the result with Eq. (2.4.2). After a bit of algebra we get

$$P_1 = \frac{n}{s_1'}, \qquad P_2 = \frac{1}{s_2'} - \frac{n}{s_1' - d},$$

$$P = \frac{1}{f'} = \frac{1}{s_2'}\left(\frac{s_1' - d}{s_1'}\right) = \left(P_2 + \frac{n}{s_1' - d}\right)\frac{s_1' - d}{s_1'}.$$

Multiplying out the preceding equation, we finally get the result sought in the form

$$P = P_1 + P_2 - (d/n)P_1 P_2. \qquad (2.4.3)$$

In the steps leading to Eq. (2.4.3), both n and d are positive. If the directions of the arrows in Fig. 2.5 are reversed, the above derivation reproduces Eq. (2.4.3), provided we again define $d > 0$ and set $n_1' = n_2 = -n$ to make the

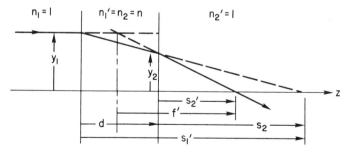

Fig. 2.5. Cross section of thick lens. See Eq. (2.4.3) for lens power. In the thin lens limit, $f' = s_2 = s_1'$.

14 2. Preliminaries: Definitions and Paraxial Optics

index of the lens positive. Thus P in Eq. (2.4.3) is the same for either direction of light.

B. THIN LENS

A thin lens is defined as one in which the separation of the two surfaces is negligible compared to other axial distances, that is, $s_2 = s_1'$ effectively. For a thin lens in air, Eqs. (2.4.1) apply directly. Letting $s_1 = s$ and $s_2' = s'$, the addition of these equations gives

$$\frac{1}{s'} - \frac{1}{s} = (n-1)\left(\frac{1}{R_1} - \frac{1}{R_2}\right) = P_1 + P_2 = P = \frac{1}{f'} = -\frac{1}{f}. \quad (2.4.4)$$

The net power of a thin lens is simply the reciprocal of its focal length and is the same as that of a thick lens with $d = 0$, as expected. Although a thin lens has two surfaces, it is of interest to note that the Gaussian relations that describe the lens are actually somewhat simpler than those for a single refracting surface.

The transverse magnification of each surface is given by Eq. (2.2.3) with the results $m_1 = s_1'/ns_1$ and $m_2 = ns_2'/s_2$. The net transverse magnification of a thin lens is then $m = m_1 m_2 = s'/s$.

As a final item for thin lenses, we note that Eq. (2.4.3) also applies to two thin lenses separated by distance d, where $n = 1$ in the space between the lenses. The simple analysis showing this is left to the reader.

C. THICK PLANE-PARALLEL PLATE

A thick plane-parallel plate, as shown in Fig. 2.6, has zero power but also has an image that is displaced laterally along the optical axis relative

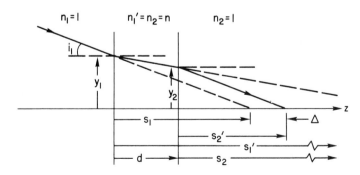

Fig. 2.6. Image shift Δ for plane-parallel plate of thickness d and index n in air. See Eqs. (2.4.5)-(2.4.7).

to the object. Applying Eq. (2.2.2) at each surface gives

$$n_1'/s_1' = n_1/s_1, \qquad n_2'/s_2' = n_2/s_2.$$

Assuming the plate of index n is in air, $n_1 = n_2' = 1$, $n_1' = n_2 = n$, and noting that $s_2 = s_1' - d$, we get

$$s_1' = ns_1, \qquad s_2' = s_1 - d/n.$$

The distance from object to image is $\Delta = s_2' - s_1 + d$, or

$$\Delta = d[1 - (1/n)]. \qquad (2.4.5)$$

Note that the displacement Δ is independent of the object distance and, as is true in all cases in the paraxial approximation, independent of height y.

In the paraxial approximation an optical system is free of any aberrations; that is, an object point is imaged precisely into an image point. When the exact form of Snell's law is used, however, most systems will have some form of aberration. A thick plate is a good example of a simple system with aberration; that is, it fails to take all rays from a single object point into a single image point. This is easily shown by applying Snell's law in its exact form at each surface. With the intermediate steps left to the reader, the geometry of Fig. 2.6 leads to

$$\Delta = d\left(1 - \frac{\cos i_1}{n \cos i_1'}\right). \qquad (2.4.6)$$

A comparison of Eqs. (2.4.5) and (2.4.6) gives

$$\Delta_{\text{exact}} - \Delta_{\text{par}} = \frac{d}{n}\left(1 - \frac{\cos i_1}{\cos i_1'}\right) \simeq \frac{y_1^2 d(n^2 - 1)}{2 s_1^2 n^3}, \qquad (2.4.7)$$

hence the image position depends on the ray height at the first surface. We consider the aberrations of a thick plate in more detail later.

V. TWO-MIRROR TELESCOPES

We now apply the results of the preceding sections to the general class of two-mirror systems. In this section we are concerned only with the paraxial properties of such systems and will limit our discussion to the case where $s_1 = \infty$. Figure 2.7 shows two examples of particular two-mirror systems, the so-called Cassegrain and Gregorian types, of which the Cassegrain is the more common type for an optical telescope.

Symbols in Fig. 2.7 are defined in the legend. Note that subscript 1 refers to the first mirror (primary) and 2 refers to the second mirror (secondary)

2. Preliminaries: Definitions and Paraxial Optics

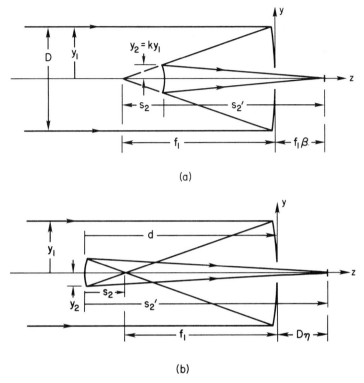

Fig. 2.7. Schematic diagrams of two-mirror reflecting telescopes: (a) Cassegrain, (b) Gregorian. Designated parameters are y_1 and y_2, height of ray at margin of primary and secondary, respectively; D, telescope diameter $= 2|y_1|$; $2|y_2|$, diameter of secondary mirror; R_1 and R_2, vertex radius of curvature of primary and secondary mirrors, respectively; s_2 and s_2', object and image distance of intermediate object (located at focal point of primary) measured from the secondary mirror vertex; f_1, focal length of primary mirror; and d, distance from primary to secondary, $d > 0$.

in the optical train. The sign of f_1 is taken positive when the primary is concave. Following the sign convention, y_1 and y_2 have the same signs for a Cassegrain and opposite signs for a Gregorian.

A. NORMALIZED PARAMETERS

It is very helpful to describe any two-mirror system in terms of a set of dimensionless or normalized parameters, defined as $k = y_2/y_1$, where y_1 and y_2 are the heights of the rays at the margin of the primary and secondary, respectively; $\rho = R_2/R_1$, where R_1 and R_2 are the vertex radii of the curvature of the primary and secondary mirror, respectively; $m = -s_2'/s_2$, transverse

V. Two-Mirror Telescopes

magnification of secondary; $f_1\beta = D\eta$, back focal distance, or distance from primary vertex to final focal point; β and η, back focal distance in units of f_1 and D, respectively; $F_1 = f_1/D$, primary mirror focal ratio; and $F = f/D$, system focal ratio, where f is the focal length.

In Fig. 2.7 we have $\beta > 0$ when the focal point lies outside the space between the primary and secondary. Depending on whether the system is Cassegrain or Gregorian, the signs of some of these dimensionless parameters differ. In particular, k and m are each positive for a Cassegrain and negative for a Gregorian.

The relationships between these parameters are obtained with the aid of Eq. (2.3.1) applied to the secondary, from which we get

$$m = \frac{\rho}{\rho - k}, \quad \rho = \frac{mk}{m-1}, \quad k = \frac{\rho(m-1)}{m}, \quad 1 + \beta = k(m+1), \quad \eta = F_1\beta. \tag{2.5.1}$$

It should be kept in mind that the relations in Eqs. (2.5.1) apply specifically to the case where the original object is at infinity. Given this caveat, we will see that it is convenient to describe telescope characteristics in term of these parameters, especially system aberrations.

The net power of a two-mirror telescope is found by using Eq. (2.4.3), which can be rewritten as

$$P = P_1[1 + (P_2/P_1) - (d/n)P_2].$$

In using this relation we recall that d and n in Eq. (2.4.3) are positive, independent of the direction of light between the surfaces. From Eq. (2.3.2) we find $P_1 = -2/R_1$, $P_2 = 2/R_2$, hence $P_2/P_1 = -1/\rho$. In using Eq. (2.3.2) note that $n = 1$ for the primary and $n = -1$ for the secondary, according to the sign convention. Substituting in terms of dimensionless parameters from Eq. (2.5.1), and noting that $d/n = (1-k)/P_1$, gives

$$P = P_1[1 - (k/\rho)] = P_1/m, \tag{2.5.2}$$

hence the telescope power is positive for a Cassegrain telescope and negative for a Gregorian. In accord with our convention for single mirrors, we take the telescope focal length positive for a Cassegrain and negative for a Gregorian. In terms of the focal lengths and focal ratios, therefore,

$$m = f/f_1 = F/F_1. \tag{2.5.3}$$

B. STOPS AND PUPILS

We now turn our attention to a brief discussion of stops and pupils. For a more complete discussion the reader should consult any of the intermediate-level texts listed in the bibliography at the end of the chapter.

The *aperture stop* is an element of an optical system that determines the amount of light reaching the image. This stop is often the boundary of a lens or mirror, although it may be a separate diaphragm. In addition to controlling the amount of light entering the system, it is one of the determining factors in the sizes of the system aberrations. For most telescopes the primary mirror serves as the aperture stop.

The *field stop* is an element that determines the angular size of the object field that is imaged by the system. In most systems the boundary of the field stop is the edge of the detector, although it may also be a separate diaphragm in the image plane ahead of the detector.

In a general optical system the image of the aperture stop formed by that part of the system preceding it in the optical train is called the *entrance pupil*. For telescopes of the type defined above, as well as for prime focus (single mirror) and refracting telescopes, it is almost always true that no imaging elements precede the aperture stop. In this case the entrance pupil coincides with the aperture stop.

The image of the aperture stop formed by that part of the system following it is called the *exit pupil*. The significance of the exit pupil is that rays from the boundary of the aperture stop approach the final image point as if coming from the boundary of the exit pupil, for all incidence angles at the aperture stop boundary.

We now apply these definitions to telescopes of the type shown in Fig. 2.7. Taking the aperture stop at the primary, the exit pupil is then the image of the primary as formed by the secondary. Figure 2.8 shows the exit pupil location for a Cassegrain telescope; for a Gregorian the exit pupil is located between the primary and secondary mirrors.

Applying Eq. (2.3.1) to the geometry in Fig. 2.8 with $f_1 \delta$ defined as the distance from the exit pupil to the telescope focal point, and converting to dimensionless parameters, gives

$$\delta = \frac{m^2 k}{m+k-1} = \frac{m^2(1+\beta)}{m^2+\beta}, \qquad (2.5.4)$$

where $\delta > 0$ when the focal surface of the system lies to the right of the exit pupil, as shown in Fig. 2.8. Although Eq. (2.5.4) was derived from the diagram for a Cassegrain, it also applies to a Gregorian telescope. Using Eqs. (2.3.3) and (2.2.3) we find that the exit pupil diameter is $D|\delta/m| = f_1|\delta/F|$.

Because the centers of the aperture stop and exit pupil are on the axis of the telescope, the so-called chief ray appears to come from the center of the exit pupil after reflection from the secondary. The *chief ray* is defined as the ray that passes through the center of the aperture stop. If the angle of incidence of the chief ray at the primary is θ, its angle with respect to

V. Two-Mirror Telescopes

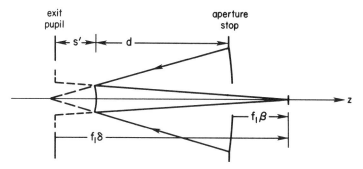

Fig. 2.8. Location of exit pupil for Cassegrain telescope. The exit pupil is closer to the secondary than is the primary focal point. See Eq. (2.5.4).

the telescope axis is ψ after reflection from the secondary. The relation between these angles is easily derived from the geometry shown in Fig. 2.9, where the focal length of a refracting telescope equivalent to a Cassegrain type is f. From Fig. 2.9

$$\psi f_1 \delta = f\theta = mf_1 \theta, \qquad (2.5.5)$$

hence $\psi/\theta = m/\delta$. Because δ is generally of order unity, the chief ray angle at the focal surface is of order m larger than the incident chief ray angle.

C. SCALE; FOCAL SURFACE SHIFT

Two other items appropriate for introduction at this point are (a) telescope scale and (b) effect of secondary mirror displacement on focal surface location.

For a telescope of focal length f, the scale is

$$S(\text{arc-sec/mm}) = \frac{206265}{f(\text{mm})}, \qquad (2.5.6)$$

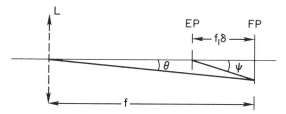

Fig. 2.9. Relation between incident and final chief ray angles, θ and ψ, respectively, in two-mirror telescope. L, Lens of equivalent refractor; EP, exit pupil; FP, focal plane. See Eq. (2.5.5).

where the units of arc-seconds per millimeter are those most often used. For conversion to radian measure the identities 0.206 arc-sec = 1 μrad and 3.44 arc-min = 1 mrad can be used.

For a given pair of primary and secondary mirrors the location of the telescope focal surface depends on the location of the secondary, as given by Eq. (2.5.1). If the secondary is moved along the optical axis, then both m and k are changed, and so also is the position of the focal surface. Let ds_2' be the displacement of the focal surface when the secondary is displaced by ds_2. Differentiating Eq. (2.3.1) while holding R_2 constant, we find that

$$ds_2' = -m^2 \, ds_2, \qquad (2.5.7)$$

where $ds_2 < 0$ when the secondary is moved closer to the primary.

The displacement given by Eq. (2.5.7) is measured relative to the secondary, which is now in a new position. Relative to the object at the focal point of the primary, or relative to the primary, the focal surface has moved by $ds_2' - ds_2 = -(m^2 + 1) \, ds_2$. This relation can easily be checked by working with the dimensionless parameters. From Eq. (2.5.1) we have

$$\beta = k(m+1) - 1 = k[(2\rho - k)/(\rho - k)] - 1,$$
$$d\beta/dk = [\rho^2/(\rho - k)^2] + 1 = m^2 + 1, \qquad (2.5.8)$$

where $dk = -ds_2/f_1$ and $d\beta$ is the shift of focal surface relative to primary mirror vertex in units of f_1.

Although Eq. (2.5.8) does not set any apparent limit on how far the secondary can be moved, there is a limit set by the onset of aberrations. Two-mirror telescopes generally have a mirror separation set to make the on-axis aberration zero. For a different secondary position the on-axis aberration is no longer zero, and its size sets a practical limit to the amount of secondary displacement. These limits will be considered during our discussion of telescope aberrations.

D. COMPARISON OF TYPES

Both Cassegrain and Gregorian telescopes, especially the former, can have a long overall focal length in a mechanical structure that is many times shorter. From Fig. 2.7 we see that a typical Cassegrain telescope has a secondary mirror–focal surface separation comparable to f_1, or about m times smaller than f. More precisely, the secondary mirror–focal surface distance is $f_1(1 + \beta - k)$, or $1 + \beta - k$ in units of f_1. Using Eq. (2.5.1) we get

$$\text{secondary–focal surface distance} = mkf_1, \qquad (2.5.9)$$

a relation that applies to both Cassegrain and Gregorian types. We show in Chapter 6 that a Gregorian telescope is significantly longer than a

Cassegrain if both telescopes have the same values of m and f_1. The advantages of a relatively short structure from an engineering point of view are obvious because there will be less flexure in a short telescope than in a long one.

Another difference between these two types of telescopes is the size of the secondary required to accept all of the light reflected from the primary. Each diagram in Fig. 2.7 shows a secondary mirror whose diameter is $|k|D$, the minimum required for a single point source. To cover a field on the sky of angular diameter 2θ without vignetting any light from the primary, the secondary must be larger by $2\theta(1-k)f_1 = 2\theta F_1(1-k)D$. Thus the full diameter of the secondary is

$$D_2 = D[|k| + 2\theta F_1(1-k)]. \qquad (2.5.10)$$

Because $k < 0$ for a Gregorian, the diameter D_2 of the Gregorian secondary is larger for the same θ and F_1, hence it blocks a larger fraction of the light headed for the primary.

These two-mirror designs have the added feature that the system focal length is easily changed simply by putting in a different secondary mirror without changing the physical length by a large factor. As an example, consider a Cassegrain telescope with parameters $F_1 = 3$, $\beta = 0.25$, and $m = 3$. Using Eq. (2.5.1) we find $k = 0.3125$ and the secondary mirror–focal surface separation is 0.9375. If we choose to increase the telescope focal length by a factor of 3, hence making $m = 9$, while keeping $\beta = 0.25$, then $k = 0.125$ and $mk = 1.125$. The modified telescope is only 1.2 times longer than the original one.

A final, and very significant, advantage of two-mirror systems is the additional freedom provided for controlling image quality. With proper choices of surface parameters it is possible to have the aberrations of the primary canceled, entirely or in part, by those of the secondary, thus giving a system with better image quality. We discuss these considerations in detail in subsequent chapters.

BIBLIOGRAPHY

Intermediate-level texts in optics
Hecht, E., and Zajac, A. (1974). "Optics." Addison–Wesley, Reading, Massachusetts.
Jenkins, F., and White, H. (1976). "Fundamentals of Optics," 4th ed. McGraw–Hill, New York.
Longhurst, R. (1967). "Geometrical and Physical Optics," 2nd ed. Wiley, New York.

Advanced texts in optics
Born, M., and Wolf, E. (1980). "Principles of Optics," 6th ed. Pergamon, Oxford.
Ghatak, A., and Thyagarajan, K. (1978). "Contemporary Optics." Plenum, New York.

Chapter 3 | Fermat's Principle: An Introduction

A very powerful method in dealing with geometric optics, the analysis of optical systems by tracing rays, is a principle ascribed to Fermat. For a single plane reflecting or refracting surface it states that the actual path that a light ray follows, from one point to another via the surface, is one for which the time required is a minimum. For this particular case, Fermat's principle can be called the *principle of least time.*

Although the principle as stated above is correct for a single surface, it must be modified for application to a general optical system. In its modern form Fermat's principle states that the actual path a ray follows is such that the time of travel between two fixed points has a stationary value with respect to small changes of that path. In other words, the path of a ray from one point to another is such that the time taken has no more than an infinitesimal difference of second order from the time taken in traveling along other closely adjacent paths between the same points. Hence, to a first approximation, the travel time of the actual ray is equal to that along a closely adjacent path.

We first look at some of the consequences of this statement from a general point of view. The discussion involving calculus of variations can be skipped on a first reading, though results derived for the atmosphere are important for observations with ground-based telescopes. In subsequent sections we look at a number of other specifics that follow from this principle.

I. FERMAT'S PRINCIPLE IN GENERAL

The simplest case illustrating Fermat's principle is shown in Fig. 3.1. A surface Σ lies between two points, P_0 and P_1, with a ray joining these points consisting of straight line segments. The solid line is the actual ray path and the dashed line some other path. If the time of travel from P_0 to P_1 is denoted by τ, then the condition that τ have a stationary value for the actual path is

$$\partial\tau/\partial x = \partial\tau/\partial y = 0, \tag{3.1.1}$$

where x and y are the generalized coordinates of the point where the ray intersects the surface.

An equivalent statement of Fermat's principle is obtained by replacing the words "time of travel" with "optical path length." If dt is an infinitesimal time of travel, then $c\,dt$ is the corresponding optical path length, where c is the velocity of light in vacuum. The optical path length (hereafter denoted by OPL) is expressed in terms of the geometric path length and index of refraction as follows:

$$d(\text{OPL}) = c\,dt = (c/v)v\,dt = n\,ds,$$
$$\text{OPL} = c\int dt = \int n\,ds, \tag{3.1.2}$$

where v is the speed of light in the medium of index n. The general statement of Fermat's pinciple is either $\delta\tau = 0$ or $\delta(\text{OPL}) = 0$, where n can be a function of all the coordinates that specify the position.

We now consider the two-dimensional case where the index of refraction $n = n(y, z)$ and $ds = \sqrt{(dy^2 + dz^2)}$. Letting $y' = dy/dz$, Fermat's principle gives

$$\delta \int_{P_0}^{P_1} n(y,z)\sqrt{(1+y'^2)}\,dz = 0, \tag{3.1.3}$$

where ds has been replaced by $dz\sqrt{(1+y'^2)}$. Letting $F(y, y', z)$ represent

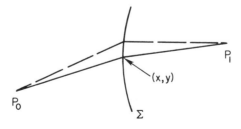

Fig. 3.1. Ray paths through interface between different optical media.

3. Fermat's Principle: An Introduction

the integrand in Eq. (3.1.3) we get

$$\delta \int_{P_0}^{P_1} F(y, y', z)\, dz = \int_{P_0}^{P_1} \delta F(y, y', z)\, dz = 0, \tag{3.1.4}$$

where

$$\delta F = \frac{\partial F}{\partial y}\delta y + \frac{\partial F}{\partial y'}\delta y' = \frac{\partial F}{\partial y}\delta y + \frac{\partial F}{\partial y'}\frac{d}{dz}(\delta y).$$

Substituting for δF in Eq. (3.1.4) and integrating the term containing y' by parts, we get

$$\int_{P_0}^{P_1} \frac{\partial F}{\partial y}\delta y\, dz + \frac{\partial F}{\partial y'}\delta y\bigg|_{P_0}^{P_1} - \int_{P_0}^{P_1} \frac{d}{dz}\left(\frac{\partial F}{\partial y'}\right)\delta y\, dz = 0. \tag{3.1.5}$$

The second term in (3.1.5) is zero because δy is zero at the endpoints. Therefore we can write Eq. (3.1.5) as

$$\int_{P_0}^{P_1}\left[\frac{\partial F}{\partial y} - \frac{d}{dz}\left(\frac{\partial F}{\partial y'}\right)\right]\delta y\, dz = 0.$$

This expression must vanish for an arbitrary δy and therefore

$$\frac{\partial F}{\partial y} - \frac{d}{dz}\left(\frac{\partial F}{\partial y'}\right) = 0, \tag{3.1.6}$$

which is the equation required to satisfy Fermat's principle.

We now take Eq. (3.1.6), replace F with the expression it represents, and carry out the differentiations indicated. As noted following Eq. (3.1.3) we have $F = n(y,z)\sqrt{(1+y'^2)}$. Noting that y' is not an explicit function of y, nor is n a function of y', we get

$$\sqrt{(1+y'^2)}\,\frac{\partial n}{\partial y} - \frac{d}{dz}\left[\frac{ny'}{\sqrt{(1+y'^2)}}\right] = 0$$

$$= \sqrt{(1+y'^2)}\,\frac{\partial n}{\partial y} - n\frac{d}{dz}\left[\frac{y'}{\sqrt{(1+y'^2)}}\right] - \frac{y'}{\sqrt{(1+y'^2)}}\frac{dn}{dz}. \tag{3.1.7}$$

Although (3.1.7) is a rather formidable equation in appearance, it is easily simplified after making some trigonometric substitutions. Figure 3.2 shows a segment of the ray path with the dashed line tangent to the path at point P. At this point

$$\tan\alpha = \frac{dy}{dz} = y', \qquad \sin\alpha = \frac{dy}{ds} = \frac{y'}{\sqrt{(1+y'^2)}},$$

$$\cos\alpha = \frac{dz}{ds} = \frac{1}{\sqrt{(1+y'^2)}}, \qquad \frac{d}{dz}\sin\alpha = \cos\alpha\,\frac{d\alpha}{dz}. \tag{3.1.8}$$

I. Fermat's Principle in General

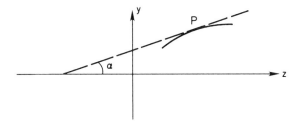

Fig. 3.2. Small segment of curved ray path in inhomogeneous medium. Dashed line is tangent to ray at point P.

Using Eqs. (3.1.8) and noting that

$$\frac{dn}{dz} = \frac{\partial n}{\partial z} + y' \frac{\partial n}{\partial y},$$

we write Eq. (3.1.7) as

$$\cos \alpha \frac{\partial n}{\partial y} - \sin \alpha \frac{\partial n}{\partial z} - n \cos \alpha \frac{d\alpha}{dz} = 0. \quad (3.1.9)$$

As a final item we note that the curvature K of a path in space is defined as

$$K = \frac{d\alpha}{ds} = \frac{d\alpha}{dz} \frac{dz}{ds} = \cos \alpha \frac{d\alpha}{dz}.$$

Substitution of this result into Eq. (3.1.9) gives

$$nK = n \cos \alpha \frac{d\alpha}{dz} = \cos \alpha \frac{\partial n}{\partial y} - \sin \alpha \frac{\partial n}{\partial z}. \quad (3.1.10)$$

The result in Eq. (3.1.10) gives the local curvature of a light ray subject to Fermat's principle in a medium in which the index of refraction is a smoothly varying function of position. Note that this relation applies to a ray in the yz plane with $n = n(y, z)$.

As a special case of Eq. (3.1.10), assume the index n is constant. In this case the partial derivatives on the right side of Eq. (3.1.10) are zero and hence the curvature is zero. Thus the path of a light ray in a homogeneous medium is a straight line.

We now apply these results to an inhomogeneous medium, the Earth's atmosphere. The items discussed in the remainder of this section include atmospheric refraction and its variation with zenith angle and wavelength and the effect of time-varying index changes on the path of a light ray.

A. ATMOSPHERIC REFRACTION

Assume that the atmosphere is a flat, layered medium with the index $n = n(z)$ only, hence the curvature of the atmosphere is neglected. In this case Eq. (3.1.10) becomes

$$nK = n \cos \alpha \frac{d\alpha}{dz} = -\sin \alpha \frac{dn}{dz}, \qquad (3.1.11)$$

where the z-axis points toward the center of the Earth. The change in the index of the atmosphere from the top ($n = 1$) to the surface ($n = 1.00029$) is small, hence the path of a ray from a star is not deviated appreciably for α not close to 90°. Integrating Eq. (3.1.11) with the assumption that α is nearly constant, hence $\cos \alpha$ and $\sin \alpha$ brought out from the integral, gives

$$\delta\alpha = -\tan \alpha_0 \, \delta n, \qquad (3.1.12)$$

where α_0 is the angle of incidence at the top of the atmosphere, or zenith angle, and δn is the change in index.

For a ray passing downward through the atmosphere $\delta n > 0$, and hence $\delta\alpha < 0$. Thus the angle the ray makes with the z axis decreases as the ray proceeds down through the atmosphere; that is, the ray is bent "toward" the z axis. That the effect is small is seen by taking, for example, $\alpha_0 = 45°$ and finding the ray deviation $\delta\alpha = 0.00029$ radians or about 1 arc-min.

The index of the atmosphere is a function of wavelength, as shown by the entries in Table 3.1, hence the deviation $\delta\alpha$ is not the same for different wavelengths. The parameter R_0 in Table 3.1 is the constant of refraction, the index difference δn expressed in units of arc-seconds.

The change $d(\delta\alpha)$ is the differential atmospheric refraction, with

$$d(\delta\alpha) = -\tan \alpha_0 \, d(\delta n), \qquad (3.1.13)$$

Table 3.1.
Index of Refraction of Atmosphere[a]

λ (nm)	$n - 1$	R_0 (arc-sec)
320	3.049E−4	62.86
400	2.982	61.48
550	2.929	60.38
700	2.907	59.93
1000	2.890	59.58

[a] Values of n from Allen (1973). Index given at $T = 0°C$, pressure = 760 mm Hg, water vapor pressure = 4 mm Hg.

and $d(\delta n)$ is the change in index between two chosen wavelengths. From the values in Table 3.1 we see that the index changes more rapidly at shorter wavelengths, hence differential refraction could adversely affect certain types of observations in the near ultraviolet at large zenith angle. As an example, using the entries in Table 3.1, $d(\delta\alpha)$ in arc-seconds is about $1.38\tan\alpha_0$ over the range from 320 to 400 nm and $2.48\tan\alpha_0$ over the range from 320 to 550 nm. With $\tan\alpha_0 \geq 1$, for example, the visible image of a star centered on a small aperture could result in no ultraviolet light passing through the aperture.

B. ATMOSPHERIC TURBULENCE

The assumption that $n = n(z)$ neglects variations in index that are present in a turbulent atmosphere at constant height due primarily to temperature fluctuations. Consider a ray that enters the atmosphere from directly overhead, with α the deviation of the ray from a vertical path. Assuming $\alpha \ll 1$ we can write Eq. (3.1.10) as

$$n(\partial\alpha/\partial z) = \partial n/\partial y, \qquad (3.1.14)$$

where the term in $\sin\alpha$ is dropped because α is small. Letting $n = 1 + \delta n$, Eq. (3.1.14) becomes (to first order)

$$\partial\alpha/\partial z = \partial(\delta n)/\partial y, \qquad (3.1.15)$$

where δn is the fluctuation in the index of refraction from the local mean. In the general case there are corresponding equations in which x replaces y. Integrating Eq. (3.1.15) from the top of the atmosphere ($z = 0$) through a distance s gives

$$\alpha_y(s) = \int_0^s [\partial(\delta n)/\partial y]_z \, dz. \qquad (3.1.16)$$

The deviation given by Eq. (3.1.16) is, of course, a function of time, with random variations in time for α_y and α_x. Because $\langle \delta n \rangle$ is zero, where $\langle \,\rangle$ denotes an average over time, the time averages of the deviations are also zero. The mean-square deviations, however, are not zero, and the net result is a ray that wanders randomly about a mean position.

The net effect of these variations leads to the phenomenon called *seeing*. In a small telescope the effect is seen as a star image in motion with excursions typically of a few arc-seconds. In a large telescope the cumulative effect of seeing is to give a blurred image with little or no motion of the image as a whole.

Although the approach using Fermat's principle shows the origin of seeing effects, the statistical processes that lead to the effects described make it impractical to proceed further with this approach. Selected results based on a statistical approach to atmospheric turbulence are given in Chapter 16.

II. EXAMPLES USING FERMAT'S PRINCIPLE

In this section we consider several additional examples and approach them from the point of view of Fermat's principle. In doing this we will rederive some of the results of Chapter 2 as well as find some new ones. For all of the cases discussed we assume that homogeneous media are separated by a surface across which the index changes abruptly.

A. LAWS OF REFRACTION AND REFLECTION

Fermat's principle can be used to derive Snell's law at a plane interface where the index changes from n to n', as shown in Fig. 3.3. For this situation the condition that the path is stationary is

$$\delta \left[n \int_{P_1}^{P_0} ds + n' \int_{P_0}^{P_2} ds \right] = 0,$$

which, on evaluation of the integrals, gives

$$\delta \{ n\sqrt{(z_1^2 + y_0^2)} + n'\sqrt{[z_2^2 + (y_2 - y_0)^2]} \} = 0. \tag{3.2.1}$$

This is, as expected, simply the sum of two optical lengths. Our variable is y_0 and differentiating Eq. (3.2.1) gives

$$\left\{ n \frac{d}{dy_0} \sqrt{(z_1^2 + y_0^2)} + n' \frac{d}{dy_0} \sqrt{[z_2^2 + (y_2 - y_0)^2]} \right\} \delta y_0 = 0.$$

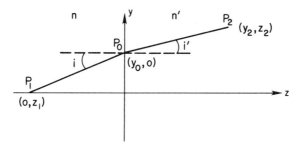

Fig. 3.3. Ray through plane interface between two homogeneous media with different indices of refraction.

The expression in braces in this relation is independent of δy_0, and therefore we set the expression equal to zero. Doing the differentiation gives

$$n \frac{y_0}{\sqrt{(z_1^2 + y_0^2)}} - n' \frac{y_2 - y_0}{\sqrt{[z_2^2 + (y_2 - y_0)^2]}} = 0. \quad (3.2.2)$$

Examination of Fig. 3.3 shows that the factors multiplying n and n' are $\sin i$ and $\sin i'$, respectively, and hence Eq. (3.2.2) is simply Snell's law of refraction, $n \sin i = n' \sin i'$. The law of reflection follows directly if we let $n' = -n$, in which case we also have $i' = -i$.

The nature of this stationary condition can be examined further by differentiating Eq. (3.2.2) with respect to y_0 and looking at the sign of the result. Because the sign is positive, the path taken by the ray in going from one fixed point to another is such that its time of travel or OPL is a minimum.

B. SPHERICAL INTERFACE

Although we already derived the paraxial equation for refraction at a spherical interface in Section 2.II, we will repeat the exercise using Fermat's principle. The spherical surface separating two homogeneous media, along with conjugate points B and B', is shown in Fig. 3.4. With due regard for signs according to the Cartesian convention, the optical length F from B to B' via point P is given by $F = -nl + n'l'$, where from the law of cosines

$$l = -\sqrt{[R^2 + (R-s)^2 - 2R(R-s)\cos\phi]},$$

$$l' = \sqrt{[R^2 + (s'-R)^2 + 2R(s'-R)\cos\phi]}.$$

Substituting l and l' into F, we have an expression in which ϕ is the variable. We apply Fermat's principle and find the stationary condition by setting $dF/d\phi = 0$. This gives

$$\frac{dF}{d\phi} = -\frac{nR(R-s)\sin\phi}{l} - \frac{n'R(s'-R)\sin\phi}{l'} = 0. \quad (3.2.3)$$

In the paraxial limit $l = s$ and $l' = s'$. Substitution of these into Eq. (3.2.3) immediately leads to Eq. (2.2.2).

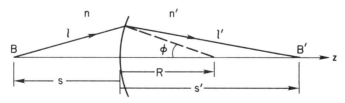

Fig. 3.4. Refraction at spherical interface.

C. FOCAL LENGTH OF THIN LENS

As an example of a slightly more complex system, we use Fermat's principle to find the focal length of a thin lens of index n with radii of curvature R_1 and R_2 as shown in Fig. 3.5.

To find the focal length we make use of the fact that Fermat's principle must apply to every ray between two conjugate points of a focusing system. For example, in Fig. 3.4 we see that a ray from B to B' along the z axis must have an OPL that is stationary with respect to closely adjacent paths. But each of these adjacent paths is itself stationary, hence the OPL is the same along all paths between two conjugate points, at least to a first approximation, provided the rays pass through the system. Stated differently, the OPL (or time of travel) between two conjugates of a perfect focusing system is neither a minimum nor a maximum.

Returning to the thin lens shown in Fig. 3.5, we find the OPL for each of two rays. For the ray coincident with the z axis we get

$$F_0 = [BO_1] + n[O_1 O_2] + f',$$

while for the ray at height y in the paraxial range

$$F_p = [BO_1] + z_1 + n[P_1 P_2] - z_2 + l,$$

where $z_2 < 0$ and l is measured from the y_2 axis. Although the distance $[BO_1]$ is infinite, this is of no consequence because on setting F_0 equal to F_p this distance drops out, and we get

$$nd + f' = z_1 + n(d - z_1 + z_2) - z_2 + l. \tag{3.2.4}$$

In Eq. (3.2.4) we have substituted $d = [O_1 O_2]$, $d - z_1 + z_2 = [P_1 P_2]$. Rearranging Eq. (3.2.4) leads to

$$l - f' = (n-1)(z_1 - z_2). \tag{3.2.5}$$

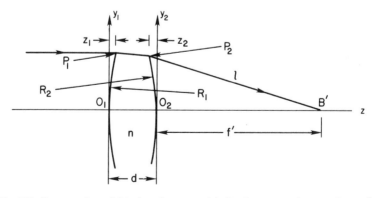

Fig. 3.5. Cross section of thin lens (not to scale). By sign convention $z_1 > 0$, $z_2 < 0$.

The radii of curvature, R_1 and R_2, are given by

$$R_1^2 = y_1^2 + (R_1 - z_1)^2 = R_1^2 + y^2 - 2R_1 z_1,$$
$$R_2^2 = y_2^2 + (-R_2 + z_2)^2 = R_2^2 + y^2 - 2R_2 z_2,$$

where $y_1 = y_2 = y$ for a thin lens in the paraxial range. In this approximation we get $z_1 = y^2/2R_1$ and $z_2 = y^2/2R_2$.

From Fig. 3.5 we see that $l^2 = y^2 + f'^2 = f'^2(1 + y^2/f'^2)$. Taking the square root and using the binominal expansion gives $l - f' = y^2/2f'$. Taking these results, substituting for z_1, z_2, and $l - f'$ in Eq. (3.2.5), and canceling common factors gives

$$1/f' = (n-1)(1/R_1 - 1/R_2),$$

a result already seen in Eq. (2.4.4).

D. DISPERSING PRISM

As our final example in this section we consider a glass prism as shown in Fig. 3.6. Because $n = n(\lambda)$ the angle of deviation θ is also a function of λ, where λ is the wavelength of light. With rays incident as shown in Fig. 3.6, there is some wavelength whose rays in the prism follow paths parallel to the prism base. For these rays the diagram is symmetric about the vertical bisector of the prism, and hence $s_1 = s_2 = s$, $\alpha_1 = \alpha_2 = \alpha$, and $a_1 = a_2 = a$.

Applying Fermat's principle to this symmetric situation we get

$$2L \cos \alpha = nt, \qquad (3.2.6)$$

where the left side of the equation is the OPL of the upper ray and the right side is the OPL of the lower ray in Fig. 3.6. We are interested in seeing how θ changes with wavelength. Differentiating Eq. (3.2.6) with respect to

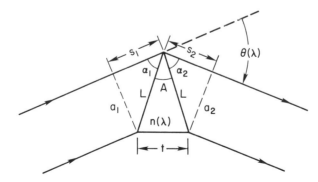

Fig. 3.6. Dispersing prism.

wavelength gives

$$t\frac{dn}{d\lambda} = -2L\sin\alpha\frac{d\alpha}{d\lambda} = -2L\sin\alpha\frac{d\alpha}{d\theta}\frac{d\theta}{d\lambda}. \quad (3.2.7)$$

From Fig. 3.6 we see that $L\sin\alpha = a$, $\theta = \pi - A - 2\alpha$, from which we get $d\alpha/d\theta = -1/2$. Substituting into Eq. (3.2.7) we get

$$d\theta/d\lambda = (t/a)\, dn/d\lambda, \quad (3.2.8)$$

where t/a is the ratio of the base length to the beam width.

The index of refraction of most optical glasses can be expressed approximately in the form

$$n(\lambda) = A' + (B/\lambda^2), \quad (3.2.9)$$

where A' and B are constants. Differentiating Eq. (3.2.9) and combining with Eq. (3.2.8) we get

$$d\theta/d\lambda = -(2t/a)(B/\lambda^3). \quad (3.2.10)$$

The negative sign indicates that θ decreases as λ increases, hence blue light is deviated more than red light. We also note that the angular dispersion (the name given to $d\theta/d\lambda$) is numerically larger for shorter wavelengths.

III. PHYSICAL INTERPRETATION OF FERMAT'S PRINCIPLE

Fermat's principle is a statement about the behavior of light rays in terms of optical path length. The statement does not in any way make use of the fact that light is an electromagnetic wave capable of undergoing destructive interference. By treating light as a wave we can give a physical interpretation of Fermat's principle in terms of destructive interference of waves following different paths. This is most easily done by means of a specific example.

If, in Fig. 3.3, we choose $n = 1$, $n' = \sqrt{(5/2)}$, then the stationary path is that for which P_2 is at $(2, 2)$ when P_1 is at $(0, -1)$ and P_0 is at $(1, 0)$. The optical path length between P_1 and P_2 is then a minimum. For another wave originating at P_1 to reach P_2 half a cycle after the wave following the minimum path, we need Δy_0 of 1475 wavelengths for light of 500 nm, when the coordinates of the points are given in meters.

Stated in another way, the extra path length introduced when y_0 is changed by 1475 wavelengths (or 0.7375 mm) is only half of a wavelength when the change is in the neighborhood of the stationary path. If, on the other hand, we choose P_0 at $(1.2, 0)$, then the path from P_1 to P_2 is not a stationary one. In this case a half-wavelength change in OPL is introduced when Δy_0 is about 2.8 wavelengths. The variation in OPL as a function of y_0 is shown in Fig. 3.7.

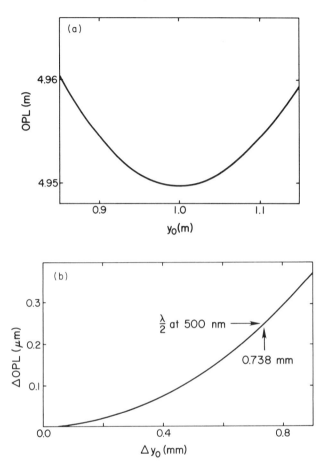

Fig. 3.7. Optical path difference for ray in Fig. 3.3. See text, Section 3.III, for defined coordinates. (b) Greatly magnified view near minimum in (a).

Fermat's principle can therefore be thought of as giving the path through which the highest transmission of light is possible. This path is the one that presents to the light waves the largest area without destructive interference for waves that pass through that area.

IV. FERMAT'S PRINCIPLE AND REFLECTING SURFACES

The application of Fermat's principle to a spherical refracting interface and thin lens in Section 3.II gives results that apply in the paraxial domain.

34   3. Fermat's Principle: An Introduction

In these examples the surface shapes were specified (all spherical), with the result that the derived equations are strictly true only for paraxial rays. In this section on reflecting surfaces we adopt a different procedure and require that rays over the entire aperture satisfy Fermat's principle. We then find the appropriate surface shape needed to satisfy this requirement.

A. CONCAVE MIRROR, ONE CONJUGATE AT INFINITY

We first consider the concave mirror shown in Fig. 3.8. Parallel rays are incident from the left with all rays focused at a distance f from the mirror vertex. For convenience we let f, l, and Δ be positive quantities. Applying Fermat's principle to a ray on the optical axis and a ray at height y, we see that equal OPLs requires $2f = l + (f - \Delta)$, or $l = f + \Delta$.

From the geometry in Fig. 3.8 we see that

$$l^2 = y^2 + (f - \Delta)^2. \tag{3.4.1}$$

Eliminating l in Eq. (3.4.1) gives $y^2 = 4f\Delta$, which in terms of z is

$$y^2 = -4fz. \tag{3.4.2}$$

Equation (3.4.2) is the equation of a parabola whose vertex is at $(0, 0)$. The paraboloidal surface of revolution is obtained by rotating the parabola about the z-axis; its equation is found by replacing y^2 by $x^2 + y^2$. Using Eq. (2.3.2) we can express f in terms of R, which, on applying the sign convention to R, gives

$$y^2 = 2Rz. \tag{3.4.3}$$

R is the radius of curvature at the mirror vertex, and both R and z are negative in Fig. 3.8.

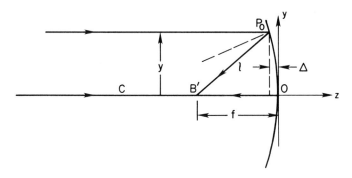

Fig. 3.8. Rays from distant point source incident on concave reflector, where l is the distance from P_0 to B'. Image at B' is point for surface given by Eq. (3.4.3).

B. CONCAVE MIRROR, BOTH CONJUGATES FINITE

Figure 3.9 shows a concave mirror with an object point at B and the corresponding image point at B', both on the z axis. Here we adopt the sign convention for s and s' at the outset, while choosing l, l', and Δ as positive quantities. Given s and $s' < 0$ in Fig. 3.9, the application of Fermat's principle to the two rays leaving B gives

$$l + l' = -(s + s'),$$

where $l^2 = y^2 + (-s - \Delta)^2$, and $l'^2 = y^2 + (-s' - \Delta)^2$. Eliminating l and l' between these relations, and letting $\Delta = -z$ as in Eq. (3.4.2), leads to the relation

$$y^2 - 4z\frac{ss'}{s+s'} + 4z^2\frac{ss'}{(s+s')^2} = 0. \tag{3.4.4}$$

This is the equation for an ellipse with center $(0, a)$, with a and b the semimajor and semiminor axes, respectively. We can easily put Eq. (3.4.4) into the standard form of an ellipse equation if we choose $2a = s + s'$, $b^2 = ss'$. The standard equation for an ellipse with center $(0, a)$ is

$$\frac{(z-a)^2}{a^2} + \frac{y^2}{b^2} = 1,$$

which can be written as

$$y^2 - 2z\frac{b^2}{a} + z^2\frac{b^2}{a^2} = 0. \tag{3.4.5}$$

The choice of a and b as given above follows directly from a comparison of Eqs. (3.4.4) and (3.4.5). It is not surprising that Fermat's principle leads

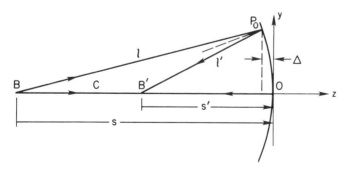

Fig. 3.9. Rays between conjugates at finite distances via concave reflector, where l (l') is the distance from P_0 to B (B'). Imagery is perfect for surface given by Eq. (3.4.4).

to an ellipse as the appropriate curve with the two conjugate points at the foci of the ellipse, considering the standard technique for drawing an ellipse with pencil, string, and two pins. A rotation of the ellipse about the z axis gives an ellipsoidal mirror, with the surface equation given by (3.4.5) after replacing y^2 by x^2+y^2.

Note that the sphere is a special case of an ellipsoid in which $s=s'$ and $a=b$. Note also that the parabola given by Eq. (3.4.2) is a special case of Eq. (3.4.4) in which $s=\infty$ and $s'=-f$.

C. CONVEX MIRROR, BOTH CONJUGATES FINITE

Figure 3.10 shows a convex mirror with a virtual object point at B and the conjugate image point at B', both on the z-axis. As with the ellipse, we adopt the sign convention for s and s' but choose l, l', and Δ as positive quantities. The dashed arc in Fig. 3.10 is a circular arc whose center is at B. Applying Fermat's principle to the two rays heading toward B gives

$$l+l'=2s',$$

while the geometry of Fig. 3.10 gives $d^2=y^2+(-s-\Delta)^2$, $l+d=s'-s$, $l'^2=y^2+(s'+\Delta)^2$. Eliminating l, l', and d between these relations and putting $\Delta=-z$ leads to

$$y^2-4z\frac{ss'}{s+s'}+4z^2\frac{ss'}{(s+s')^2}=0, \qquad (3.4.6)$$

an equation identical to Eq. (3.4.4). There is, however, an important difference between Eq. (3.4.4) and Eq. (3.4.6). In the former equation s and

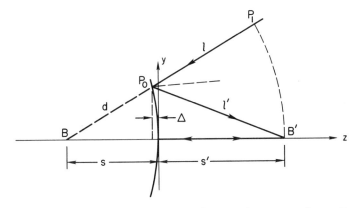

Fig. 3.10. Rays between conjugates at finite distances via convex reflector. Imagery is perfect for surface given by Eq. (3.4.6).

s' have the same sign because both conjugates are on the same side of the mirror vertex; in the latter equation s and s' have opposite signs. As is easily demonstrated, Eq. (3.4.6) is the equation of a hyperbola.

The standard equation for a hyperbola with a vertex at $(0, 0)$ is

$$\frac{(z-a)^2}{a^2} - \frac{y^2}{b^2} = 1,$$

which can be rewritten as

$$y^2 + 2z\frac{b^2}{a} - z^2\frac{b^2}{a^2} = 0. \quad (3.4.7)$$

Equations (3.4.6) and (3.4.7) agree if we choose $b^2 = -ss'$ and $2a = s + s'$. As before, the replacement of y^2 by $x^2 + y^2$ gives a hyperboloid of revolution about the z-axis.

The case of a convex mirror with one conjugate at infinity is left as an exercise for the reader. The appropriate surface for this situation is a paraboloid.

V. CONIC SECTIONS

Each of the surface cross sections derived in the preceding section is a conic section, and it is therefore appropriate to find a single equation describing the family of such curves with the vertex at the origin. We proceed by working with Eq. (3.4.4) for an ellipse. From Eq. (2.3.1) we get

$$ss'/(s+s') = R/2, \quad (3.5.1)$$

where this relation applies in the paraxial region, hence R is the vertex radius of curvature. For an ellipse the eccentricity e is defined as $e = c/a$, where c is the distance from one of the foci to the center of the ellipse and $c^2 = a^2 - b^2$. Substituting in terms of s and s' we get

$$1 - e^2 = \frac{4ss'}{(s+s')^2}, \quad e^2 = \frac{(s-s')^2}{(s+s')^2}. \quad (3.5.2)$$

Substituting Eqs. (3.5.1) and (3.5.2) into Eq. (3.4.4) gives

$$y^2 - 2Rz + (1-e^2)z^2 = 0. \quad (3.5.3)$$

Although derived from the ellipse equation, the relation in Eq. (3.5.3) describes the family of conic sections, provided we choose e appropriately. In the literature one often sees a conic section described in terms of a *conic constant* K, where $K = -e^2$. In terms of both e and K the various conic sections are as follows:

Prolate ellipsoid:	$e^2 < 0$,	$K > 0$
Sphere:	$e = 0$,	$K = 0$
Oblate ellipsoid:	$0 < e < 1$,	$-1 < K < 0$
Paraboloid:	$e = 1$,	$K = -1$
Hyperboloid:	$e > 1$,	$K < -1$

In all of the discussion to follow, we use K to describe the conic sections. Transforming Eq. (3.5.3) to get the equation for the surface of revolution gives

$$\rho^2 - 2Rz + (1+K)z^2 = 0, \qquad (3.5.4)$$

where $\rho^2 = x^2 + y^2$. Although we are also using ρ as one of the normalized parameters to describe a two-mirror telescope, it will always be clear from context which ρ is being used.

At this point it is instructive to calculate r, the local radius of curvature at a point (ρ, z) on the mirror surface. The relation for radius of curvature is

$$r = (1 + z'^2)^{3/2} / z'',$$

where $z' = dz/d\rho$, $z'' = d^2z/d\rho^2$. Solving Eq. (3.5.4) for z and carrying out the calculation gives

$$r = R[1 - K(\rho^2/R^2)]^{3/2}. \qquad (3.5.5)$$

For $K = 0$ we get $r = R$, as expected. As we go through the family of conic surfaces from sphere to ellipsoid to paraboloid to hyperboloid, we see that r becomes progressively larger for a given ρ and R. Alternatively, the local curvature, defined as $1/r$, becomes progressively less. As the point on the surface approaches the vertex, hence $\rho \to 0$, we see that $r \to R$. Near the vertex all of the surfaces have nearly the same shape and, in the paraxial approximation, are identical.

In summary, then, we see that conic surfaces used as mirrors provide perfect imagery for a single pair of conjugates. A given conic mirror, however, will not strictly satisfy Fermat's principle at any other pair of conjugates. As we will see, this failure to image a point into a point implies the presence of aberrations, a subject we explore in detail in subsequent chapters. In spite of this apparent limitation, the family of conic surfaces is the basis for most multi-mirror systems.

VI. CONCLUDING COMMENTS

A. RAYS AND WAVEFRONTS

The application of Fermat's principle to find conic surfaces that are perfect mirrors makes use of rays and optical path lengths. A different way of looking at what a focusing system does is in terms of wavefronts. A wavefront is simply a surface on which every point has the same optical path distance from a point source of light. In a homogeneous medium this surface is obviously a sphere whose center is the point object. In the same medium rays are radial lines directed outward, and at each point on a wavefront a ray is perpendicular to the wavefront.

A perfect system that satisfies Fermat's principle is therefore one that converts a spherical wavefront centered on the object to a spherical wavefront centered on the image. Conversely, if Fermat's principle is not satisfied for all rays over a large aperture, then the wavefront converging toward the image is no longer spherical and the image has aberrations. The connection between ray and wavefront aberrations is established in the discussion in Chapter 5.

B. HOW PERFECT IS "PERFECT"?

Fermat's principle as used thus far is concerned only with rays and ignores the wave nature of light. Because of the wave character of light, no image is perfect in the sense that it is a point of infinitesimal size. The question to be addressed therefore, albeit not from a rigorous point of view here, is the minimum size of an image given an otherwise perfect optical system.

Consider an optical system that is perfect according to Fermat's principle, as shown schematically in Fig. 3.11. Light from two distant point sources, A and B, with angular separation θ fills the aperture of diameter D. According to wave theory, two image points cannot be resolved or separated if the difference in light travel time of rays to them from opposite edges of the aperture is less than approximately one period of the wave. Equivalently, the points cannot be resolved if the optical path difference between these rays is less than approximately one wavelength. In Fig. 3.11 the optical path difference between these rays is Δ with the resolution limit set by $\Delta \sim \lambda$. From the geometry we see that

$$\theta_{min} \sim \lambda/D. \quad (3.6.1)$$

From the angular resolution limit in Eq. (3.6.1) we can infer that the individual images A' and B' must each have an angular diameter $\theta \approx \lambda/D$,

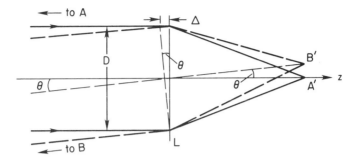

Fig. 3.11. Schematic of perfect optical system from which approximate diffraction limit is derived. See Eq. (3.6.1) and Section 3.VI.

as seen from L. If the angular diameter of each image was substantially smaller than λ/D, then the images would be resolved, contrary to the limit set by Eq. (3.6.1).

The reasoning used to arrive at Eq. (3.6.1) and an estimate of the minimum possible image size is not a rigorous procedure, nor does it tell how the light is distributed within the image. A more rigorous approach requires analysis using diffraction theory, a topic we consider in Chapter 10.

REFERENCES

Sources of information on atmospheric refraction and seeing
Allen, C. (1973). "Astrophysical Quantities," 3rd ed., Chap. 6. Athlone, London.
Stock, J., and Keller, G. (1960). "Stars and Stellar Systems I, Telescopes," Chap. 9. University of Chicago Press, Chicago.

BIBLIOGRAPHY

Discussion of Fermat's principle
Born, M., and Wolf, E. (1980). "Principles of Optics," 6th ed., Chap. 3. Pergamon, Oxford.
Hecht, E., and Zajac, A. (1974). "Optics," Chap. 4. Addison-Wesley, Reading, Massachusetts.

Chapter 4 Introduction to Aberrations

Thus far our discussion of optical systems has proceeded along two different lines. In Chapter 2 we developed the paraxial equations for spherical refracting and reflecting surfaces and noted that in the paraxial limit there is a one-to-one correspondence between object and image point. In Chapter 3 we turned our attention to Fermat's principle and reflecting surfaces of conic cross section. Our analysis led to the result that for a given pair of conjugate object and image points there is a conic surface that gives a perfect image, independent of the paraxial approximation.

In this chapter we begin to examine what happens when Fermat's principle is not strictly satisfied in the range outside the paraxial approximation. We will see that the geometrical image in this case is no longer a point but becomes a blur. An optical system that produces a blurred image, where the blur is in addition to the diffraction blur noted in Section 3.VI, is a system with aberrations.

To illustrate the onset of aberrations we consider a very simple optical system, a single conic mirror. After calculating the aberrations of several such mirrors for selected object points, we introduce the topic of aberration compensation. By this we mean that the aberrations of one optical element can be offset, wholly or partially, by those of another element. The two systems considered in this chapter are the Schmidt camera and the family of Cassegrain telescopes. This discussion is only an introduction; a more complete description of aberrations and compensation follows in Chapter 5.

I. REFLECTING CONICS AND FOCAL LENGTH

We begin by calculating the focal length of a concave mirror or, more specifically, the distance from the mirror vertex to the point where a reflected ray from a distant object intersects the optical axis. Figure 4.1 shows a ray parallel to the optical axis striking a mirror at height ρ, where ρ is defined by Eq. (3.5.4). Contrary to our normal convention, the light from the object proceeds from right to left, a choice made for convenience. With this choice distances to the right of the mirror vertex measured along the z axis are positive. We also take angle ϕ in Fig. 4.1 positive when $\rho > 0$.

From the geometry of Fig. 4.1 we find $f = z + z_0$, where

$$z_0 = \frac{\rho}{\tan 2\phi} = \frac{\rho(1 - \tan^2 \phi)}{2 \tan \phi}. \quad (4.1.1)$$

From Fig. 4.1 we also note that $\tan \phi$ is simply $dz/d\rho$, the negative of the slope of the normal to the mirror.

From Eq. (3.5.4) we find the relation

$$\frac{dz}{d\rho} = \frac{\rho}{R - (1+K)z} = \tan \phi.$$

Substituting this into Eq. (4.1.1) gives

$$z_0 = \frac{\rho}{2}\left[\frac{R - (1+K)z}{\rho} - \frac{\rho}{R - (1+K)z}\right]. \quad (4.1.2)$$

Putting Eq. (4.1.2) into $f = z + z_0$ gives

$$f = \frac{R}{2} + \frac{(1-K)z}{2} - \frac{\rho^2}{2(R - (1+K)z)}. \quad (4.1.3)$$

Because the surface equation (3.5.4) is quadratic in z, the solution for z contains a square root, as will Eq. (4.1.3) when z is eliminated. We proceed,

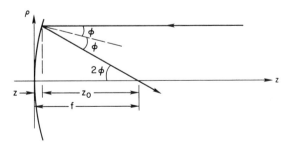

Fig. 4.1. Geometry of ray from distant object reflected from concave mirror.

therefore, by expanding the square root as a power series in small quantities before substituting into Eq. (4.1.3). Solving Eq. (3.5.4) for z gives

$$z = \frac{R}{1+K}\left[1 - \left(1 - \frac{\rho^2}{R^2}(1+K)\right)^{1/2}\right]$$

$$= \frac{\rho^2}{2R} + (1+K)\frac{\rho^4}{8R^3} + (1+K)^2\frac{\rho^6}{16R^5} + \cdots. \quad (4.1.4)$$

Substituting Eq. (4.1.4) into Eq. (4.1.3) gives

$$f = \frac{R}{2} - \frac{(1+K)\rho^2}{4R} - \frac{(1+K)(3+K)\rho^4}{16R^3} - \cdots. \quad (4.1.5)$$

Examination of Eq. (4.1.5) shows that $f = R/2$ for $K = -1$, a paraboloid. Although higher-power terms are not included in Eq. (4.1.5), this statement about a paraboloid is true when all terms are included. This is easily verified by setting $K = -1$ prior to making the above substitutions.

For a sphere or ellipsoid the conic constant $K > -1$ and $f < R/2$, while for a hyperboloid $f > R/2$. As expected, Fermat's principle is strictly satisfied, hence f = constant for any ρ, only for a paraboloid when the object is at infinity. For any other conic the change in focal length Δf as a function of ρ is

$$\Delta f = f(\rho) - f(\text{paraxial}) = -\frac{(1+K)\rho^2}{4R} - \frac{(1+K)(3+K)\rho^4}{16R^3} - \cdots. \quad (4.1.6)$$

Thus for any conic surface other than a paraboloid the image of a distant object on the optical axis is blurred. Examination of Eq. (4.1.6) shows that Δf is independent of the sign of ρ, hence the blur is symmetric about the z axis.

II. SPHERICAL ABERRATION

We now examine in detail the nature of the aberration for the case of an object on the optical axis, an aberration called *spherical aberration*. Of particular interest is the size of the blur, measured perpendicular to the optical axis, at or near the paraxial focus.

A. TRANSVERSE

We define TSA, the *transverse spherical aberration*, as the intersection of a ray from height ρ on the mirror with the paraxial focal plane, as shown in Fig. 4.2. From similar triangles there we find

$$\mathrm{TSA}/\Delta f = \rho/(f-z),$$

where both TSA and Δf are negative in Fig. 4.2. Using Eqs. (4.1.4)–(4.1.6), applying the binomial expansion, and retaining all terms through fifth order gives

$$\mathrm{TSA} = -(1+K)\frac{\rho^3}{2R^2} - 3(1+K)(3+K)\frac{\rho^5}{8R^4} + \cdots. \quad (4.2.1)$$

Each term is designated according to the power of ρ. The first term is the third-order transverse spherical aberration (TSA3); the second term is fifth-order transverse spherical aberration (TSA5). For $K = 0$, a spherical surface, each term in Eq. (4.2.1) is negative for $\rho > 0$ and positive for $\rho < 0$. The sign of TSA indicates where a given ray crosses the paraxial focal plane, in accord with the sign convention established for distances measured perpendicular to the z axis. Because of the presence of the factor $(1+K)$ in Eq. (4.2.1), the sign of TSA for a hyperboloid is opposite to that for a sphere or ellipsoid.

The relative size of the two terms in Eq. (4.2.1) for rays from the edge of the aperture is given by

$$\frac{\mathrm{TSA5}}{\mathrm{TSA3}} = \frac{3(3+K)\rho^2}{4R^2} = \frac{3(3+K)}{64F^2},$$

where F is the focal ratio. For a sphere TSA5 is 10% of TSA3 when $F = 1.19$. Thus it is sufficient to neglect the TSA5 term for all but very fast mirrors.

We can also find the spherical aberration for the case shown in Fig. 4.2 by working directly with surface equations, one for a paraboloid and one for a surface with conic constant K. From Eq. (4.1.4) we find the difference between the surfaces, through terms in ρ^4, given by

$$\Delta z = z_p - z(K) = -(\rho^4/8R^3)(1+K) + \cdots. \quad (4.2.2)$$

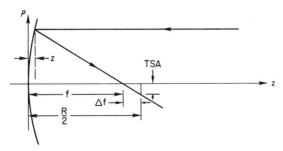

Fig. 4.2. Transverse spherical aberration (TSA) at paraxial focus. See Eqs. (4.1.6) and (4.2.1).

II. Spherical Aberration

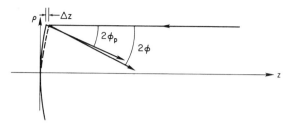

Fig. 4.3. Path difference between ray reflected from paraboloid (solid curve) and conic (dashed curve). Size of Δz, given in Eq. (4.2.2), is greatly exaggerated in the diagram.

where the subscript p denotes the paraboloid. From Fig. 4.3 we see that the path difference between two rays, one incident on the paraboloid and one on the other surface at the same height, is approximately $2\Delta z$, provided the angles ϕ and ϕ_p are small.

We also see from Fig. 4.3 that the directions of the reflected rays differ by $2(\phi_p - \phi)$, where $\phi = dz/d\rho$, $\phi_p = dz_p/d\rho$. From Eq. (4.1.4) we find

$$2(\phi_p - \phi) = (d/d\rho)(2\Delta z) = -(1+K)(\rho^3/R^3). \tag{4.2.3}$$

B. ANGULAR

This difference in direction between the reflected rays is ASA, the *angular spherical aberration*. Because Eq. (4.2.3) is taken only to third order, it is more precise to say ASA3 for the difference in direction given in Eq. (4.2.3).

From the geometry in Fig. 4.4 we see directly the relation between transverse and angular aberration, which is

$$\text{TSA3} = (R/2)(\text{ASA3}) = -(1+K)(\rho^3/2R^2), \tag{4.2.4}$$

where, as before, we assume the angle ϕ_p is not too large. This result is the same as the first term in Eq. (4.2.1).

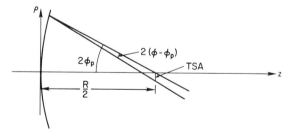

Fig. 4.4. Relation between TSA and angular difference between ray paths after reflection. See Eq. (4.2.3).

We need to review briefly the steps taken in arriving at Eq. (4.2.4) and the approximations used. In stating that the path difference between the two rays is $2\Delta z$, we replaced any cosine factors present in the exact difference by unity. In effect, we used the paraxial approximation in making this statement. The same approximation was made in writing the relation between TSA3 and ASA3 in Eq. (4.2.4). This approximation is quite good, even for a mirror as fast as $f/2$. In this case we find $\tan \phi = 0.25$ and $\cos \phi = 0.97$, hence our third-order result is accurate to a few percent. If we wanted to use the same method to find the fifth-order term, we could not use the paraxial approximation but would need to retain higher-power terms in the expansions of tangents and cosines of angles. But, as noted above, third-order aberration results suffice for reflecting surfaces in most optical systems.

The procedure followed to get the third-order result in Eq. (4.2.4) can be generalized to any pair of object and image conjugates. All that is needed is Δz, the difference between the reflecting surface that images the object without aberration and the actual surface, with the paraxial approximation used in the same way as above. The relations are

$$\text{angular aberration} = \frac{d}{d\rho}(2\Delta z),$$

$$\text{transverse aberration} = s'\frac{d}{d\rho}(2\Delta z). \qquad (4.2.5)$$

Equations (4.2.5) apply specifically to reflecting surfaces in air in which the mirror is oriented as shown in Fig. 4.1. In general the optical path difference, which includes the index of refraction, is required by Fermat's principle and the calculation of Δz. The more general discussion of Eqs. (4.2.5), including the index of refraction, is given in Chapter 5.

The importance of the procedure outlined in the paragraph preceding Eqs. (4.2.5) lies in its utility when applied to optical systems with more than one surface. In Chapter 5 we develop the method by which Δz can be determined in a general way for any optical system. Once Δz is known, it is then a straightforward matter to calculate the angular and transverse aberrations.

C. EXAMPLE: SPHERE WITH FINITE CONJUGATES

As an illustration of the utility of Eqs. (4.2.5) we consider an object point at a finite distance and an ellipsoid with the correct conic constant needed to form a perfect image. If a sphere is used in place of the ellipsoid, aberration is present in the image. Following the prescription above, we find the difference Δz between these two surfaces. From Eq. (4.1.4) we get

$$\Delta z = z_e - z_s = K\rho^4/8R^3,$$

through the terms of interest. Therefore Eqs. (4.2.5) give

$$\text{ASA3} = K(\rho^3/R^3), \qquad \text{TSA3} = K(\rho^3/R^3)s', \qquad (4.2.6)$$

where the range of K is $-1 < K < 0$ for real conjugates. It is convenient to rewrite Eqs. (4.2.6) in terms of the magnification m. Substituting Eq. (2.3.3) into Eqs. (3.5.1) and (3.5.2) gives

$$\frac{s'}{R} = \frac{1-m}{2}, \qquad K = -\left(\frac{m+1}{m-1}\right)^2.$$

Therefore

$$\text{ASA3} = -\left(\frac{m+1}{m-1}\right)^2 \frac{\rho^3}{R^3}, \qquad (4.2.7a)$$

$$\text{TSA3} = +\frac{(m+1)^2}{m-1} \frac{\rho^3}{2R^2}. \qquad (4.2.7b)$$

These relations give the spherical aberration of a sphere used at magnification m. Note that $m < 0$ for real conjugates, hence TSA for a concave spherical mirror always has the same sign for a given ρ, independent of the sign of R. This sign is such that the focus for marginal rays, the rays reflected from the edge of the mirror, is closer to the vertex than the paraxial focus, as shown in Fig. 4.2. Note that the substitution of $m = 0$ into Eqs. (4.2.7) gives the same ASA3 and TSA3 as when $K = 0$ is substituted into Eqs. (4.2.3) and (4.2.4), as expected. As a final comment, note that the spherical aberration is zero when $m = -1$. For this magnification $s = s'$, and the sphere is the perfect surface according to Fermat's principle.

TSA as given by the relations above is a measure of the image size at the paraxial focus. The distribution of rays near the paraxial focus is such that the image has a minimum size between the paraxial focus and the focus for marginal rays. A cross section of the image near paraxial focus is shown in Fig. 4.5 for a spherical mirror with $m = 0$ and focal ratio $F = 2$. The paraxial focus is at the origin of the (ρ', z') coordinate frame. Each ray is drawn so that

$$\rho' = -\rho^3/2R^2, \qquad \text{at} \quad z' = 0, \qquad (4.2.8a)$$

$$z' = -\rho^2/4R, \qquad \text{at} \quad \rho' = 0, \qquad (4.2.8b)$$

where $\rho' = \text{TSA3}$ from Eq. (4.2.1) and $z' = f - R/2$ from Eq. (4.1.5). Note that the vertical scale in Fig. 4.5 is stretched relative to the horizontal scale.

4. Introduction to Aberrations

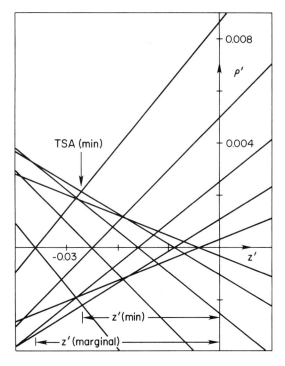

Fig. 4.5. Ray distribution near paraxial focus for image with spherical aberration. Paraxial focus is at $(0, 0)$. See Eq. (4.2.8) for definition of parameters.

It is easy to see in Fig. 4.5 that the image width at the minimum blur is four times smaller than TSA3 at the paraxial focus. The location of the minimum blur is given by $z'(\min) = 0.75 z'(\text{marginal})$. Analytical calculations such as those given by Welford support these graphical conclusions. The image at its minimum size is called the circle or disk of least confusion.

Using Eq. (4.2.8b) we can find ρ for those rays which cross the optical axis at $z'(\min)$. For such rays let $\rho = \rho'$ and let ρ_0 be the marginal ray coordinate at the mirror. Then

$$\frac{z'(\min)}{z'(\text{marginal})} = 0.75 = \frac{\rho'^2}{\rho_0^2}. \tag{4.2.9}$$

Solving Eq. (4.2.9) gives $\rho' = 0.866\rho_0$. We use this result in our analysis of a Schmidt camera in Section IV.

From our discussion we see that the diameter of the circle of least confusion is $|\rho^3/4R^2|$ when the object is at infinity. At the mirror this blur subtends an angle α where

$$\alpha = \rho^3/2R^3 = 1/128F^3. \tag{4.2.10}$$

As the focal ratio F increases, the subtended angle α decreases. A point is reached, however, where the image diameter no longer decreases but reaches the limit set by diffraction according to Eq. (3.6.1). The smallest F for which a spherical mirror used to image a distant object is approximately diffraction-limited is found by equating α in Eq. (4.2.10) to θ in Eq. (3.6.1). The result is

$$D \sim 128\lambda F^3,$$
$$D \sim 0.007 F^3, \quad \text{for} \quad \lambda = 550 \text{ nm}.$$

As examples we find $F \sim 11$ for $D = 10$ cm and $F \sim 24$ for $D = 1$ m. Thus, in spite of spherical aberration, a spherical mirror in collimated light is effectively diffraction-limited, provided the focal ratio is large enough.

An interesting exercise left to the reader is to take $D \sim 128\lambda F^3$, solve for λ, and substitute the result into Eq. (4.2.2). It turns out that Δz_0, the difference between a paraboloid and a "diffraction-limited" sphere at the edges, is approximately $\lambda/8$. Hence the path difference between two marginal rays to the two mirrors is $\sim \lambda/4$. Alternatively, the wavefront emerging from the spherical mirror is no longer spherical but differs from the spherical wavefront emerging from the paraboloid by $\lambda/4$ at the margin. Although this limit is found here in a special case, it turns out that this is a useful criterion for establishing when any optical system gives images that are approximately diffraction-limited.

III. OFF-AXIS ABERRATIONS

We now turn our attention to off-axis aberrations, those aberrations present when the object point does not lie on the optical axis. We consider here only a special case, a paraboloid in collimated light, reserving the general discussion for the next chapter.

Figure 4.6 shows a cross section of a paraboloid with optical axis z and vertex at 0. The image at B is, of course, a perfect one geometrically. The same paraboloid images a distant point object at angle θ from the z axis at point B', where the distance BB' is approximately $f\theta$.

To determine the kinds of aberrations present in the image at B', we first find a system that takes the rays at angle θ and forms a perfect image at B'. This system is obviously a paraboloid whose optical axis is parallel to the incident beam and passes through B', with its vertex at distance f from B'. The coordinate system for this paraboloid is shown in Fig. 4.6; the optical axis is denoted by z' and the vertex is at $0'$. We then find the distance between these two paraboloids and use Eqs. (4.2.5) to find the aberrations.

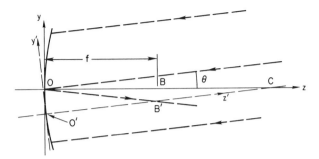

Fig. 4.6. Collimated beam at angle θ with optical axis incident on paraboloid. $0'$ is the origin of the rotated coordinate system; point C lies on both the z and z' axes.

Carrying out the task outlined in the previous paragraph is not a simple matter and, besides, the method used in Chapter 5 gives the results in a much more direct way. Thus we will only outline the procedure briefly and give a few significant results. First, the unprimed coordinate system is translated to put its origin at $0'$; then it is rotated through angle θ. The equation for the original parabola in this coordinate frame is then solved for z' in terms of y', R, and θ. Second, the equation for the parabola whose vertex is at $0'$ is solved for z. The difference between these values of z gives the Δz of interest, with the result of the form

$$2\Delta z = a_1 \frac{y^3 \theta}{R^2} + a_2 \frac{y^2 \theta^2}{R} + a_3 y \theta^3. \qquad (4.3.1)$$

The numerical coefficients of these terms are not of interest here because we are looking only at the difference between two parabolas, not two paraboloids. The importance of Eq. (4.3.1) is in the combination of y and θ in each term, and we proceed to find the angular aberration using Eqs. (4.2.5) with y replacing ρ. Letting AA represent angular aberration we get

$$\text{AA} = 3a_1 \frac{y^2 \theta}{R^2} + 2a_2 \frac{y \theta^2}{R} + a_3 \theta^3. \qquad (4.3.2)$$

The terms in Eq. (4.3.2) represent different aberrations: the first is called *coma*, the second is *astigmatism*, and the last is *distortion*. The character of each aberration is quite different because of the different way in which each depends on y and θ. Our following description of each aberration is limited to the yz plane and is, therefore, incomplete. A complete description, based on rays over the entire aperture, is given in Chapter 5.

Coma is proportional to $y^2 \theta$ and hence is changed in sign when θ changes sign. Coma is invariant to the sign of y and therefore rays from opposite sides of the mirror are on the same side of the central ray in the vicinity of the paraxial focus. Selected rays for an image in which coma is the only aberration are shown in Fig. 4.7a, from which we see that the image is asymmetric. Analysis using rays over the full aperture shows that a comatic image resembles a comet, hence the name, in which the paraxial focus is at the "head" of the comet.

Astigmatism is proportional to $y\theta^2$ and hence is unchanged by a sign change in θ. A change in the sign of y changes the sign of the astigmatism and therefore rays from opposite sides of the mirror are on opposite sides of the central ray near the paraxial focus. Selected rays for an image in which astigmatism is the only aberration are shown in Fig. 4.7b, from which we see that the image is laterally symmetric about the central ray. Although it appears from Fig. 4.7b that a good image can be obtained by shifting from B' to the point where the marginal rays intersect the central ray, analysis using marginal rays around the full aperture shows that this is not the case.

Distortion is proportional to θ^3 and does not depend on y. Thus this aberration, if it is the only one present, does not affect the image quality, only its position. For a set of point objects equally spaced perpendicular

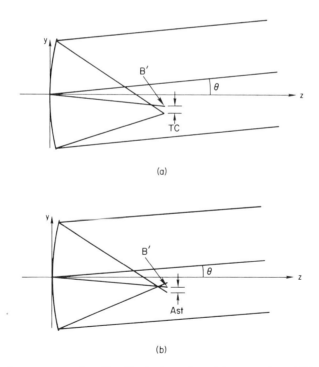

Fig. 4.7. Image cross sections for off-axis aberrations: (a) coma only and (b) astigmatism only. Dependence of angular aberrations on mirror parameters is given in Eq. (4.3.2).

to the optical axis, the set of images would not be equally spaced if distortion was present.

There is one final aberration that is present in Fig. 4.6, though it does not appear in Eq. (4.3.2), the aberration called *curvature of field*. Its character is most easily seen by noting that the transformation that takes the origin from 0 to 0' is essentially one of a rotation about the center of curvature C. Thus the motion of point B in Fig. 4.6 to B' is along a circular arc whose center is C. The foci for various θ, in the absence of other aberrations, are located on a curved surface, hence the name curvature of field.

In the special case considered here the radius of curvature of the image surface is the focal length f and the surface is concave as seen from point C. In Chapter 5 we show that the radius of this surface, called the Petzval surface, is independent of the object location and depends only on the lens or mirror surface radius of curvature and the indices of refraction on each side.

At this stage five aberrations have been identified: spherical aberration, coma, astigmatism, distortion, and curvature of field. The first of these is independent of field angle, but all others depend on some power of θ. The first three aberrations in this list affect image quality, while the last two affect only image position.

From Eqs. (4.2.7b) and (4.3.2) we see that the transverse aberrations depend on aperture radius y and field angle θ according to the relation

$$\text{aberration} \propto y^n \theta^m, \tag{4.3.3}$$

where $n + m = 3$. Hence each of these aberrations is called a third-order aberration. The main task in the analysis of the image quality in any optical system is to determine how much of each of these aberrations is present and to eliminate or reduce the amount of each by proper selection of system parameters.

IV. ABERRATION COMPENSATION

In Section 3.VI we noted that a perfect optical system is one in which the wavefront emerging from the final surface is spherical. From the discussion in this chapter it is evident that there is a close relation between deviations from a spherical wavefront and the appearance of aberrations. Along any ray the actual wavefront may be behind or in front of the ideal wavefront, depending on whether that portion of the wavefront has been retarded or advanced.

IV. Aberration Compensation

Although the analysis so far has been limited to aberrations of a single-surface optical system, it should be evident that compensation of aberrations, wholly or in part, should be possible in systems with more than one surface. In terms of Fermat's principle exact compensation means that a wavefront advance introduced by one or more surfaces is canceled by an equal wavefront retardation introduced by other surfaces. As far as the final wavefront is concerned, it is only the net advance or retardation that determines the size of any image defect.

In this section we examine two systems, a Cassegrain telescope and a Schmidt camera, for each of which the net spherical aberration is zero. Each system is composed of two optical elements chosen so that a wavefront advance due to one element is balanced by an equal retardation introduced by the other. The object point for each is a distant point source on the optical axis.

A. CASSEGRAIN TELESCOPE

The configuration of mirrors in a Cassegrain telescope is shown in Fig. 2.7a. Based on the discussion in Section 3.IV, one pair of mirrors for which the spherical aberration is zero is a paraboloidal primary and a hyperboloidal secondary. The former is the perfect mirror for a point object at infinity, while the latter is perfect for finite conjugate points on opposite sides of the mirror. The conic constants are $K_1 = -1$ for the primary and $K_2 = -(m+1)^2/(m-1)^2$ for the secondary, where $m = -s_2'/s_2$ in Fig. 2.7a. The relation for K_2 follows from Eq. (3.5.2) with the substitution $K = -e^2$. Choosing a value of $k = y_2/y_1$ sets the position of the secondary mirror relative to the primary. With the values of m and k chosen, the paraxial relations in Eq. (2.5.1) are used to find the values of ρ and β. If, for example, $m = 5$ and $k = 0.2$, then $\rho = 0.25$, $\beta = 0.2$, and $K_2 = -2.25$. The telescope specification is completed when the primary diameter and focal length are chosen.

The paraboloid–hyperboloid combination is called a *classical Cassegrain*. We now show how this configuration can be changed into a different one by changing the conic constants of both the primary and secondary mirrors. This is done in a way that keeps the net spherical aberration zero, hence a change in K_1 is accompanied by a change in K_2 such that the wavefront advance at one mirror is equal to the wavefront retardation at the other.

Starting with the classical Cassegrain configuration in Fig. 4.8, each surface is changed into a different conic by "bending" the original surface. If the new surfaces lie to the left of the original surfaces, as shown in Fig. 4.8, then the wavefront has been advanced at the primary and retarded at

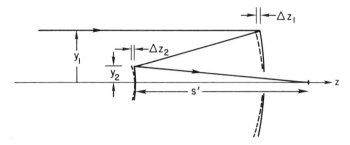

Fig. 4.8. Cassegrain telescope schematic. Classical Cassegrain has mirrors shown by solid curves; modified Cassegrain has mirrors shown by dashed curves. R_1, R_2, m, and k are the same for both telescopes. Surface differences are given in Eqs. (4.4.1).

the secondary. The advance and retardation are $2\Delta z_1$ and $2\Delta z_2$, where Δz_1 and Δz_2 are the surface differences at the primary and secondary, respectively. Using Eq. (4.1.4), each surface has z as follows:

$$z_1(\text{original}) = \frac{y_1^2}{2R_1},$$

$$z_1(\text{new}) = \frac{y_1^2}{2R_1} + (1+K_1)\frac{y_1^4}{8R_1^3},$$

$$z_2(\text{original}) = \frac{y_2^2}{2R_2} + \left[1 - \left(\frac{m+1}{m-1}\right)^2\right]\frac{y_2^4}{8R_2^3},$$

$$z_2(\text{new}) = \frac{y_2^2}{2R_2} + (1+K_2)\frac{y_2^4}{8R_2^3}.$$

Therefore

$$2\Delta z_1 = -(1+K_1)\frac{y_1^4}{4R_1^3}, \tag{4.4.1a}$$

$$2\Delta z_2 = -\left[K_2 + \left(\frac{m+1}{m-1}\right)^2\right]\frac{y_2^4}{4R_2^3}, \tag{4.4.1b}$$

where R_1 and R_2 are held constant. Applying the condition that the advance equals the retardation requires $2\Delta z_1 = 2\Delta z_2$. Note that this is equivalent to stating that the optical distance from object to image is unchanged, hence Fermat's principle is satisfied for all rays from a distant point source. Applying this condition gives

$$K_1 + 1 = \frac{y_2^4}{y_1^4}\frac{R_1^3}{R_2^3}\left[K_2 + \left(\frac{m+1}{m-1}\right)^2\right] = \frac{k^4}{\rho^3}\left[K_2 + \left(\frac{m+1}{m-1}\right)^2\right]. \tag{4.4.2}$$

As an example consider the paraxial parameters for the classical Cassegrain and choose $K_2 = 0$. From Eq. (4.4.2) we get $K_1 = -0.7696$. This combination of an ellipsoidal primary and a spherical secondary is called a *Dall–Kirkham* configuration.

The solutions of Eq. (4.4.2) represent the family of Cassegrain telescopes for which spherical aberration of a distant point source is zero. For a given set of k, m, and p there is an infinity of combinations of K_1 and K_2 that satisfy Eq. (4.4.2). In practice, the choice of K_1 and K_2 from this set depends on other considerations, such as off-axis aberrations and the ease with which the mirrors can be made and tested. In the case of a Dall–Kirkham, for example, the separate mirrors are easily tested but large coma results in a small usable field. Discussion of all of the aberrations of Cassegrain and other two-mirror telescopes and comments on testing are given in Chapter 6.

B. SCHMIDT CAMERA

A Schmidt camera is composed of three elements: a concave spherical mirror, an aperture stop whose center is located at the center of curvature of the mirror, and a refracting plate in the plane of the stop, as shown schematically in Fig. 4.9. For the moment ignore the refracting plate and consider only the mirror and stop. Placement of the stop at the center of curvature gives a system that is effectively axis-free. Rays through the stop from an off-axis object point "see" an optical system, a portion of the spherical mirror, which is the same as that for rays from an on-axis point. In effect, any line through C from an off-axis point is equivalent to the z-axis. This axis-free character is, of course, true only for the sphere because of its constant curvature. Therefore the aberrations of an image of any object point, for this arrangement of mirror and stop, are just those of an on-axis image, namely spherical aberration. Because of the symmetry about point C the image surface is spherical and curvature of field is present but, as noted above, this aberration does not affect image quality.

Because this very simple optical system is free of astigmatism and coma, it is the basis for cameras and telescopes that are designed for wide-field

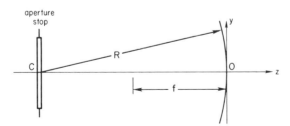

Fig. 4.9. Schmidt camera configuration. Center of curvature C of spherical mirror is at center of aperture stop. Surface figure on glass plate at stop is given by Eq. (4.4.4).

applications. Note that freedom from these aberrations is true for objects at any point to the left of the stop in Fig. 4.9.

The remaining optical element, the refracting plate or corrector, serves the function of correcting the spherical aberration due to the mirror, where the wavefront advance at the mirror is compensated by an equal wavefront retardation by the corrector. To find the required wavefront advance we take a distant object point, the only case considered here, and find Δz between the sphere and a reference paraboloid. The latter surface is, of course, the surface that would give a perfect on-axis image. From Eq. (4.2.2) we find for $K = 0$ that

$$2\Delta z = -\rho^4/4R^3. \qquad (4.4.3)$$

Consider a plane-parallel plate of thickness t and index n. At any height y near one surface of this plate we remove a layer of air of thickness τ and replace it with a layer of glass of optical thickness $n\tau$. The net change in optical path due to this change is $(n-1)\tau$ for a ray parallel to the z-axis. Because the light is "slowed down" in the glass, this optical path difference is the required retardation and

$$(n-1)\tau = 2\Delta z = -\rho^4/4R^3. \qquad (4.4.4)$$

Defining $\eta = \rho/\rho_0$ where ρ_0 is the radius of the aperture stop, and noting that $f = -R/2$ gives

$$\tau = \frac{\eta^4 \rho_0^4}{32(n-1)f^3} = \frac{f\eta^4}{512(n-1)F^4}. \qquad (4.4.5)$$

For an otherwise flat plate Eq. (4.4.5) defines the surface figure on one face required to correct the spherical aberration of the mirror. From the point of view of Fermat's principle it does not matter whether the figured surface faces the mirror or the incident light. In either orientation rays at the edge of the aperture are retarded relative to those near the axis. Note also that Eq. (4.4.5) applies only to the case where the perfect reference surface is a paraboloid. For objects whose distance is not effectively at infinity the appropriate reference surface is an ellipsoid, but the procedure for finding τ is the same.

We now consider how the corrector plate has effected the compensation of the spherical aberration of the mirror. Rays through the corrector near its center are essentially undeviated and hence are focused at the paraxial focal point of the mirror. Rays farther out on the corrector are deviated away from the z-axis because in cross section the corrector is a thin prism. If the effective prism angle is α, as shown in Fig. 4.10, the ray deviation in the paraxial approximation is $(n-1)\alpha$. Because of this deviation both the point at which the ray strikes the mirror and its angle of incidence are

IV. Aberration Compensation

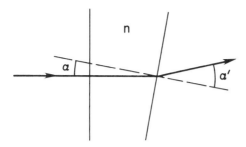

Fig. 4.10. Small section of corrector plate near edge, in cross section. Net ray deviation is $(n-1)\alpha$.

changed, and the reflected ray is directed toward the paraxial focus. From the center of the corrector outward the angle α increases as ρ^4 and is a maximum at the edge; thus the marginal rays are deviated by the largest amount.

If the corrector had a constant index of refraction it would affect rays of any wavelength in the same way but, of course, this is not the case. Because n is not constant the deviation is also a function of wavelength. Denoting the deviation by δ, a simple differentiation gives

$$d\delta/d\lambda = \alpha dn/d\lambda. \qquad (4.4.6)$$

Thus rays of different wavelength are directed in slightly different directions with the effect largest at the edge of the corrector. A system corrected at one wavelength is no longer corrected at other wavelengths and the image now has the aberration called *chromatic spherical aberration*. This image defect is always present when the corrector is a single element, but it can be minimized by selecting a different focal point for the system.

Looking at Fig. 4.5 we see that the blur is least at the circle of least confusion. At this distance from the mirror neither the paraxial rays nor the marginal ones intersect the z axis but, as given in Eq. (4.2.9), the rays from the zone at $0.866\rho_0$ are in focus on the axis. It is also evident from Fig. 4.5 that the maximum deviation necessary to bring all rays to this same focus is less than that required to bring the marginal rays to the paraxial focus. The reference surface needed to minimize the overall ray deviation is a paraboloid whose focal point is at the circle of least confusion in Fig. 4.5, hence has a radius of curvature different from the one used to derive Eq. (4.4.3).

The required change in R is found by substituting Eq. (4.2.9) into Eq. (4.1.6) to get the change in focal length needed to move the focus to the circle of least confusion. For $K = 0$ the result is $\Delta f = -3\rho_0^2/16R$, and

therefore
$$R' - R = 2\Delta f = -3\rho_0^2/8R, \quad (4.4.7)$$

where R' is the radius of curvature of the modified paraboloid. Substituting Eq. (4.4.7) into Eq. (4.1.4) gives

$$z' - z = \frac{\rho^2}{2}\left(\frac{1}{R'} - \frac{1}{R}\right) = \frac{3\rho_0^2\rho^2}{16R^3}. \quad (4.4.8)$$

With reference to the new paraboloid the wavefront advance at the spherical mirror is

$$2\Delta z = \frac{3\rho_0^2\rho^2}{8R^3} - \frac{\rho^4}{4R^3}. \quad (4.4.9)$$

Equating the wavefront advance and retardation gives the surface figure as

$$\tau = \frac{3\rho_0^2\rho^2}{8(n-1)R^3} - \frac{\rho^4}{4(n-1)R^3} = \frac{f\eta^4}{512(n-1)F^4}\left(1 - \frac{3}{2\eta^2}\right), \quad (4.4.10)$$

where $\tau < 0$ over the entire corrector aperture. In contrast to the corrector whose profile is given by Eq. (4.4.5), this corrector is thickest at its center. Comparing Eq. (4.4.10) to (4.4.5) we also see that an additional term has been introduced into the surface figure, one that amounts to including a radius term. Rewriting Eq. (4.4.10) we get

$$\tau = \frac{\rho^2}{2R_c} - \frac{\rho^4}{4(n-1)R^3}, \quad (4.4.11)$$

where R_c, the radius of curvature of the modified corrector, is

$$R_c = [4(n-1)R^3]/3\rho_0^2. \quad (4.4.12)$$

Throughout the analysis leading to Eq. (4.4.11), we carefully followed the sign conventions established in Chapter 2. From the diagram in Fig. 4.9 we see that $R < 0$, hence $R_c < 0$ as well. In absolute terms τ has its largest value when $\rho = 0.866\rho_0$, as is easily verified by setting $d\tau/d\rho = 0$ and solving for ρ. The rays for which the deviation is a maximum are those for which $d\tau/d\rho$ is largest in an absolute sense, which occurs at $\eta = 0.5$ and $\eta = 1$. The shapes of the corrector profile and the emerging wavefront, much exaggerated, are shown in Fig. 4.11.

As a final item for the Schmidt camera we calculate the chromatic TSA present when the index n differs from the one used in the design profile. The starting point is the wavefront retardation of the corrector, $(n'-1)\tau$, where n' is the variable. The change in retardation as a function of a change in index is $\tau\delta n$, where $\delta n = n' - n$ and n is the design index. Equivalently, this is the optical path difference for index n'.

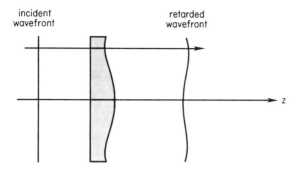

Fig. 4.11. Profiles, greatly exaggerated, of corrector and emerging wavefront. Ray shown at height $\sqrt{3}/2$ of full aperture is undeviated.

Substitution of the OPD for $2\Delta z$ in Eqs. (4.2.5) and setting $s' = f$ gives the chromatic aberration. The result is

$$\text{TSA} = f\frac{d}{d\rho}(\tau\delta n) = f\,\delta n\frac{d\tau}{d\rho} = \frac{f\eta^3\,\delta n}{64(n-1)F^3}\left(1 - \frac{3}{4\eta^2}\right). \quad (4.4.13)$$

Putting in $\eta = 0.5$ or 1 gives the chromatic TSA for the rays that have the maximum deviation or largest effective prism angle. The two values of η give the same TSA value in magnitude and the result in absolute terms is

$$\text{TSA} = \frac{f}{256F^3}\frac{\delta n}{n-1}, \quad (4.4.14)$$

which is effectively the radius of the chromatic image.

The prescription of the corrector profile required to correct the spherical aberration of a spherical mirror used in collimated light is correct through fourth-order terms in ρ. As noted in Eq. (4.2.1) there are higher-order terms in the expression for spherical aberration and Eq. (4.4.11) can be extended to eliminate their presence. Such terms are significant for cameras of small focal ratio, but these terms are not considered here. For details on aberrations of higher order, the reader should consult the books by Bouwers and Linfoot.

The introduction of the refracting corrector element gives an optical system with excellent image quality over a large field, with chromatic aberration setting the limit on the image quality. It should also be noted that the corrector does have an axis and therefore the camera is no longer axis-free. As a result there will be off-axis aberrations, though these aberrations are relatively small because the corrector is nearly a plane-parallel plate and is generally quite thin. For details on the magnitude of these off-axis effects the reader should consult an excellent article by Bowen. This article is also of interest in showing how the aberrations of a Schmidt camera can be calculated without recourse to Fermat's principle.

The two systems treated in this section have only been examined in part. Our intent here has been to use Fermat's principle as a tool to facilitate the analysis of optical systems and to demonstrate its power in the process, at least for on-axis aberrations. The full capability of this tool, including analysis of off-axis aberrations, will be available after the treatment in the following chapter.

REFERENCES

Bowen, I. (1960). "Stars and Stellar Systems I, Telescopes," Chap. 4. Univ. of Chicago Press, Chicago, Illinois.
Welford, W. (1974). "Aberrations of the Symmetrical Optical System," Chap. 6. Academic Press, New York.

BIBLIOGRAPHY

For an introduction to aberrations see the intermediate-level texts in optics listed in the bibliography in Chapter 2.

For a discussion of ray-tracing and aberrations in a form suitable for constructing a computer program
"Optical Design" (1962). MIL-HDBK-141. U.S. Government Printing Office, Washington, D.C.

Telescopes
Bowen, I. (1967). "Annual Review of Astronomy and Astrophysics," Vol. 5, p. 45. Annual Reviews, Palo Alto, California.
Gascoigne, S. (1973). *Appl. Opt.* **12**, 1419.
Learner, R. (1981). "Astronomy Through the Telescope." Van Nostrand-Reinhold, Princeton, New Jersey.
Meinel, A. (1960). "Stars and Stellar Systems I, Telescopes," Chap. 3. Univ. of Chicago Press, Chicago, Illinois.
Meinel, A. (1976). "Applied Optics and Optical Engineering," Vol. 5, Chap. 6. Academic Press, New York.
Wetherell, W., and Rimmer, M. (1972). *Appl. Opt.* **11**, 2817.

Schmidt cameras
Bouwers, A. (1946). "Achievements in Optics," Chap. 1. Elsevier, Amsterdam.
Bowen, I. (1960). "Stars and Stellar Systems I, Telescopes," Chap. 4. Univ. of Chicago Press, Chicago, Illinois.
Linfoot, E. (1955). "Recent Advances in Optics," Chap. 4. Oxford Univ. Press, London and New York.

Chapter 5 | Fermat's Principle and Aberrations

At this point the stage is set for a general application of Fermat's principle to a surface of revolution and the derivation of its aberrations. The theory of aberrations, generally called the Seidel theory, is a classical subject and has been treated in detail by many authors, including Born and Wolf, and Welford. An excellent introduction to the theory of aberrations is given by Longhurst. The treatment here leads to nothing new, but the approach is one which leads to results that can easily be applied to optical systems of specific interest to astronomers, such as telescopes, cameras, and spectrometers. Rather than simply citing results derived from the Seidel theory, we start with Fermat's principle and derive the desired relations in a systematic way. These aberration relations are then reduced to specific forms appropriate to given surface types, such as conic mirrors, spherical refracting surfaces, and aspheric plates as used in Schmidt cameras and telescopes.

I. APPLICATION TO SURFACE OF REVOLUTION

A sketch of a general surface of revolution about the z axis is shown in Fig. 5.1, with the origin of the coordinate system at the vertex of the surface. The homogeneous medium to the left of the surface has index n; the medium to the right has index n'. The object and image points are at Q and Q', respectively, and an arbitrary ray from Q intersects the surface at $B(x, y, z)$.

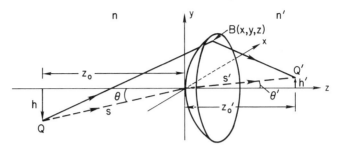

Fig. 5.1. Path of arbitrary ray through refracting surface. Points Q and Q' lie in the yz plane; point B is on the surface. The chief ray passes through the origin of the coordinate system.

Because the surface is symmetric about the z axis, there is no loss of generality in placing Q and Q' in the yz plane. The surface equation is a generalized form of Eq. (4.1.4) through fourth-power terms of ρ as follows:

$$\begin{aligned} z &= \frac{\rho^2}{2R} + (1+K)\frac{\rho^4}{8R^3} + \frac{b\rho^4}{8(n'-n)} \\ &= \frac{\rho^2}{2R} + \frac{\rho^4}{8}\left[\frac{1+K}{R^3} + \frac{b}{n'-n}\right] = \frac{\rho^2}{2R} + \frac{\alpha\rho^4}{8}. \end{aligned} \quad (5.1.1)$$

where α represents the quantity in brackets, and $\rho^2 = x^2 + y^2$. The term in b explicitly includes the type of aspheric term required for corrector plates, one example of which is discussed in Section 4.IV. The form of the b term in Eq. (5.1.1) is chosen to simplify its appearance in the aberration relations. In Fig. 5.1, the center of the aperture stop is located at the origin of the coordinate system. The case for an aperture stop displaced along the z axis is considered later in this chapter.

Applying Eq. (3.1.2) to the ray shown in Fig. 5.1 gives

$$\text{OPL} = n[QB] + n'[BQ'], \quad (5.1.2)$$

where $[QB]$ and $[BQ']$ are the segments of the ray to the left and right of point B, respectively. From the geometry in Fig. 5.1 we have

$$\begin{aligned} [QB] &= [(y-h)^2 + (z_0-z)^2 + x^2]^{1/2} \\ [BQ'] &= [(y-h')^2 + (z_0'-z)^2 + x^2]^{1/2}. \end{aligned} \quad (5.1.3)$$

The dashed line in Fig. 5.1 from Q to Q' passes through the center of the aperture stop and intersects the surface at its vertex. This line represents the chief ray through the system with angles θ and θ' with respect to the z axis. The OPL measured along the chief ray is simply $-ns + n's'$, where s is negative and s' is positive by the sign convention.

I. Application to Surface of Revolution

The signs of other quantities in Fig. 5.1 are as follows: θ, θ', h', and z_0' are positive, and h and z_0 are negative. With due regard for these signs, we have

$$h = s \sin \theta, \qquad z_0 = s \cos \theta, \qquad s^2 = z_0^2 + h^2,$$
$$h' = s' \sin \theta', \qquad z_0' = s' \cos \theta', \qquad s'^2 = z_0'^2 + h'^2. \qquad (5.1.4)$$

Substituting the relations in the first line of Eq. (5.1.4) into $[QB]$ gives

$$[QB] = -s\left[1 - \frac{2y}{s}\sin\theta + \frac{\rho^2}{s}\left(\frac{1}{s} - \frac{\cos\theta}{R}\right) + \frac{\rho^4}{4s^2}\left(\frac{1}{R^2} - \alpha s \cos\theta\right)\right]^{1/2}.$$

The relation for $[BQ']$ is similar in form except that θ' replaces θ, s' replaces s, and the leading minus sign is dropped. The expression for $[QB]$ is now transformed by applying the binomial expansion and retaining all terms through fourth order, with the result

$$[QB] = -\left\{ s - y\sin\theta + \frac{y^2}{2}\left(\frac{\cos^2\theta}{s} - \frac{\cos\theta}{R}\right) + \frac{x^2}{2}\left(\frac{1}{s} - \frac{\cos\theta}{R}\right) \right.$$
$$+ \frac{x^2 y}{2s}\sin\theta\left(\frac{1}{s} - \frac{\cos\theta}{R}\right) + \frac{y^3}{2s}\sin\theta\left(\frac{\cos^2\theta}{s} - \frac{\cos\theta}{R}\right)$$
$$\left. + \frac{\rho^4}{8}\left[\frac{1}{R^2}\left(\frac{1}{s} - \frac{(1+K)\cos\theta}{R}\right) - \frac{1}{s}\left(\frac{1}{s} - \frac{\cos\theta}{R}\right)^2 - \frac{b\cos\theta}{n'-n}\right]\right\}.$$

A similar relation follows for $[BQ']$ once the changes noted above are made. Substitution of these relations for $[QB]$ and $[BQ']$ into Eq. (5.1.2) then gives the OPL for this general ray, as follows:

$$\text{OPL} = (-ns + n's') - y(n'\sin\theta' - n\sin\theta)$$
$$+ \frac{y^2}{2}\left[\frac{n'\cos^2\theta'}{s'} - \frac{n\cos^2\theta}{s} - \frac{n'\cos\theta' - n\cos\theta}{R}\right]$$
$$+ \frac{x^2}{2}\left[\frac{n'}{s'} - \frac{n}{s} - \frac{n'\cos\theta' - n\cos\theta}{R}\right]$$
$$- \frac{x^2 y}{2}\left[\frac{n\sin\theta}{s}\left(\frac{1}{s} - \frac{\cos\theta}{R}\right) - \frac{n'\sin\theta'}{s'}\left(\frac{1}{s'} - \frac{\cos\theta'}{R}\right)\right]$$
$$- \frac{y^3}{2}\left[\frac{n\sin\theta}{s}\left(\frac{\cos^2\theta}{s} - \frac{\cos\theta}{R}\right) - \frac{n'\sin\theta'}{s'}\left(\frac{\cos^2\theta'}{s'} - \frac{\cos\theta'}{R}\right)\right]$$
$$+ \frac{\rho^4}{8}\left[\frac{1}{R^2}\left(\frac{n'}{s'} - \frac{n}{s} - \frac{(1+K)}{R}\right)(n'\cos\theta' - n\cos\theta) + \frac{n}{s}\left(\frac{1}{s} - \frac{\cos\theta}{R}\right)^2\right.$$
$$\left. - \frac{n'}{s'}\left(\frac{1}{s'} - \frac{\cos\theta'}{R}\right)^2 - \frac{b}{n'-n}(n'\cos\theta' - n\cos\theta)\right]. \qquad (5.1.5)$$

Although Eq. (5.1.5) is a formidable equation in appearance, the application of Fermat's principle simplifies it considerably. We begin by noting that the first set of parentheses denotes the OPL for the chief ray. Because Fermat's principle is concerned with optical path differences and stationary values, as given in Eq. (3.1.1), it is appropriate to remove this term by defining G as the OPD between the general ray and the chief ray. Given this definition, we have

$$G = \text{OPL} - \text{OPL(chief ray)}$$
$$= A_0 y + A_1 y^2 + A_1' x^2 + A_2 y^3 + A_2' x^2 y + A_3 \rho^4, \qquad (5.1.6)$$

where the A_i's are the multiplying factors in Eq. (5.1.5). Applying Fermat's principle in the form $\delta(\text{OPL}) = 0$ to Eq. (5.1.6) gives

$$\frac{\partial}{\partial x}(\text{OPL}) = \frac{\partial G}{\partial x} = 0, \qquad \frac{\partial}{\partial y}(\text{OPL}) = \frac{\partial G}{\partial y} = 0. \qquad (5.1.7)$$

Equation (5.1.7) is satisfied for $x = y = 0$ only if $A_0 = 0$, hence $n' \sin \theta' = n \sin \theta$, which is simply Snell's law for the chief ray.

Returning to Eq. (5.1.6) we see that one or the other of the terms proportional to the square of the distance from the surface vertex can be made zero by a proper choice of s'. The term in y^2 is zero if

$$\frac{n' \cos^2 \theta'}{s_t'} - \frac{n \cos^2 \theta}{s} = \frac{n' \cos \theta' - n \cos \theta}{R}, \qquad (5.1.8)$$

where s_t' is the location of the *tangential* astigmatic image. As we will see, this image is a line image oriented perpendicular to the plane defined by the z axis and the chief ray.

Alternatively the term in x^2 is zero if

$$\frac{n'}{s_s'} - \frac{n}{s} = \frac{n' \cos \theta' - n \cos \theta}{R}, \qquad (5.1.9)$$

where s_s' is the location of the *sagittal* astigmatic image. This image is also a line, but lying in the plane containing the z axis and the chief ray. A sketch of these images as defined by selected rays is shown in Fig. 5.2. Note that $s_t' = s_s'$ in the paraxial approximation where $\cos \theta = \cos \theta' = 1$, and both Eqs. (5.1.8) and (5.1.9) reduce to Eq. (2.2.2), as expected.

The separation between the two astigmatic images is found by solving Eqs. (5.1.8) and (5.1.9) for $1/s_t'$ and $1/s_s'$, respectively, and taking the difference between the two expressions. The result is

$$\frac{\Delta s'}{s_s' s_t'} = \frac{\tan^2 \theta'}{n'} \left[\frac{n' \cos \theta' - n \cos \theta}{R} + \frac{n}{s}\left(1 - \frac{n'^2}{n^2}\right) \right],$$

where $\Delta s' = s_s' - s_t'$. To terms through θ^2 this expression reduces to

$$\frac{\Delta s'}{s'^2} = \frac{n^2 \theta^2}{n'}\left(\frac{1}{n's'} - \frac{1}{ns}\right). \qquad (5.1.10)$$

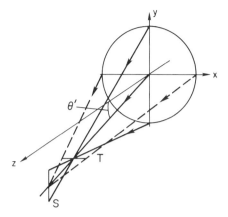

Fig. 5.2. Location and orientation of astigmatic line images. T and S denote tangential and sagittal images, respectively.

In Eq. (5.1.10) s' differs little from either s'_s or s'_t if $\Delta s'$ is small. Note that the separation of the astigmatic images, and thus also the lengths of the line images, is proportional to θ^2. Derivation of the image length follows in the next section.

At this point our analysis of Eq. (5.1.6) by means of Fermat's principle has given Snell's law and the locations of the astigmatic images. In the next section we use these results to evaluate the remaining coefficients in Eq. (5.1.6), which, if nonzero, determine the magnitude and type of aberration present in an image.

II. EVALUATION OF ABERRATION COEFFICIENTS

For an optical system that satisfies Eq. (5.1.7) for any (x, y) within the aperture, each of the coefficients in Eq. (5.1.6) must be zero and the system is perfect. If one or more of these coefficients is nonzero, then aberrations are present. Not surprisingly, the size of a given aberration is directly proportional to the corresponding coefficient in Eq. (5.1.6).

Before we evaluate each coefficient it is important to note that this analysis is limited to finding third-order angular and transverse aberrations only. For the terms in Eq. (5.1.6) this means retaining only those terms for which the sum of powers of θ and ρ, x, or y is not greater than four. Because $A_3\rho^4$ already has a fourth-power factor in ρ, each of the cosine factors can be replaced by one, with the result that A_3 is independent of θ. Following this procedure we see that $A_2 = A_2'$ with each proportional to θ, and A_1 and A_1' are each proportional to θ^2.

Using Snell's law and Eq. (2.2.2) to simplify the terms in Eq. (5.1.5) we get

$$A_3 = -\frac{1}{8}\left[\frac{K}{R^3}(n'-n) + b\right] - \frac{n^2}{8}\left(\frac{1}{s} - \frac{1}{R}\right)^2\left(\frac{1}{n's'} - \frac{1}{ns}\right), \quad (5.2.1)$$

$$A_2 = \theta\frac{n^2}{2}\left(\frac{1}{s} - \frac{1}{R}\right)\left(\frac{1}{n's'} - \frac{1}{ns}\right), \quad (5.2.2)$$

where the first term in brackets in Eq. (5.2.1) represents the contribution from the nonspherical part of the surface. Note that there is no term in Eq. (5.2.2) involving K or b and thus any nonspherical surface component does not contribute to the aberration associated with this coefficient, provided the aperture stop is at the surface. We will see later that this statement about Eq. (5.2.2) is not true when the aperture stop is displaced from the surface.

The evaluation of the remaining coefficients, A_1 and A_1', depends on the image distance chosen. For example, choosing $s' = s_s'$ makes $A_1' = 0$, and the coefficient A_1 is evaluated by substituting $s' = s_s'$ into A_1 in Eq. (5.1.5). The result to second order in θ is

$$A_1 = -\theta^2\frac{n^2}{2}\left(\frac{1}{n's'} - \frac{1}{ns}\right). \quad (5.2.3)$$

If, on the other hand, the choice were $s' = s_t'$, then $A_1 = 0$ and

$$A_1' = \theta^2\frac{n^2}{2}\left(\frac{1}{n's'} - \frac{1}{ns}\right), \quad (5.2.4)$$

hence $A_1 = -A_1'$. In either case terms involving K or b are absent but, as with A_2, they will be present when the aperture stop does not coincide with the surface.

The difference in sign between A_1 and A_1' is a measure of the differences between the marginal rays at the ends of each of the line images. As seen in Fig. 5.2, the marginal rays in the yz plane intersect the chief ray before reaching the sagittal image, while the marginal rays in the xz plane reach the chief ray after passing through the tangential image. In terms of transverse aberrations at the two images, the magnitudes are the same but the signs are opposite.

Although the details are not given here, it is worth noting that choosing s' as the midpoint between the line images leads to the result that $A_1 = -A_1'$ with each one-half as large as the values given in Eqs. (5.2.3) and (5.2.4). A glance at Fig. 5.2. shows that this result is expected. The image blur at this position is circular in cross section.

Because of the close relationship between A_1 and A'_1, it does not matter which s' is chosen to characterize the astigmatism present in an image. Our choice is $s' = s'_s$ or $A'_1 = 0$, hence Eq. (5.2.3) is the relation used in subsequent discussions.

The direct way in which A_1 is a measure of the astigmatism is seen in a comparison of Eqs. (5.2.3) and (5.1.10), from which it follows that

$$\Delta s'/s'^2 = -2A_1/n'. \tag{5.2.5}$$

Using Eq. (5.2.5) it is a simple matter to derive an expression for the transverse astigmatism at the sagittal image. Defining the transverse astigmatism (abbreviated TAS) as one-half the length of the line image, we find from the geometry of Fig. 5.2 that

$$\text{TAS} = -(\Delta s'/s')y = 2A_1 ys'/n', \tag{5.2.6}$$

where TAS < 0 at the sagittal image in Fig. 5.2 when $y > 0$, as required by the sign convention. The diameter of the astigmatic blur circle midway between the line images is $|\text{TAS}|$.

At this point let us summarize our findings. We have relations for the aberration coefficients and, in the case of A_1, have its relationship to a transverse aberration. The next step is to find the connection between A_2 and A_3 and their respective transverse aberrations. This is done by first establishing the connection between nonzero terms in Eq. (5.1.6) and deviations of the wavefront converging on the image point from the spherical shape produced by a perfect system.

III. RAY AND WAVEFRONT ABERRATIONS

An optical system free of aberrations takes light from an object point Q, for which the wavefront is a sphere with center at Q, and images it at the Gaussian image point Q'. The wavefront of the light converging toward Q' is a sphere whose center is at Q', and the OPL along any ray through the system is constant. Thus G in Eq. (5.1.6) is zero. This spherical wavefront is taken as our reference and designated Σ_r.

For a system with aberrations the wavefront converging toward Q' is no longer spherical and, depending on the sign of G, is either advanced or retarded at each point on the wavefront. A schematic cross section of an aberrated wavefront, designated Σ_a, is shown in Fig. 5.3, with it and the ideal wavefront Σ_r in contact at their centers where $G = 0$. At any other point on the actual wavefront G is the OPD between Σ_r and Σ_a. The geometric distance along any ray between the wavefronts is G/n' and

5. Fermat's Principle and Aberrations

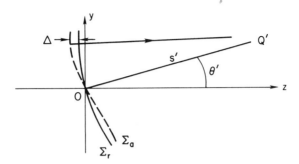

Fig. 5.3. Cross sections of reference and aberrated wavefronts, Σ_r and Σ_a, respectively. Radius of curvature of the reference wavefront is s'.

designating this distance as Δ we have

$$\Delta = \frac{G(x, y)}{n'} = \frac{1}{n'}[\Sigma_r(x, y) - \Sigma_a(x, y)]. \tag{5.3.1}$$

For the situation shown in Fig. 5.3 we have $n' > 0$, hence Δ and G have the same signs. When $\Delta > 0$ the actual wavefront is retarded with respect to the reference wavefront; the actual wavefront is advanced when $\Delta < 0$.

Differentiating Eq. (5.3.1) gives

$$\frac{\partial \Delta}{\partial y} = \frac{1}{n'}\frac{\partial G}{\partial y} = \frac{1}{n'}\left(\frac{\partial \Sigma_r}{\partial y} - \frac{\partial \Sigma_a}{\partial y}\right), \tag{5.3.2}$$

with a similar relation in which x replaces y. The quantity in parentheses in Eq. (5.3.2) is the difference in slopes between the reference and aberrated wavefronts in a slice parallel to the yz plane. Because rays are perpendicular to wavefronts, this is also the difference between the slopes of the ray for a perfect system and the actual ray, each at point (x, y) on the respective wavefronts. Given this difference in slopes, there is a consequent transverse aberration in the y direction at the image, as shown in Fig. 5.4. A similar result follows in the x direction from Eq. (5.3.2) with x in place of y. From the geometry in Fig. 5.4 we get

$$TA_y = s'\frac{\partial \Delta}{\partial y} = \frac{s'}{n'}\frac{\partial G}{\partial y}, \quad TA_x = s'\frac{\partial \Delta}{\partial x} = \frac{s'}{n'}\frac{\partial G}{\partial x}, \tag{5.3.3}$$

where the subscripts x and y on TA denote transverse aberrations in the x- and y-direction, respectively.

Substituting Eq. (5.1.6) into Eqs. (5.3.3) gives

$$TA_y = \frac{s'}{n'}[2A_1 y + A_2(x^2 + 3y^2) + 4A_3 y\rho^2], \tag{5.3.4}$$

III. Ray and Wavefront Aberrations

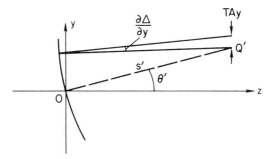

Fig. 5.4. Geometry of the wavefront slope difference and transverse aberration in the image. See Eq. (5.3.3).

$$TA_x = \frac{s'}{n'}[2A_1'x + 2A_2xy + 4A_3x\rho^2]. \quad (5.3.5)$$

We have now established the connection between the aberration coefficients and the geometric transverse aberrations. The corresponding angular aberrations, denoted by AA, are given by

$$TA_y = s'AA_y, \qquad TA_x = s'AA_x. \quad (5.3.6)$$

We could proceed by discussing each of the factors in Eqs. (5.3.4) and (5.3.5) in detail, but for our purposes that is not necessary. Instead the reader is referred to any of the references cited at the beginning of this chapter for discussions of both ray and wavefront aberrations. We simply note here that A_1 is a measure of astigmatism, as discussed in the previous section, and A_3 a measure of spherical aberration, with the discussion of it in Section 4.II sufficient for our needs.

The remaining coefficient, A_2, is a measure of coma. A sketch showing the asymmetric form of this aberration is given in Fig. 5.5. Note that the marginal rays on the y axis meet at a point three times farther from the Gaussian image than the corresponding point for the marginal rays on the x axis. The source of this difference between the tangential and sagittal coma is evident by inspection of Eq. (5.3.4). In discussions that follow, coma is specified by giving the sagittal coma.

The distribution of light rays over the comatic image is not uniform, there being a much greater density of rays near the point of the comatic image. About 80% of the energy is within a distance equal to the transverse sagittal coma from the Gaussian focus. Unlike the case of spherical aberration, a shift along the chief ray does not improve the image quality. For more details on the nature of a comatic image, the reader should consult Born and Wolf, and Welford.

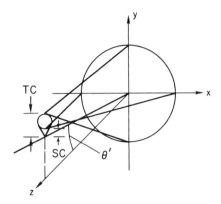

Fig. 5.5. Sketch of comatic image profile. TC and SC are tangential and sagittal coma, respectively.

Following our discussion in Section 4.III, we might have expected a term in Eq. (5.3.4) indicating the presence of distortion. This term is not present because we set $n' \sin \theta' = n \sin \theta$. An alternative approach is to set $n'\theta' = n\theta$, correct in the paraxial limit; thus cubic terms in θ and θ' remain with $A_0 \propto \theta^3$. For a mirror there are no cubic terms because the paraxial equation is exact, and hence there is no distortion when the stop is at the surface. Distortion appears when the stop is displaced from the surface.

Finally, we have not yet considered curvature of field, the final aberration noted in Section 4.III. The characteristics of this aberration are given in Section VII.

IV. SUMMARY OF ABERRATION RESULTS, STOP AT SURFACE

It is now appropriate to bring together all of the important results on aberrations and present them in a set of tables for convenient reference. Following are two tables of aberration coefficients: Table 5.1 for a general refracting surface and Table 5.2 for a reflecting surface. Because many of the applications considered in subsequent chapters involve single- or multiple-mirror systems, it is convenient to include a separate table for mirrors. The next two tables summarize the results for transverse aberrations, Table 5.3 for a general refracting surface and Table 5.4 for mirrors. Explanations and definitions are included as needed.

All of the results in this section apply specifically to the case where the aperture stop is at the surface. Because there are no optical surfaces either preceding or following this surface, the entrance and exit pupils are also located at the surface.

IV. Summary of Aberration Results, Stop at Surface

Table 5.1
Aberration Coefficients for General Surface[a,b]

$$A_1 = -\theta^2 \frac{n^2}{2}\left(\frac{1}{n's'} - \frac{1}{ns}\right), \qquad A_0 = 0$$

$$A_2 = \theta\frac{n^2}{2}\left(\frac{1}{s} - \frac{1}{R}\right)\left(\frac{1}{n's'} - \frac{1}{ns}\right)$$

$$A_3 = -\frac{1}{8}\left[n^2\left(\frac{1}{s} - \frac{1}{R}\right)^2\left(\frac{1}{n's'} - \frac{1}{ns}\right) + \frac{K}{R^3}(n'-n) + b\right]$$

[a] Entrance pupil is at surface.
[b] For a spherical refracting surface with no aspheric component, the last two terms in A_3 are absent.

Table 5.2
Aberration Coefficients for Mirror Surface[a,b]

$$A_0 = 0, \qquad A_1 = \frac{n\theta^2}{R}, \qquad A_2 = -\frac{n\theta}{R^2}\left(\frac{m+1}{m-1}\right)$$

$$A_3 = \frac{n}{4R^3}\left[K + \left(\frac{m+1}{m-1}\right)^2\right] - \frac{b}{8}$$

[a] Entrance pupil is at surface.
[b] The following relations apply to a mirror:

$$\frac{1}{s} - \frac{1}{R} = \frac{1}{R}\left(\frac{m+1}{m-1}\right), \qquad \frac{1}{n's'} - \frac{1}{ns} = -\frac{2}{nR}.$$

Table 5.3
Transverse Aberrations for General Surface[a,b]

$$\text{TAS} = -n^2\left(\frac{1}{n's'} - \frac{1}{ns}\right)\frac{y\theta^2 s'}{n'}, \qquad \Delta s' = -\frac{s'}{y}\text{TAS}$$

$$\text{TSC} = \frac{n^2}{2}\left(\frac{1}{s} - \frac{1}{R}\right)\left(\frac{1}{n's'} - \frac{1}{ns}\right)\frac{y^2\theta s'}{n'}$$

$$\text{TSA} = -\frac{1}{2}\left[n^2\left(\frac{1}{s} - \frac{1}{R}\right)^2\left(\frac{1}{n's'} - \frac{1}{ns}\right) + \frac{K}{R^3}(n'-n) + b\right]\frac{y^3 s'}{n'}$$

[a] Entrance pupil is at surface.
[b] Angular aberrations are given by the above relations with the final s' divided out.

Table 5.4
Transverse Aberrations for Mirror Surface[a]

$$\text{TAS} = -\frac{2y}{R}\theta^2 s', \qquad \text{TSC} = \frac{y^2}{R^2}\left(\frac{m+1}{m-1}\right)\theta s'$$

$$\text{TSA} = -\frac{y^3}{R^3}\left[K + \left(\frac{m+1}{m-1}\right)^2\right]s' + \frac{by^3}{2n}s'$$

[a] Entrance pupil is at surface.

A. APLANATIC CONDITION

For a spherical surface with $b = 0$, it is clear from Table 5.1 that all of the aberration coefficients are zero when $n's' = ns$. For a mirror $n' = -n$ and this condition is satisfied with $s' = -s$. Using the paraxial mirror equation (2.3.2) we find $R = \infty$, hence the surface is a plane mirror.

For a refracting surface the condition $n's' = ns$, together with the Gaussian equation (2.2.2), gives

$$ns = n's' = R(n + n'). \qquad (5.4.1)$$

This defines the object and image positions for a so-called aplanatic sphere, where the term *aplanatic* means the system has zero spherical aberration and coma. A lens of this type is often used as the first element in high-power microscope objectives. It has also been used as an element near the focus of a Schmidt camera in a spectrograph to shorten the camera focal length, as noted by Bowen. In this application its chromatic aberration is not a serious constraint in getting good image quality.

For a conic mirror with $b = 0$, the condition for zero spherical aberration fixes K in terms of m; the relation was used in Section 4.IV.A to establish the conic constant for the secondary mirror of a classical Cassegrain telescope.

From the entries in Table 5.2 we also see that a sphere ($K = 0$) in a configuration with $m = -1$, thus $s' = s = R/2$, has zero spherical aberration and coma, hence is an aplanat.

B. DEFINITIONS, CHARACTER OF ABERRATIONS

Measures of the transverse aberrations are taken from Eq. (5.3.4). Before writing these out some definitions are needed. Each aberration is designated by two letters as follows: SA, spherical aberration; SC, sagittal coma; TC, tangential coma; AS, astigmatism; and DI, distortion.

IV. Summary of Aberration Results, Stop at Surface

If the aberration is transverse, a prefix T is attached; if the aberration is angular, a prefix A is attached. With these designations the transverse aberration expressions are

$$\text{TAS} = \frac{2A_1 y s'}{n'}, \qquad \text{TSC} = \frac{A_2 y^2 s'}{n'}, \qquad \text{TSA} = \frac{4A_3 y^3 s'}{n'}. \qquad (5.4.2)$$

Note that TSC is simply one-third of the tangential coma, where the latter is based on the rays from the y axis. All of the aberrations are computed using rays from the y axis, with the full aberration given by Eq. (5.4.2) when the radius of the surface is substituted for y.

It is not necessary to use Eq. (5.3.5) to find transverse aberrations in the x direction because A'_1 is zero, given our choice $s' = s'_s$, and the extent of the blur in the x direction is known from results of Eq. (5.4.2). Thus all measures of transverse aberrations given below are in the y direction, that is, measured in the plane defined by the chief ray and z axis.

Results obtained by substituting from Tables 5.1 and 5.2 into Eq. (5.4.2) are given in Tables 5.3 and 5.4. Although $\Delta s'$ is actually a longitudinal aberration, its relation to TAS is included in Table 5.3 for completeness. All of the relations in Table 5.3 include the sign convention and thus these equations give information about the character of the aberrations as well as their magnitudes. A brief summary of the relation between aberration sign and image character follows, where the choice of sign is in accord with the figures illustrating each of the aberrations. Assuming $y > 0$:

TAS < 0: tangential line image closer to surface than sagittal image.
TSC > 0: coma flare is directed away from z axis, or Gaussian focus between flare and axis.
TSA < 0: marginal rays cross chief ray between surface and Gaussian focus.

For some purposes the sign of the aberration is of no consequence and the magnitude is all that matters. In terms of magnitudes, each of the aberrations has the following interpretation:

$$|\text{TAS}| = \begin{cases} \text{Half-length of astigmatic line image} \\ \text{Diameter of astigmatic blur circle} \end{cases}$$

$$3|\text{TSC}| = \begin{cases} \text{Length of comatic flare} \\ 1.5 \times \text{width of comatic flare} \end{cases}$$

$$|\text{TSA}| = \begin{cases} \text{radius of blur at paraxial focus} \\ 2 \times \text{diameter of circle of least confusion} \end{cases}$$

All of these results assume, of course, that only a single aberration is nonzero. If more than one aberration is present in an image, there is no simple way to characterize the blur character or dimensions.

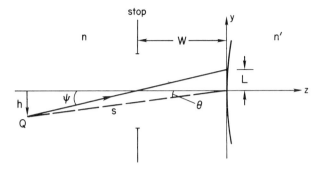

Fig. 5.6. Portion of Fig. 5.1 with aperture stop displaced from surface. The chief ray makes angle ψ with the z axis and intersects the surface at height L. The entrance pupil is at the stop. The relation between parameters is given in Eq. (5.5.2).

Inspection of the entries in Tables 5.3 and 5.4 shows that the transverse aberrations are independent of the direction of light incident on a given surface. Reversing the direction of incident light is equivalent to taking Fig. 5.1 and reversing it left for right, thus changing the sign of n, n', s', R, and θ. This result is expected because the direction of the incident light, left to right or vice versa, cannot change the image character.

V. ABERRATIONS FOR DISPLACED STOP

We now determine the aberration coefficients for a single surface with the aperture stop displaced from the surface, as shown in Figs. 5.6 and 5.7. In Fig. 5.6 the stop defines the light bundle before refraction at the surface, and the entrance pupil coincides with the stop. In Fig. 5.7 the stop follows

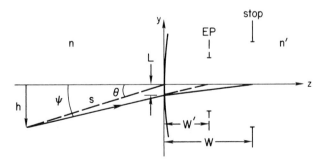

Fig. 5.7. Repeat of Fig. 5.6 with stop to right of surface. The stop is reimaged as the entrance pupil EP. Equation (5.5.2) applies when W' replaces W.

the surface and the entrance pupil, or image of the stop, is separate from the stop. In both figures the chief ray is directed toward the center of the pupil at angle ψ with the z axis and intersects the surface at height L.

Comparing these figures with Fig. 5.1 we see that an arbitrary ray that met the surface at (x, y) in Fig. 5.1 now meets the surface at $(x, y + L)$. Because a different portion of the surface refracts the light bundle from Q when the stop is displaced, it is expected that the image aberrations will differ from those derived for Fig. 5.1.

An example of this difference was previously noted in Section 4.IV.B, where the absence of coma and astigmatism for a sphere with stop at the center of curvature of the sphere was the basis for the Schmidt camera. These aberrations are not zero when the stop coincides with the surface, as is evident from Table 5.2. This one example illustrates the importance of the stop position in controlling or eliminating aberrations.

A. GENERAL FORMULATION

We now proceed to find the general aberration relations for a displaced stop. The procedure is simply one of putting $y + L$ in place of y in Eq. (5.1.6). Collecting terms in various powers of x, y, or ρ, and dropping all constants independent of these variables, we get

$$\begin{aligned} G &= y(A_0 + 2LA_1 + 3L^2A_2 + 4L^3A_3) \\ &+ y^2(A_1 + 3LA_2 + 6L^2A_3) + x^2(A_1' + LA_2' + 2L^2A_3) \\ &+ y^3(A_2 + 4LA_3) + x^2y(A_2' + 4LA_3) + A_3\rho^4 \\ &= B_0y + B_1y^2 + B_1'x^2 + B_2(x^2y + y^3) + B_3\rho^4, \end{aligned} \quad (5.5.1)$$

where $A_2' = A_2$ is used to combine the cubic terms.

Even before deriving the explicit form of the B_i, there are several important statements that follow from Eq. (5.5.1). These *stop-shift relations* are

(1) L does not appear in the ρ^4 term and thus the spherical aberration coefficient is independent of stop position.

(2) If $A_3 = 0$, the coma coefficient B_2 is independent of the stop position and the coma is that given in Section IV.

(3) If both A_2 and A_3 are zero, B_1 and B_1' are independent of the stop position and reduce to A_1 and A_1', respectively.

The importance of these statements will become evident when specific systems are discussed in following chapters. Although they are deduced here for a single surface, it turns out these statements also apply to a system made up of many surfaces.

The process required to evaluate each B_i is a straightforward one, though the algebra is a bit messy at times and therefore omitted in the discussion to follow. When the evaluation is more than simple substitution, as for astigmatism, a brief outline of the procedure is given.

To begin, we note the following relations, valid in the paraxial approximation, derived from the geometry in Fig. 5.6:

$$L = -W\psi, \qquad \theta = \psi[1 - (W/s)], \qquad (5.5.2)$$

where W is the distance from the surface to the entrance pupil, and both L and W are governed by the same sign convention as other distances. Figure 5.7 shows the geometry for the case where the stop follows the surface. In this case Eq. (5.5.2) applies if W is replaced by W'. A glance at Eq. (5.5.2) and Fig. 5.6 shows that, for a given θ, ψ increases in size as s approaches W. There will come a point where ψ is large enough to make the paraxial result in Eq. (5.5.2) invalid, and results derived using (5.5.2) will be incorrect. Unfortunately, no simple statement can be made about where this breakdown occurs, and one must check on a case-by-case basis as to the validity of results derived from third-order aberration theory. Exact ray tracing is generally used to check the results.

A final point to be made is in the choice of parameters used in presenting the aberration coefficients. The choice made here is to eliminate L and θ and to give the results in terms of W, the entrance pupil position, and ψ, the chief ray angle.

With these preliminaries behind us, we proceed with the results; Table 5.5 gives the coefficients for a general surface, while Table 5.6 gives the results for a mirror. The spherical term B_3 is taken from Tables 5.1 and 5.2 and is included here for completeness. The terms B_2 and B_0 are derived from Eq. (5.5.1) and the entries in Tables 5.1 and 5.2 by direct substitution.

The derivation of B_1 is not one of direct substitution of the A_i but involves going through steps analogous to those used in Sections I and II. First we note that

$$B_1 = A_1 + 3L\Omega, \qquad B_1' = A_1' + L\Omega, \qquad (5.5.3)$$

where $\Omega = A_2 + 2LA_3$, and A_1 and A_1' are the multiplying factors of y^2 and x^2, respectively, in Eq. (5.1.5).

The choice of $B_1 = 0$ or $B_1' = 0$ locates the tangential or sagittal images, respectively. Corresponding to Eqs. (5.1.8) and (5.1.9) we find

$$\frac{n' \cos^2 \theta'}{s_t'} - \frac{n \cos^2 \theta}{s} = \frac{n' \cos \theta' - n \cos \theta}{R} - 6L\Omega, \qquad (5.5.4)$$

$$\frac{n'}{s_s'} - \frac{n}{s} = \frac{n' \cos \theta' - n \cos \theta}{R} - 2L\Omega. \qquad (5.5.5)$$

V. Aberrations for Displaced Stop

Table 5.5
General Aberration Coefficients, Centered Pupil[a,b]

$$B_0 = \frac{(W\psi)^3}{2}\left[\Gamma\left(\frac{1}{W}-\frac{1}{R}\right)\left(\frac{1}{W}-\frac{1}{R}+\frac{1}{W}-\frac{1}{s}\right)+\frac{K}{R^3}(n'-n)+b\right]$$

$$B_1 = -\frac{(W\psi)^2}{2}\left[\Gamma\left(\frac{1}{W}-\frac{1}{R}\right)^2+\frac{K}{R^3}(n'-n)+b\right]$$

$$B_2 = \frac{(W\psi)}{2}\left[\Gamma\left(\frac{1}{s}-\frac{1}{R}\right)\left(\frac{1}{W}-\frac{1}{R}\right)+\frac{K}{R^3}(n'-n)+b\right]$$

$$B_3 = A_3 = -\frac{1}{8}\left[\Gamma\left(\frac{1}{s}-\frac{1}{R}\right)^2+\frac{K}{R^3}(n'-n)+b\right]$$

$$\Gamma = n^2\left(\frac{1}{n's'}-\frac{1}{ns}\right)$$

[a] Entrance pupil is at distance W from surface.
[b] The coefficient B_0 is based on $A_0 = 0$, not on the paraxial relation $n'\theta' = n\theta$.

The relation analogous to Eq. (5.1.10), through terms in θ^2, is

$$\frac{\Delta s'}{s'^2} = \left(\frac{\Delta s'}{s'^2}\right)_{W=0} - \frac{4L\Omega}{n'}, \qquad (5.5.6)$$

where the first term on the right is given in Eq. (5.1.10).

The next step is to choose an image at which to evaluate the astigmatism and, as in Section II, the choice is the sagittal image. Solving Eq. (5.5.5) for s', substituting the result into B_1 in Eq. (5.5.3), and evaluating to second

Table 5.6
Mirror Aberration Coefficients, Centered Pupil[a]

$$B_0 = -n(W\psi)^3\left[\frac{K}{R^3}+\frac{1}{R}\left(\frac{1}{W}-\frac{1}{R}\right)\left(\frac{1}{W}-\frac{1}{R}+\frac{1}{W}-\frac{1}{s}\right)\right]+\frac{b}{2}(W\psi)^3$$

$$B_1 = n\frac{(W\psi)^2}{R}\left[\frac{K}{R^2}+\left(\frac{1}{W}-\frac{1}{R}\right)^2\right]-\frac{b}{2}(W\psi)^2$$

$$B_2 = -n\frac{(W\psi)}{R^2}\left[\frac{K}{R}+\left(\frac{m+1}{m-1}\right)\left(\frac{1}{W}-\frac{1}{R}\right)\right]+\frac{b}{2}(W\psi)$$

$$B_3 = A_3 = \frac{n}{4R^3}\left[K+\left(\frac{m+1}{m-1}\right)^2\right]-\frac{b}{8}$$

[a] Entrance pupil is at distance W from surface.

order in θ, gives

$$B_1 = A_1 + 2L\Omega = -\frac{n'}{2}\left(\frac{\Delta s'}{s'^2}\right)_{W=0} + 2L\Omega. \tag{5.5.7}$$

Note that A_1 in the first part of Eq. (5.5.7) is the value at the sagittal image, hence different from A_1 in Eq. (5.5.3). From a comparison of Eqs. (5.5.6) and (5.5.7) we find

$$\Delta s'/s'^2 = -2B_1/n', \tag{5.5.8}$$

a relation corresponding to Eq. (5.2.5). Thus B_1 is a measure of the astigmatism when the entrance pupil is not at the surface. The entry for B_1 in Table 5.5 is obtained by evaluation of Eq. (5.5.7).

The relations between the aberration coefficients and the transverse aberrations are similar to those given above.

$$\text{TDI} = \frac{B_0 s'}{n'}, \qquad \text{TAS} = \frac{2B_1 y s'}{n'},$$

$$\text{TSC} = \frac{B_2 y^2 s'}{n'}, \qquad \text{TSA} = \frac{4B_3 y^3 s'}{n'}. \tag{5.5.9}$$

The full aberration is obtained from Eqs. (5.5.9) when y is replaced by the height at which a marginal ray from Q on the z axis intersects the surface. Tables analogous to 5.3 and 5.4 are left to the reader.

B. EXAMPLES

At this point it is appropriate to illustrate the relations with two examples. Our choices are a sphere and a paraboloid, each illuminated with collimated light, hence $m = 0$. We also assume no aspheric component and set $b = 0$.

For the sphere we find $A_3 = B_3 = n/4R^3$, hence we expect that both B_2 and B_1 will depend on the stop position. Inspection of the coefficients in Table 5.6 shows that this is the case, where both B_2 and B_1 are zero when $W = R$, as expected.

For the parabola $B_3 = 0$. Setting $m = 0$ and $K = -1$ in B_2 we see that the coma is independent of W, as expected based on our discussion following Eq. (5.5.1). Because coma is not zero the astigmatism coefficient depends on W, hence a proper choice of W will mean zero astigmatism. From B_1 in Table 5.6 we see that this choice is $W = R/2$, hence the stop is at the focal surface.

Returning to the sphere in collimated light, but allowing for an aspheric component, we find $B_3 = 0$ when $b = 2n/R^3$. This aspheric component could be put directly on the mirror, but this does little good because now the mirror has nonzero coma independent of W. The solution, of course, is to put the aspheric component on another optical element located at $W = R$, as already shown in our discussion of the Schmidt camera. How aberrations are calculated for systems with many surfaces is the topic of the next section.

VI. ABERRATIONS FOR MULTISURFACE SYSTEMS

The real power of the approach to aberrations using Fermat's principle is particularly evident when systems with many surfaces are analyzed. For any surface in such a system, say the ith one, the object and image are at Q_i and Q'_i located at distances s_i and s'_i, respectively, from the surface. Between the object and image the OPD between an arbitrary ray and the chief ray is given by Eq. (5.5.1), where W_i is the position of the entrance pupil for the surface. If this same ray is followed from the original object to the final image, then the OPD for the system is

$$G_s = G_1 + G_2 + \cdots + G_f = \sum G_i, \qquad (5.6.1)$$

where the subscript f denotes the last surface. Each term in Eq. (5.6.1) can be replaced by Eq. (5.5.1), with the appropriate (x, y) at each surface. Before making this substitution, note that a complete description of the aberrations according to Eq. (5.4.2) is obtained from rays in the yz plane only. Therefore we set $x = 0$ in Eq. (5.5.1), and the system OPD for rays in the yz plane is

$$G_s = \sum (B_{0i} y_i + B_{1i} y_i^2 + B_{2i} y_i^3 + B_{3i} y_i^4)$$

$$= \sum_k \left(\sum_i B_{ki} y_i^{k+1} \right), \qquad k = 0, 1, 2, 3, \qquad (5.6.2)$$

with the geometric distance along an arbitrary ray between the actual and reference wavefronts at the final surface given by

$$\Delta = G_s(y_i)/n'_f. \qquad (5.6.3)$$

To find the transverse aberration at the final image, we proceed along the lines followed in going from Eq. (5.3.1) to Eq. (5.3.3), but do so with only one of the aberration terms, say the kth one. With reference to the last surface we get

$$\frac{\partial \Delta}{\partial y_f} = \frac{1}{n'_f} \frac{\partial G_s}{\partial y_f} = \frac{1}{n'_f}(k+1) \sum_i B_{ki} y_i^k \frac{\partial y_i}{\partial y_f}. \qquad (5.6.4)$$

The partial derivatives in Eq. (5.6.4) are easily evaluated with the aid of Fig. 5.8, where two rays from an intermediate axial object point are shown

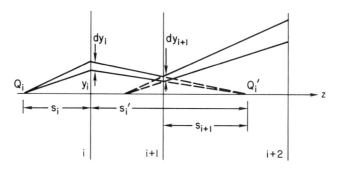

Fig. 5.8. Paths of two adjacent rays from Q_i through several surfaces, where Q_i is an intermediate axial object point. Differential heights $dy_i \propto y_i$ at each surface.

passing through several surfaces. Because each Q_i is imaged to Q_i', the ratio of the differential change in y_i to that of any other y, say the fth one, is simply the ratio y_i/y_f. Substituting this into Eq. (5.6.4) and multiplying by s_f', we can write the transverse aberration in the y direction as

$$\mathrm{TA}_y = \frac{s_f'}{n_f'}\frac{\partial G_s}{\partial y_f} = \frac{s_f'}{n_f'}(k+1)\left[\sum_i B_{ki}\left(\frac{y_i}{y_f}\right)^{k+1}\right]y_f^k \qquad (5.6.5)$$

or

$$\mathrm{TA}_y = \frac{s_f'}{n_f'}(k+1)B_{ks}y_f^k. \qquad (5.6.6)$$

In Eq. (5.6.6) the representation of the sum in brackets in Eq. (5.6.5) is reduced to a single symbol, with the subscript s denoting a system aberration coefficient. Note the close correspondence between Eq. (5.6.6) and each of the terms containing y in Eq. (5.3.4).

Calculation of the transverse aberration using Eq. (5.6.6) is based on the marginal ray height at the last surface. In many cases it is convenient to express the transverse aberration in terms of the marginal ray height at some other surface, such as at the system entrance pupil. This is easily done by multiplying and dividing Eq. (5.6.5) by the ray height at another surface, say the ιth one, raised to the $k+1$ power. The results are

$$B_{ks} = \sum_i B_{ki}\left(\frac{y_i}{y_\iota}\right)^{k+1}, \qquad (5.6.7)$$

$$\mathrm{TA}_y = \frac{s_f'}{n_f'}\left(\frac{y_\iota}{y_f}\right)(k+1)B_{ks}y_\iota^k. \qquad (5.6.8)$$

Note that the terms in Eq. (5.6.7) depend on the choice of ι but TA_y is, of course, independent of this choice.

VI. Aberrations for Multisurface Systems

The formalism needed to calculate third-order aberrations for a multisurface system is now complete. The necessary aberration coefficients are in Tables 5.5 and 5.6, and it is simply a matter of computing each one surface by surface and substituting into Eqs. (5.6.7) and (5.6.8).

A. EXAMPLE: ABERRATION COEFFICIENTS OF TWO-MIRROR TELESCOPES

As an example of the procedure, consider the Cassegrain telescope shown in Fig. 2.7a. We assume the stop is at the primary, thus its aberration coefficients can be taken from Table 5.2. Setting $n = 1$, $m = 0$, and $b = 0$ gives

$$B_0 = 0, \quad B_1 = \frac{\theta^2}{R_1}, \quad B_2 = \frac{\theta}{R_1^2}, \quad B_3 = \frac{1}{4R_1^3}(K_1 + 1).$$

For the secondary $n = -1$, $b = 0$, and $\psi = -\theta$, and from Table 5.6 we get

$$B_0 = -(W\theta)^3 \left[\frac{K_2}{R_2^3} + \frac{1}{R_2}\left(\frac{1}{W} - \frac{1}{R_2}\right)\left(\frac{1}{W} - \frac{1}{R_2} + \frac{1}{W} - \frac{1}{s_2}\right) \right],$$

$$B_1 = -\frac{(W\theta)^2}{R_2} \left[\frac{K_2}{R_2^2} + \left(\frac{1}{W} - \frac{1}{R_2}\right)^2 \right],$$

$$B_2 = -\frac{W\theta}{R_2^2} \left[\frac{K_2}{R_2} + \left(\frac{m+1}{m-1}\right)\left(\frac{1}{W} - \frac{1}{R_2}\right) \right],$$

$$B_3 = -\frac{1}{4R_2^3} \left[K_2 + \left(\frac{m+1}{m-1}\right)^2 \right].$$

With $y_t = y_1$, the marginal ray height at the primary, we get

$$B_{ks} = B_{k1} + B_{k2}(y_2/y_1)^{k+1}, \tag{5.6.9}$$

where the subscripts 1 and 2 on the B_k refer to the primary and secondary, respectively.

In terms of the normalized parameters defined in Chapter 2 for a two-mirror telescope we have $k = y_2/y_1$ and $\rho = R_2/R_1$, where this k is not to be confused with the index k in Eq. (5.6.9). For spherical aberration we then get

$$B_{3s} = \frac{1}{4R_1^3}\left\{ K_1 + 1 - \frac{k^4}{\rho^3}\left[K_2 + \left(\frac{m+1}{m-1}\right)^2 \right] \right\}. \tag{5.6.10}$$

Note that spherical aberration is zero when the expression in braces is zero, a result previously given in Eq. (4.4.2). It was derived there by starting with a classical Cassegrain and "bending" the mirrors subject to the requirement that Fermat's principle be satisfied.

Given Eq. (5.6.10) it is now a simple matter to find TSA. Using Eq. (2.5.8) we find $s_2' = mkf_1 = kf$. Putting this and $n_2' = 1$ into Eq. (5.6.8) gives

$$\text{TSA} = \frac{fy_1^3}{R_1^3}\{\quad\} = \frac{f}{64F_1^3}\{\quad\}, \tag{5.6.11}$$

where the quantity in braces is that in Eq. (5.6.10), and F_1 is the focal ratio of the primary mirror. Expressions for the other aberrations are determined using the same procedure, but this development is left for Chapter 6, where the characteristics of telescopes are explored in detail.

VII. CURVATURE OF FIELD

The remaining third-order aberration to be considered is that of curvature of field. As noted in Chapter 4 this aberration does not affect the image quality, but given the usual case of a flat detector it can adversely affect the image definition over an extended field. For the Schmidt camera, for example, the focal surface has a radius of curvature equal to the camera focal length. Matching the detector to the focal surface requires either deforming it to the proper radius or using another optical element to "flatten" the field. The former method is used with most large Schmidt telescopes by bending photographic plates. The use of a field-flattener lens is discussed later in this section.

We now consider the situation shown in Fig. 5.9, where an optical surface whose vertex is at the origin of the (x, y, z) coordinate system images the curved object surface Σ into a curved image surface Σ'. The surfaces Σ and Σ' have radii of curvature r and r', respectively, with the sign convention for each the same as for a surface radius of curvature, thus $r > 0$ and $r' < 0$ in Fig. 5.9. As a final definition, let κ denote the curvature of the image

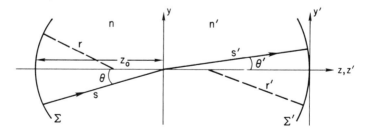

Fig. 5.9. Curved object surface Σ imaged to surface Σ'. Radii of curvature of the object and image surfaces are r and r', respectively. The adopted sign convention has $r > 0$ and $r' < 0$ in the diagram.

VII. Curvature of Field   83

surface, with $\kappa = 1/r'$. It should be noted here that our sign convention for r and r' is opposite that of Born and Wolf, but we choose to preserve its universal character.

The diagram in Fig. 5.9 can apply to any individual optical surface within a multisurface system, where Σ is an intermediate object surface and Σ' its conjugate surface. In the discussion to follow we will not designate these surfaces with specific indices, but will include them at the end as needed.

The procedure is one of finding s' in terms of θ' and applying a general relation between κ and s' to get the curvature of Σ'. A glance at Eqs. (5.5.4) and (5.5.5) shows that s' can contain only even powers of θ and θ' when expanded in a power series. Thus we can solve these equations for s' and, after substituting $n\theta = n'\theta'$, get s' through second order in the form

$$1/s' = a_0 + a_2 \theta'^2. \qquad (5.7.1)$$

With this form of s', it can be shown that the curvature κ to zeroth order is given by

$$\kappa = -(a_0 + 2a_2), \qquad (5.7.2)$$

hence κ is constant and to this approximation the image surface is a section of a sphere. [The relation between κ and s' in polar coordinates from which Eq. (5.7.2) is derived can be found in the "Mathematics Manual"—see list of references at the end of the chapter.]

A. PETZVAL SURFACE

We first determine the curvature of a special surface called the Petzval surface. This surface is the image surface in the special case where the astigmatism is zero, hence $s'_s = s'_t = s'_p$, where s'_p is the distance from the origin in Fig. 5.9 to the Petzval surface. We can find s'_p from either Eq. (5.5.4) or Eq. (5.5.5).

Given the condition that $\Delta s' = 0$ in Eq. (5.5.6) we find $2L\Omega = -A_1$, and substituting this into Eq. (5.5.5) gives

$$\frac{n'}{s'_p} = \frac{n}{s} + \frac{n' \cos \theta' - n \cos \theta}{R} + A_1. \qquad (5.7.3)$$

To put Eq. (5.7.3) into the form required by Eq. (5.7.1) means the usual power series substitutions and using $n\theta = n'\theta'$ to eliminate θ. In addition to the angles that appear explicitly in Eq. (5.7.3), the distance s depends on θ. The relation between s and θ is found using the geometry in Fig. 5.10, where the sag u of Σ is

$$u = \frac{y^2}{2r} = \frac{s^2 \sin^2 \theta}{2r} = s \cos \theta - z_0,$$

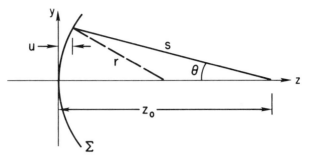

Fig. 5.10. Sag u of the object surface, with $u = y^2/2r$. See the discussion following Eq. (5.7.3).

and, solving for s, leads to

$$\frac{1}{s} = \frac{1}{z_0} - \frac{n'^2 \theta'^2}{2n^2}\left(\frac{1}{z_0} + \frac{1}{r}\right).$$

Substituting for s in Eq. (5.7.3) and collecting terms gives

$$\frac{1}{s'_p} = \frac{n}{n'z_0} + \frac{n'-n}{n'R} + \frac{\theta'^2}{2}\left[\frac{(n'-n)^2}{n'nR} - \frac{n}{n'z_0} - \frac{n'}{nr}\right],$$

which on substitution into Eq. (5.7.2) gives

$$\frac{1}{n'r'} - \frac{1}{nr} = -\left(\frac{n'-n}{n'nR}\right). \tag{5.7.4}$$

The importance of Eq. (5.7.4) lies in the fact that the curvature of the Petzval surface does not depend on the distances s and s' or on the position of the entrance pupil for this surface. This relation applies to each surface in a system and, given that $n'r'$ for the ith surface is nr for the $(i+1)$st surface, leads to a sum over all surfaces given by

$$\frac{1}{n'_f r'_f} - \frac{1}{n_1 r_1} = -\sum_i \left(\frac{n'-n}{n'nR}\right)_i, \tag{5.7.5}$$

where 1 and f refer to the first and last surfaces, respectively. For a flat object field, the most common situation, we get

$$\kappa_p = -n'_f \sum_i \left(\frac{n'-n}{n'nR}\right)_i. \tag{5.7.6}$$

Thus for any optical system for which the object field is flat, the Petzval surface is an invariant surface. If the system has astigmatism, each of the astigmatic image surfaces will have its own curvature. But, as we now show, there are definite relations between these curvatures, the amount of astigmatism, and the Petzval curvature.

B. CURVATURES OF ASTIGMATIC SURFACES

The procedure to find these curvature relations starts with the substitution of Eq. (5.2.5) into Eq. (5.5.6), giving

$$\frac{\Delta s'}{s'^2} = \frac{s'_s - s'_t}{s'_s s'_t} = -\frac{2}{n'}(A_1 + 2L\Omega).$$

Solving for $2L\Omega$ and substituting into Eq. (5.5.5) gives

$$\frac{n'}{s'_s} = \frac{n}{s} + \frac{n'\cos\theta' - n\cos\theta}{R} + A_1 + \frac{n'}{2}\left(\frac{s'_s - s'_t}{s'_s s'_t}\right),$$

where we see by comparison with Eq. (5.7.3) that the first three terms to the right of the equals sign are simply n'/s'_p. With this substitution we get

$$2(s'_p - s'_s) = s'_s - s'_t, \tag{5.7.7}$$

where the factors in the denominator cancel because of their near equality. Equation (5.7.7) can also be written as

$$s'_p - s'_t = 3(s'_p - s'_s). \tag{5.7.8}$$

The geometric interpretation of Eq. (5.7.8) is a simple one; at a given height y the distance between the Petzval and tangential surfaces is three times the distance between the Petzval and sagittal surfaces, with the sagittal surface always between the other two. Because astigmatism is zero on-axis, the image surfaces are in contact where they intersect the z axis.

Note that Eq. (5.7.8) holds for any surface in an optical system and does not depend on the object distance for that surface, nor does it depend on the entrance pupil location. Hence it must also hold for the final image surfaces of a system with many surfaces, and the relations to follow are taken at the final surfaces.

It is a simple matter to write Eq. (5.7.7) and (5.7.8) in terms of surface curvatures. From the geometry in Fig. 5.11 we get

$$u_\alpha - u_\beta = s'_\alpha - s'_\beta, \tag{5.7.9}$$

where u is the surface sag and α and β denote any pair of image surfaces. We also see that $u = y^2\kappa/2$ and thus

$$\kappa_\alpha - \kappa_\beta = \frac{2}{\theta'^2}\left(\frac{s'_\alpha - s'_\beta}{s'^2}\right), \tag{5.7.10}$$

from which it follows, using Eqs. (5.7.7) and (5.7.8), that

$$\kappa_s - \kappa_t = 2(\kappa_p - \kappa_s), \qquad \kappa_p - \kappa_t = 3(\kappa_p - \kappa_s). \tag{5.7.11}$$

Choosing $\alpha = s$ and $\beta = t$ in Eq. (5.7.10) and substituting Eq. (5.5.8) we get

$$\kappa_s - \kappa_t = -4B_{1s}/n'\theta'^2, \tag{5.7.12}$$

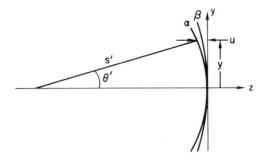

Fig. 5.11. Sags of image surfaces α and β, where $u_\alpha - u_\beta = s'_\alpha - s'_\beta$ in the paraxial approximation.

where B_{1s} is the system astigmatism coefficient from Eq. (5.6.7) with $\iota = f$. Note that B_{1s} is referenced to the last surface in the system because θ' is specified at the last surface.

It is now a simple matter to combine Eqs. (5.7.12) and (5.7.11) and solve for the curvatures of the individual surfaces. The results are given in Table 5.7. Included is an entry for the curvature κ_m of the surface midway between the S- and T-surfaces, that surface on which the astigmatic images are circular.

C. EXAMPLES

A few simple examples are now in order. Consider first a spherical mirror in collimated light, hence $m = 0$. From Table 5.6 we get

$$B_1 = \frac{n\theta^2}{R}\left(1 - \frac{W}{R}\right)^2,$$

Table 5.7

Image Surface Curvatures

$$\kappa_p = -n'_f \sum_i \left(\frac{n'-n}{n'nR}\right)_i$$

$$\kappa_s = \kappa_p + \frac{2B_{1s}}{n'\theta'^2}, \qquad \kappa_t = \kappa_p + \frac{6B_{1s}}{n'\theta'^2}$$

$$\kappa_m = \tfrac{1}{2}(\kappa_s + \kappa_t), \qquad B_{1s} = \sum B_{1i}\left(\frac{y_i}{y_f}\right)^2$$

VII. Curvature of Field

where $\psi = \theta$ from Eq. (5.5.2). For a single reflecting surface we have $n' = -n$, $\theta' = -\theta$. Substituting into the entries in Table 5.7 leads to

$$\kappa_p = \frac{2}{R}, \quad \kappa_s = \frac{2}{R} - \frac{2}{R}\left(1 - \frac{W}{R}\right)^2, \quad \kappa_t = \frac{2}{R} - \frac{6}{R}\left(1 - \frac{W}{R}\right)^2.$$

For $W = R$ each of the curvatures is $2/R$, as expected because the astigmatism is zero. For $W = 0$ we find $\kappa_s = 0$, $\kappa_t = -4/R$. Thus the Petzval and tangential surfaces have opposite curvatures with a flat sagittal surface between them. Although collimated light was specified above, note that these results hold for any object distance because B_1 is independent of the magnification m.

As a second example consider a Schmidt camera. As noted in Section 4.IV, a spherical mirror with $W = R$ has zero astigmatism but a curved image surface with curvature $2/R$. One way to flatten the surface is to introduce another element whose astigmatism is zero, to a first approximation, and to choose its characteristics to make the Petzval curvature zero for the system. This is done with a thin lens located near the image surface, as shown in Fig. 5.12. The contribution of the corrector plate is ignored in the analysis to follow because $R_c \gg R$ for any practical focal ratio, a result that is evident from Eq. (4.4.12).

The Petzval curvature for the mirror-lens combination, derived from the relation in Table 5.7, is

$$\kappa_p = \frac{2}{R} - \left(\frac{n-1}{n}\right)\left(\frac{1}{R_1} - \frac{1}{R_2}\right). \tag{5.7.13}$$

where R_1 and R_2 are the radii of curvature of the first and second surfaces, respectively, of the lens, and the index n of the lens is positive. Setting $R_2 = \infty$ gives a lens whose flat surface faces the image surface and, though this is not required, we make this choice for convenience. Therefore the

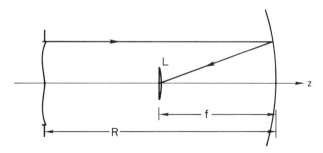

Fig. 5.12. Schmidt camera with lens L near the focal surface to give zero Petzval curvature. See Eq. (5.7.13).

Petzval surface is flat if

$$(n-1)/nR_1 = 2/R. \qquad (5.7.14)$$

Because R of the mirror is negative, so also is R_1, hence the lens is plano-convex in cross section and has positive power. Though the combination now has a flat Petzval field, the thin lens may introduce some astigmatism. We show in Chapter 9 that the amount introduced is small if the lens is close to the image surface.

Our final example is that of a two-mirror telescope, either Cassegrain or Gregorian. For light incident on the primary according to our usual convention we have $n_1 = 1$, $n_2 = -1$. The Petzval curvature is then

$$\kappa_p = 2\left(\frac{1}{R_2} - \frac{1}{R_1}\right) = \frac{2}{R_1}\left(\frac{1-\rho}{\rho}\right). \qquad (5.7.15)$$

where $\rho = R_2/R_1$. For a Gregorian $\rho < 0$ and κ_p is opposite in sign to R_1. Thus the Petzval surface for a Gregorian is convex as seen from the secondary. For a Cassegrain the Petzval surface is concave as seen from the secondary, provided $\rho < 1$. Discussion of the curvatures of the other astigmatic surfaces of two-mirror telescopes is left to Chapter 6.

VIII. ABERRATIONS FOR DECENTERED PUPIL

The aberration results given to this point are correct for an optical system in which all of the elements, including the aperture stop and pupils, are centered. By centered we mean there is a single axis, designated the z axis, about which the system can be rotated without change, where the z-axis passes through the center of each element. If one or more of these elements is displaced laterally from the z axis or rotated about a line perpendicular to the z axis, the system is no longer rotationally symmetric and aberrations are introduced. The lateral displacement is commonly referred to as a decenter and the rotation as a tilt. Decenter and/or tilt of, for example, the secondary mirror in a two-mirror telescope is one important case of this loss of symmetry and is discussed in Chapter 6.

A. GENERAL FORMULATION

In this section we find the aberration coefficients for a general surface with its associated pupil where the center of the pupil is displaced from the z axis of the surface. A cross section of this situation is shown in Fig. 5.13, where the pupil is displaced in the y direction by L'. The chief ray

VIII. Aberrations for Decentered Pupil

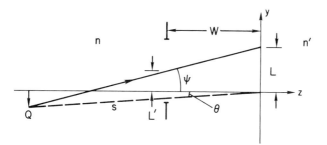

Fig. 5.13. Sketch of chief ray through center of stop displaced by L' from the z axis of the surface. The relation between parameters is given in Eq. (5.8.1).

passes through the center of the pupil and makes angle ψ with the z axis, the axis of symmetry of the surface. In the paraxial approximation we find the following relations from the geometry in Fig. 5.13:

$$L = L' - W\psi, \qquad \theta = \psi\left(1 - \frac{W}{s}\right) + \frac{L'}{s}, \qquad (5.8.1)$$

where W is the distance from the surface to the entrance pupil, and the sign of each distance and angle is set by the sign convention. The relations in Eq. (5.8.1) are a generalization of those in Eq. (5.5.2).

The procedure for finding the aberration coefficients is the same as that followed in Section 5.V, except that Eq. (5.8.1) is used instead of Eq. (5.5.2) when substituting into Eq. (5.5.1). The results of carrying out these substitutions for coma and astigmatism are given in Table 5.8 for a general surface and Table 5.9 for a mirror surface. Distortion is not included because the change in it is too small to be significant. The spherical aberration is independent of L', hence $B_3 = B_3(\text{cen})$.

Table 5.8

General Aberration Coefficients, Decentered Pupil[a]

$$B_1 = B_1(\text{cen}) - \frac{L'^2}{2}\left[\frac{\Gamma}{R^2} + \frac{K}{R^3}(n' - n) + b\right]$$

$$- L'(W\psi)\left[\frac{\Gamma}{R}\left(\frac{1}{W} - \frac{1}{R}\right) - \frac{K}{R^3}(n' - n) - b\right]$$

$$B_2 = B_2(\text{cen}) + \frac{L'}{2}\left[\frac{\Gamma}{R}\left(\frac{1}{s} - \frac{1}{R}\right) - \frac{K}{R^3}(n' - n) - b\right]$$

$$\Gamma = n^2\left(\frac{1}{n's'} - \frac{1}{ns}\right)$$

[a] $B_i(\text{cen})$ are entries in Table 5.5.

Table 5.9

Mirror Aberration Coefficients, Decentered Pupil[a]

$$B_1 = B_1(\text{cen}) + nL'^2 \left(\frac{K+1}{R^3} - \frac{b}{2n} \right)$$

$$+ \frac{2nL'(W\psi)}{R^2} \left(\frac{1}{W} - \frac{K+1}{R} + \frac{bR^2}{2n} \right)$$

$$B_2 = B_2(\text{cen}) + \frac{nL'}{R^3} \left[K - \left(\frac{m+1}{m-1} \right) - \frac{bR^3}{2n} \right]$$

[a] $B_i(\text{cen})$ are entries in Table 5.6.

Examination of the entry for B_2 in Table 5.8 shows that the part of the coma coefficient that results from the decentering is not dependent on the angle of the chief ray. Hence the effect of the decentering is to introduce constant coma over the entire image field, in addition to any angle-dependent coma that is present. The effect of the decentering on B_1 is to introduce a constant term and a term that depends linearly on the angle of the chief ray. Hence astigmatism is also present over the entire image field.

Calculation of the system aberration coefficients is carried out following the procedure in Section 5.VI. In all cases of interest, it turns out that the effect of a decentered stop is much greater on coma than on astigmatism, hence the image surface curvatures are not significantly affected and the results of Table 5.7 can be used with $B_{1s}(\text{cen})$.

B. EXAMPLE: SCHMIDT CAMERA

At this point it is instructive to give an example of a system with decentered stop. The example discussed is a Schmidt camera in which the axis of the corrector plate is displaced from the mirror axis, as shown in Fig. 5.14. The aperture stop of the system is the corrector plate, with collimated light incident.

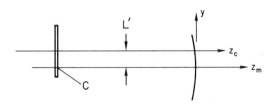

Fig. 5.14. Schmidt camera with axis of corrector z_c displaced from mirror axis z_m by L'.

VIII. Aberrations for Decentered Pupil

Because $W = 0$ for the corrector, its aberration coefficients are

$$B_1 = 0, \quad B_2 = 0, \quad B_3 = -b/8, \tag{5.8.2}$$

where $R = \infty$ is the choice for the radius of curvature of the corrector. (For a corrector profile that minimizes chromatic aberration R is finite, as shown in Section 4.IV. However, as we show in Chapter 7, the aberration coefficients are dominated by the term in b.)

The parameters for the spherical mirror are $W = R$, $m = 0$, $b = 0$, and $n = 1$, in which case $B_1(\text{cen})$ and $B_2(\text{cen})$ are zero. From Table 5.9 we then find

$$B_1 = \frac{L'^2}{R^3}, \quad B_2 = \frac{L'}{R^3}, \quad B_3 = \frac{1}{4R^3}. \tag{5.8.3}$$

With the ray heights at the corrector and mirror equal, the system aberration coefficients according to Eq. (5.6.7) are simply the sums of corresponding terms in Eqs. (5.8.2) and (5.8.3). Putting these sums into Eq. (5.6.6), and dividing by s' to get the angular aberration, gives

$$\text{AAS} = \frac{1}{2F}\left(\frac{L'}{R}\right)^2, \quad \text{ATC} = \frac{3}{16F^2}\left(\frac{L'}{R}\right), \tag{5.8.4}$$

where ATC is the angular tangential coma. Spherical aberration is zero provided $b = 2/R^3$.

The relation for ATC in Eqs. (5.8.4) can be used to find the largest permissible L' for a given ATC. If we choose a blur limit of 1 arc-second, then the reader can verify that L'/R, expressed in arc-seconds, cannot exceed $16F^2/3$. The value of AAS for this value of L'/R is about 1000 times smaller and thus is negligible.

It is also instructive to take this same system and tilt the mirror with respect to the corrector, as shown in Fig. 5.15. From the geometry in Fig. 5.15 we see that $L' = -\alpha R$ and $\psi = \theta - \alpha$, where α is the tilt angle of the mirror and ψ is the angle of the chief ray relative to the mirror axis. The fact that ψ depends on α is of no consequence here because $B_1(\text{cen})$ and

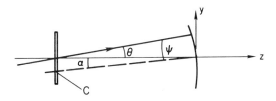

Fig. 5.15. Schmidt camera with axis of mirror, denoted by dashed line, tilted by angle α with respect to the z axis of the corrector.

$B_2(\text{cen})$ for the mirror are zero, independent of ψ. Hence Eqs. (5.8.2) through (5.8.4) are the same for this system, with α replacing L'/R.

It is not surprising that the systems in Figs. 5.14 and 5.15 have the same aberrations because they are, in fact, equivalent. The tilt, in effect, has offset the center of curvature of the mirror by a distance L' from the center of the corrector and, because the sphere has no preferred axis, the systems are the same. Note that this equivalence between a tilt and decenter does not hold for any surface that has a unique axis.

Before leaving this system, it is worth examining Fig. 5.15 from the point of view of Fermat's principle. For $\theta = 0$ rays through the upper half of the corrector are advanced at the mirror, while those through the bottom half are retarded. Hence an asymmetry is introduced into the reflected wavefront and the dominant aberration in the image is coma.

IX. CONCLUDING REMARKS

All of the results needed to calculate the aberrations of a general centered optical system to third order are now in place. By centered we mean there is a single axis of symmetry passing through the vertices of all the optical surfaces. It is well to remember that these results are not exact, but for most systems used in optical astronomy they are sufficient. Exact image characteristics derived from ray tracing can, of course, be used to supplement the third-order results.

A comparison of the form of the coefficients in this chapter with those in, for example, Born and Wolf shows a significant difference in notation. The results given in Chapter 5 of Born and Wolf are derived in terms of Seidel variables, while our results are given in terms of actual variables. Though the two approaches give the same final system aberrations, the representation we have chosen is more convenient to use in practice.

With the availability of sophisticated computer ray-tracing programs, the reader may question the necessity of a detailed development of these analytical results. From the point of view of an optical designer starting from scratch to choose a suitable system for a particular application, the analytical results are preferred because one can usually determine rather quickly whether a given type of system is appropriate. Once a basic arrangement of optical elements has been selected, a computer can be used to optimize the system and check image characteristics.

Getting the required aberration relations has been a lengthy process. It would have been sufficient to simply present the final results without the derivations, but for the reader who is venturing into this field for the first time it is useful to see the source of the results. Discussions in subsequent chapters are directed toward finding the characteristics of systems, with the results above available for reference.

We also have the results needed to calculate the aberrations introduced when one or more of the optical elements in a system is decentered. The general treatment is complicated when more than one element is decentered, and we limit our following discussion to those cases in which one element is decentered.

There is one more topic of aberration theory, which is covered in a later chapter. In Chapter 14 we use Fermat's principle as a starting point to derive the characteristics of diffraction grating surfaces. These results, when combined with those above, will allow us to discuss the characteristics of a variety of spectrographic instruments.

REFERENCES

Bowen, I. (1960). "Stars and Stellar System I, Telescopes," Chap. 4. Univ. of Chicago Press, Chicago, Illinois.
Merritt, F. (1962). "Mathematics Manual." McGraw-Hill, New York.

BIBLIOGRAPHY

For an introduction to aberrations see the intermediate-level texts in optics listed in the bibliography in Chapter 2.

Advanced theory of aberrations
Born, M., and Wolf, E. (1980). "Principles of Optics," 6th ed. Pergamon, Oxford.
Welford, W. (1974). "Aberrations of the Symmetric Optical System." Academic Press, New York.

Chapter 6 | Reflecting Telescopes

Reflecting telescopes and their associated instrumentation are the principal tools of the observational astronomer. In this chapter we consider the characteristics of the reflecting telescope in its various forms. Although refracting telescopes are still in use, they are relatively few in number and do not compete in light-gathering power with the large reflectors. We choose to consider reflecting telescopes only.

In the discussions to follow we consider the various kinds of reflectors, their inherent aberrations for a distant object field, and their advantages and limitations. Because of the aberrations there are definite field limitations, which are noted for each type. The aberration calculations are based on the results of Chapter 5, with the results of the calculations presented in terms of angular measure as seen on the sky (or object field). These measures are given in both analytical and graphical form, with the latter given in units of arc-seconds. Although close attention is given to the sign convention in deriving the aberration formulas, the final angular results are given without regard for sign. The one exception to this is the field curvature, for which the sign is essential.

Descriptions of many of the types discussed here appear in the literature, and references are given at the end of the chapter. In our discussion we cover a large number of telescope types with a common notation to facilitate comparison between them. In this chapter and suceeding ones our discussion assumes the reader has digested the main themes in the preceding chapters. If this is not the case, then, at a minimum, Section 2.V, Chapter 4, and

Sections 5.IV, 5.VI, and 5.VII should be reviewed. Only pure mirror systems are considered in this chapter, including a brief discussion of mirror testing in Section 6.V. Schmidt telescopes and systems with refracting corrector systems are the subjects of Chapters 7-9.

I. PARABOLOID

The single-mirror paraboloid is the simplest telescope that is free from spherical aberration, a result noted in Chapter 5. The paraboloid is almost always used with the aperture stop at the mirror and thus the remaining aberrations can be taken directly from Table 5.4. Because we are interested primarily in the angular aberrations, we divide each of the transverse aberrations by s' to get the desired results. Setting $m = 0$ we get

$$\text{ASC} = \theta(y^2/R^2) = \theta/16F^2, \qquad (6.1.1)$$

$$\text{AAS} = \theta^2(2y/R) = \theta^2/2F, \qquad (6.1.2)$$

where y is the height of a marginal ray at the mirror and the telescope focal ratio $F = |R/4y|$. Note that we have not carried any minus signs forward from Table 5.4 because the absolute size of the aberration is usually the item of primary concern. If we had paid attention to the signs, we would have found a coma flare directed away from the center of the field, where $\theta = 0$, and a tangential astigmatic image lying closer to the mirror than the sagittal line image.

Results from Eqs. (6.1.1) and (6.1.2) are shown in Fig. 6.1 for three focal ratios. The principal item to notice in Fig. 6.1 is the dominance of coma for small field angles, which sets the limit to the radius of the field over which the image quality can be considered "good." By "good" we mean an angular blur size that is less than or equal to the blur given an otherwise perfect image by atmospheric distortion. In our discussions we take the typical blur due to atmospheric effects as 1 arc-sec.

Figure 6.2 shows the distribution of rays at the Gaussian image for an $f/10$ paraboloid at $\theta = 9$ arc-min off-axis, with the asymmetric character of the comatic image clearly evident. A total of 112 rays were traced, hence each number in Fig. 6.2 is approximately the percent of the total energy in that part of the image. The full extent of the image in this case is set by the tangential coma, which is 3 times the sagittal coma. With the atmospheric blur set at 1 arc-sec, we can use Fig. 6.1 and the relation between tangential and sagittal coma to determine the limiting field radius for good images.

Angular aberration (arc-sec)
1.5
1.0
0.5
0
5
10
15
20
θ (arc-min)
4
8
10
4
8
10

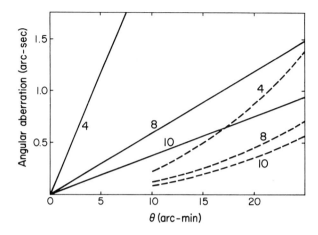

Fig. 6.1. Angular aberrations of paraboloid in collimated light at selected focal ratios. Solid lines: sagittal coma; dashed curves: astigmatism. The number on each curve is the focal ratio. See Eqs. (6.1.1) and (6.1.2).

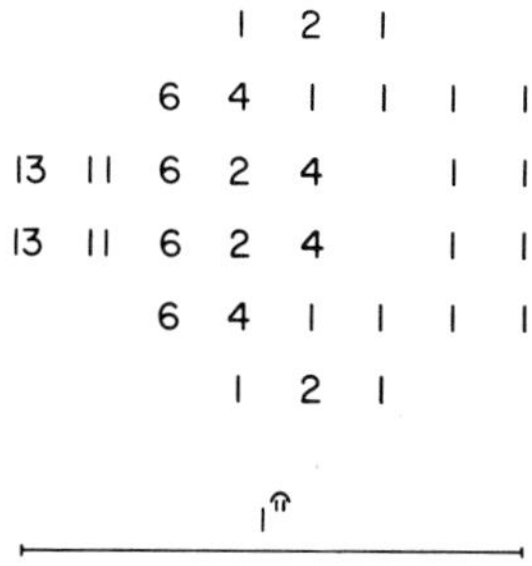

1 2 1
6 4 1 1 1 1
13 11 6 2 4 1 1
13 11 6 2 4 1 1
6 4 1 1 1 1
1 2 1
1″

Fig. 6.2. Distribution of rays over image of $f/10$ paraboloid at a field angle of 9 arc-min.

Table 6.1

Limiting Field Radius for Good Images: Paraboloid Telescope

F	θ (arc-min)
4	1.42
8	5.69
10	8.89

The results are given in Table 6.1, from which it is clear that the paraboloid is limited to small fields, especially for small focal ratios.

With coma as the aberration that limits the size of the field, the placement of the aperture stop at any other position does not improve the image quality. Coma is independent of stop position when spherical aberration is zero, as noted in Section 5.V, and reducing astigmatism has no effect on the overall blur size.

The remaining item to consider is the field curvature. Using Table 5.7 we find

$$\kappa_p = 2/R, \quad \kappa_m = -2/R, \quad u_m = f\theta^2/2, \quad (6.1.3)$$

where u_m is the sag of the median surface midway between the tangential and sagittal focal surfaces. Given the limiting field radii in Table 6.1, the reader can verify that for any practical focal length f the median image surface is essentially flat.

In summary, then, the paraboloid telescope is limited to small fields with coma setting the field limit. All other aberrations are negligible over this field.

II. TWO-MIRROR TELESCOPES

We introduced the topic of two-mirror telescopes in Chapter 2 with schematic diagrams of two types, Cassegrain and Gregorian, in Fig. 2.7, as well as a set of definitions of normalized parameters with which to describe any two-mirror telescope. Selected items from Section 2.V are summarized in Table 6.2 for convenient reference.

It is instructive to study the relations between the normalized parameters in Table 6.2 because they define the bounds on the parameters for each of the possible telescope types. For all types we require that the final image of the telescope is real.

Table 6.2

Normalized Parameters for Two-Mirror Telescopes

$k = y_2/y_1$
$\rho = R_2/R_1$
$m = f/f_1 = F/F_1$
$mkf_1 =$ distance from secondary to focal surface
$\beta f_1 =$ back focal distance

$$m = \frac{\rho}{\rho - k}, \quad \rho = \frac{mk}{m-1}, \quad k = \frac{1+\beta}{m+1}$$

If the primary is concave, hence f_1 positive, the requirement of a real final image means $mk>0$. If m and k are positive the telescope type is Cassegrain; if their signs are negative the telescope type is Gregorian. In both cases $|k|<1$ to ensure that some light reaches the primary.

If the primary is convex, hence f_1 negative, then a real final image requires that mk is negative. In this case the secondary must be larger than the primary, hence $k>1$, and m is negative. This type of telescope, with its concave secondary, is the so-called inverse Cassegrain.

The different combinations of m, k, and ρ are summarized in Table 6.3. It is worth noting here that among the Cassegrains with concave secondary and the inverse Cassegrains are the so-called Couder and Schwarzschild designs discussed later in this chapter. The Cassegrain with flat secondary is not included in the analysis and discussion to follow.

We now proceed to find the aberration relations for two-mirror telescopes by using Eq. (5.6.9) and the aberration coefficients of the primary and secondary which precede it. Before writing the system aberration coefficients, W is written in terms of the normalized parameters: $W/R_2 = (k-1)/2\rho$, $W/s_2 = (1-k)/k$. We also express K_2 in terms of K_1 by using the condition for zero spherical aberration, which, from Eq. (5.6.10), is

$$K_2 = -\left(\frac{m+1}{m-1}\right)^2 + \frac{m^3}{k(m-1)^3}(K_1+1). \qquad (6.2.1)$$

With these substitutions, and after straightforward but tedious algebra, the two-mirror aberrations given in Table 6.4 are found.

The choice of parameters used in Table 6.4 is arbitrary and different combinations may be more convenient, depending on the application. The advantage of expressing the system coefficients in terms of m and β is that certain important conclusions are more easily deduced.

Getting from the system coefficients in Table 6.4 to the angular aberrations requires substituting each B_s in turn into Eq. (5.6.8), where, as noted in

Table 6.3

Parameter Combinations for Two-Mirror Telescopes[a]

m	k	ρ	Type	Secondary
>1	>0	>0	Cassegrain	Convex
$=1$	>0	∞	Cassegrain	Flat
0 to 1	>0	<0	Cassegrain	Concave
<0	<0	<0	Gregorian	Concave
<0	>1	>0	Inverse Cassegrain	Concave

[a] For $m=1$, $k=(1+\beta)/2$.

II. Two-Mirror Telescopes

Table 6.4

Aberration Coefficients for Two-Mirror Telescopes[a]

$$B_{2s} = \frac{\theta}{m^2 R_1^2}\left[1 + \frac{m^2(m-\beta)}{2(1+\beta)}(K_1+1)\right] = \frac{\theta}{4f^2}[\ ]$$

$$B_{1s} = \frac{\theta^2}{mR_1}\left[\frac{m^2+\beta}{m(1+\beta)} - \frac{m(m-\beta)^2}{4(1+\beta)^2}(K_1+1)\right] = -\frac{\theta^2}{2f}[\ ]$$

$$B_{0s} = \frac{\theta^3(m-\beta)(m^2-1)}{4m^2(1+\beta)^2}\left[m + 3\beta + \frac{m^2(m-\beta)^2}{2(1+\beta)(m^2-1)}(K_1+1)\right]$$

[a] In terms of m and β, spherical aberration is zero according to the following relation:

$$K_1 + 1 = \frac{(m-1)^3(1+\beta)}{m^3(m+1)}\left[K_2 + \left(\frac{m+1}{m-1}\right)^2\right].$$

Section 5.VI, $s'_2/n'_2 = kf$ and $y_1/y_2 = 1/k$. Take care to note that Eq. (5.6.8) gives tangential coma; our result below is for sagittal coma. To get the angular aberration projected on the sky, the transverse aberration is divided by f. Final results are given in Table 6.5, with quantities in brackets taken from Table 6.4. At this point it is important to point out that the relations in Tables 6.4 and 6.5 apply to any type of two-mirror telescope for which the spherical aberration is zero.

Before discussing the characteristics of specific telescope types, we give the general relations for the image surface curvatures based on the results in Section 5.VII. As noted following Eq. (5.7.12), the coefficient B_{1s} and the angle θ' are referenced to the last surface in the system, hence the secondary. The relation between this θ' and the field angle θ is derived by noting that the focal-surface-to-secondary distance is k times smaller than f. Hence a point on the image surface, which subtends angle θ on the sky, subtends angle θ/k at the secondary.

Table 6.5

Angular Aberrations for Two-Mirror Telescopes[a]

$$\text{ASC} = \frac{\theta y^2}{4f^2}[\ ] = \frac{\theta}{16F^2}[\ ]$$

$$\text{AAS} = \frac{\theta^2 y}{f}[\ ] = \frac{\theta^2}{2F}[\ ] \quad \text{ADI} = B_{0s}$$

[a] $\text{ATC} = 3 \cdot \text{ASC}$.

The coefficient B_{1s} referenced to the secondary is calculated using the relation in Table 5.7. The relation between this result, denoted $B_{1s}(\text{sec})$, and that given in Table 6.4, denoted $B_{1s}(\text{pri})$, is

$$k^2 B_{1s}(\text{sec}) = B_{1s}(\text{pri}).$$

Therefore $B_{1s}(\text{sec})/\theta'^2 = B_{1s}(\text{pri})/\theta^2$, and we find

$$\kappa_p = \frac{2}{R_1} \left[\frac{m(m-\beta) - (m+1)}{m(1+\beta)} \right], \quad (6.2.2)$$

$$\kappa_m = \frac{2}{mR_1} \left[\frac{(m^2-2)(m-\beta) + m(m+1)}{m(1+\beta)} - \frac{m(m-\beta)^2}{2(1+\beta)^2}(K_1+1) \right].$$

With all of the necessary relations now in hand, we turn our attention to a discussion of the characteristics of specific telescope types. The categories considered are the so-called classical telescopes, those for which the primary mirror is a paraboloid, and the aplanatic telescopes, those with zero coma. Brief mention is also made of some less important types, such as the Dall-Kirkham with its spherical secondary.

A. CLASSICAL TYPE

This category of two-mirror telescopes is one for which $K_1 = -1$. The condition for zero spherical aberration then requires that the conic constant of the secondary is

$$K_2 = -\left(\frac{m+1}{m-1}\right)^2. \quad (6.2.3)$$

For the Cassegrain, $m > 0$ and the secondary is a hyperboloid; for the Gregorian and inverse Cassegrain, $m < 0$ and the secondary is an oblate ellipsoid.

With the substitution of $K_1 = -1$ in Eqs. (6.2.2) and the formulas in Tables 6.4 and 6.5, the expressions are much simplified. For convenient reference, these relations are given in Table 6.6

The first thing to note about the relations in Table 6.6 is that the coma is exactly the same as that of a paraboloid of the same focal ratio, as given in Eq. (6.1.1). Note also that this is true for either a Cassegrain or a Gregorian, hence neither type has an advantage with respect to this aberration. The image of an $f/10$ Cassegrain is essentially that shown in Fig. 6.2.

To evaluate the astigmatism, we note that β is typically a small positive number of the order of a few tenths, while $|m|$ is typically 10 or more times larger. A good measure of the astigmatism is thus obtained by setting $\beta = 0$,

II. Two-Mirror Telescopes

Table 6.6

Aberrations of Classical Two-Mirror Telescopes

$$\text{ASC} = \frac{\theta}{16F^2}$$

$$\text{AAS} = \frac{\theta^2}{2F}\left[\frac{m^2+\beta}{m(1+\beta)}\right]$$

$$\kappa_m = \frac{2}{R_1}\left[\frac{(m^2-2)(m-\beta)+m(m+1)}{m^2(1+\beta)}\right]$$

with the result that $\text{AAS} = m\theta^2/2F$. Comparison of this result with Eq. (6.1.2) shows that a classical telescope whose focus is at the primary vertex has astigmatism $|m| \times$ larger than that of a paraboloid of the same F. As in the case of coma, there is no discriminant due to astigmatism between Cassegrain and Gregorian types.

The astigmatism can also be put in the form $\text{AAS} = \theta^2/2F_1$. Thus astigmatism for $\beta = 0$ depends only on the focal ratio of the primary mirror. Comparison of the tangential coma blur size with the astigmatic blur diameter shows that coma is almost always the aberration that sets the limiting field size for good images. Verification of this statement is left to the reader.

Looking at the curvature of the median image surface in the case $\beta = 0$, we see it is approximately $2(m+1)/R_1$. This relation is not exact, but it illustrates two features of the image surface. First, the sign of κ_m is opposite for the Cassegrain and Gregorian types; the surface of best images for the Cassegrain (Gregorian) is concave (convex) as seen from the secondary. Second, this surface is more strongly curved for larger $|m|$. This, however, is rarely a limitation because the field covered is usually smaller in angle when m is larger.

In summary, the classical two-mirror telescope is limited to small fields with coma setting the field limit for good images. The astigmatism is larger than that of a paraboloid, but for small fields this is rarely a limiting factor. Because of the small field size, distortion is typically a few hundredths of an arc-second at most and thus much smaller than the atmospheric blur. Compared to the asymmetry of a comatic image, distortion is not important. The only differences between the Cassegrain and Gregorian of the classical type are the sign and magnitude of the image surface curvature but, given the relatively small usable fields, these differences are of little consequence.

B. APLANATIC TYPE

The classical telescope is clearly limited in field coverage by the presence of coma in the off-axis images. In this section we consider the category of telescopes for which, to third order, coma is zero. Any optical system in which both spherical aberration and coma are absent is called an aplanat. In recent years the aplanatic Cassegrain telescope, or Ritchey-Chretien as it is commonly called, has been the overwhelming choice of builders of large telescopes of 2-m aperture or larger, including the 2.4-m Hubble Space Telescope. Thus this class of telescope has been carefully studied and merits our close attention. An extensive article by Wetherell and Rimmer is an additional source of information on aplanatic telescopes.

It should not be surprising that both spherical aberration and coma can be eliminated in a system with two conic mirrors. A glance at Eq. (6.2.1) shows that there are two free parameters, the conic constants of the mirrors. One conic constant is chosen to make B_{2s} in Table 6.4 zero, after which the condition for zero spherical aberration sets the other. Thus the conic constants for an aplanat are

$$K_1 = -1 - \frac{2(1+\beta)}{m^2(m-\beta)}, \qquad (6.2.4)$$

$$K_2 = -\left(\frac{m+1}{m-1}\right)^2 - \frac{2m(m+1)}{(m-\beta)(m-1)^3}. \qquad (6.2.5)$$

For the Ritchey-Chretien (hereafter called RC) the primary is now a hyperboloid, as is the secondary. The conic constant for the secondary of the RC is more negative than for the classical Cassegrain. For the aplanatic Gregorian (hereafter called AG) the primary is now an ellipsoid. The conic constant for the secondary of the AG is more negative than that of the classical Gregorian, provided $|m| > 1$, but the conic is still ellipsoidal. In each case the two mirrors have been "bent" in the same direction in the manner shown in Fig. 4.8. However, the direction of deformation for the mirrors of the RC is opposite that for the AG, as the reader can easily verify.

Substitution of Eq. (6.2.4) into Eqs. (6.2.2) and the coefficients in Table 6.4 gives the aberrations for the aplanatic telescopes, with the results given in Table 6.7.

As with the classical telescopes, we let $\beta = 0$ to determine the approximate magnitudes of the aberrations. The results are

$$\text{AAS} = \frac{\theta^2}{2F}\left(m + \frac{1}{2}\right), \qquad \text{ADI} = \frac{\theta^3}{4}(m^2 - 2), \qquad \kappa_m = \frac{2}{R_1}(m+1).$$

Table 6.7

Aberrations of Aplanatic Two-Mirror Telescopes

$$\text{AAS} = \frac{\theta^2}{2F}\left[\frac{m(2m+1)+\beta}{2m(1+\beta)}\right]$$

$$\text{ADI} = \theta^3 \frac{(m-\beta)}{4m^2(1+\beta)^2}[m(m^2-2)+\beta(3m^2-2)]$$

$$\kappa_m = \frac{2}{R_1} \cdot \frac{(m+1)}{m^2(1+\beta)}[m^2-\beta(m-1)]$$

Compared to the classical type at the same focal ratio, the astigmatism for the RC is larger and that of the AG is smaller. At a given R_1, the curvature of the median image surface is larger for the RC than for the AG, with the curvatures again of opposite sign. A comparison of κ_m with κ_p in Eqs. (6.2.2) shows a median surface more strongly curved than the Petzval surface for the RC, but less strongly curved for the AG. The distortion is the same for both types of aplanatic telescope and is slightly less than for a classical type with $\beta = 0$.

In summary, the aplanatic two-mirror telescope has a field limit for good images set by astigmatism with a smaller field for the RC than for the AG. Note that the image blur due to astigmatism is symmetric and therefore the centers of the images can be located more accurately. In terms of field curvature, the AG is also better than the RC. At the edge of the usable field of an aplanatic telescope the distortion is usually a few hundredths of an arc-second and may need to be taken into account in certain types of observations.

C. COMPARISON OF CLASSICAL AND APLANATIC TYPES

It is now appropriate to take a specific set of parameters and show all of the characteristics of each type of telescope. The parameters chosen are characteristic of those for a typical telescope with $\beta = 0.25$ selected to give an accessible focal surface. Each telescope has the same primary mirror and overall focal ratio. A listing of these parameters, including the conic constants, is given in Table 6.8.

The important characteristics of each telescope type, with the parameters of Table 6.8, are given in Table 6.9. In addition to the angular aberrations,

Table 6.8
Parameters for Two-Mirror Telescopes[a]

Parameter	CC	CG	RC	AG
K_1	−1.000	−1.000	−1.0417	−0.9632
K_2	−2.778	−0.360	−3.1728	−0.4052

[a] CC, classical Cassegrain; CG, classical Gregorian; RC, Ritchey-Chretien; AG, aplanatic Gregorian; $F_1 = 2.5$, $|F| = 10$, $\beta = 0.25$, $|m| = 4$.

entries are included that provide a normalized measure of the size of each telescope type.

From the results in Table 6.9 we can deduce the approximate field angle at which the dominant angular aberration is equal to the diameter of a star image blurred by atmospheric effects or "seeing." If the blur diameter is 1 arc-sec, the field angle at which the aberration blur equals the "seeing" blur is about 9 arc-min for the classical telescopes dominated by coma, about 18 arc-min for the RC, and about 20 arc-min for the AG. Thus the field diameter for the parameters chosen is roughly a factor of two larger for the aplanatic type of telescope.

From the astigmatic surface curvatures in Table 6.9 we find that, in absolute value, the median surfaces have greater curvature and the Petzval surfaces have smaller curvature for the Cassegrain types than for the Gregorian types. The median surface curvature is also larger for the aplanatic

Table 6.9
Characteristics of Two-Mirror Telescopes[a,b]

Parameter	CC	CG	RC	AG
m	4.00	−4.00	4.00	−4.00
k	0.25	−0.417	0.25	−0.417
$1 - k$	0.75	1.417	0.75	1.417
mk	1.000	1.667	1.000	1.667
ATC	2.03	2.03	0.00	0.00
AAS	0.92	0.92	1.03	0.80
ADI	0.079	0.061	0.075	0.056
$\kappa_m R_1$	7.25	−4.75	7.625	−5.175
$\kappa_p R_1$	4.00	−8.00	4.00	−8.00

[a] Parameters are those of telescopes in Table 6.8. Aberrations are given at a field angle of 18 arc-min in units of arc-seconds.

[b] Coma is given in terms of tangential coma.

type than for its classical counterpart. Because $R_1 < 0$, the astigmatic surfaces as seen from the secondary are concave and convex for the Cassegrain and Gregorian types, respectively.

If aberrations were the only discriminant of the four telescope types in Table 6.9, the aplanatic Gregorian would emerge as the preferred choice. Other factors, however, strongly favor the RC, and it is this type that has been the overwhelming choice for new large telescopes over the past two decades. The reason for this choice lies in rows 2-4 in Table 6.9.

Recall that k is the ratio of diameters, secondary over primary, for an on-axis light bundle, and thus k^2 is the minimum fractional area of the primary obscured by the secondary. The parameter $(1-k)$ is the separation of the primary and secondary in units of f_1, while mk is the distance from the secondary to the final focal surface in the same units.

The obstruction of the light by the secondary in the Gregorian is clearly larger than in the Cassegrain, hence the latter is slightly favored. Comparing values of $(1-k)$ for the Cassegrain and Gregorian types, we find that the primary-secondary separation is almost 1.9 times larger for the Gregorian. We also find that the distance from the secondary to the focal surface is nearly 70% larger for the Gregorian. Thus for a given focal length and primary and final focal ratio, the physical length of the Gregorian is substantially greater.

This greater length has two very significant impacts on the choice of a telescope and the cost of an observatory facility. First, the cost of a building and dome needed to house the telescope is significantly greater for a larger telescope. For a large telescope the building costs are usually comparable to the cost of the telescope. Second, the cost of the Gregorian telescope itself is greater because the framework supporting the mirrors is longer and more massive. This framework must keep the mirrors in proper alignment if the image quality is to be held to the values given in Table 6.9. In Section III to follow we consider the effects of misalignment of the primary and secondary and show that significant aberrations can be introduced if the mirrors are not properly aligned.

For the same overall focal length, it is possible to make a Gregorian with the same physical length as a Cassegrain. To do so, however, requires a primary with a smaller focal ratio, hence the magnification m is also larger. A consequence of these changes is a primary mirror that is more difficult to make and a telescope that is significantly more sensitive to misalignment than the Cassegrain.

Thus for most large telescopes intended for stellar observations, the Ritchey-Chretien is the favored type. The one feature of the Gregorian that might be important for certain observations is its real exit pupil. A physical stop located here will act to suppress stray light scattered from the support structure of the telescope. Unless this specific feature of the Gregorian is an essential one, the less expensive RC is preferred.

D. HYBRID TYPES

Most large telescopes are provided with interchangeable secondaries with each primary-secondary combination giving a different telescope focal length and focal ratio. With a secondary other than the one designed for the Cassegrain focus, the telescope focus is usually located at a different physical position. For a telescope in an equatorial mount plane mirrors redirect the light, with the final beam directed along the polar axis of the telescope to the so-called coudé focus. For a telescope in an altitude-azimuth (alt-az) mount a plane mirror directs the light along the altitude axis to the so-called Nasmyth focus. In both cases m and β are larger than at the Cassegrain focus.

For a classical telescope the conic constant of the secondary is given by Eq. (6.2.3) for each m selected, and the telescope is still of the classical type. The relations in Table 6.6 apply, with the parameters for the Cassegrain replaced by the new values.

A Ritchey-Chretien primary in combination with a different secondary, on the other hand, is no longer aplanatic, and the results in Table 6.7 do not apply. This type of telescope is not a Ritchey-Chretien and we choose to call it a *hybrid* telescope. The aberration coefficients given in Table 6.4 apply to hybrid telescopes, provided K_1 for the original RC primary is used. Denoting the parameters for the RC as m_c and β_c, the conic constant of the primary is, according to Eq. (6.2.4), given by

$$K_1 = -1 - \frac{2(1+\beta_c)}{m_c^2(m_c - \beta_c)}.$$

The conic constant of the secondary is set by the condition that the spherical aberration of the hybrid telescope is zero. Substituting K_1 above into the last relation in Table 6.4 gives

$$K_2 = -\left(\frac{m+1}{m-1}\right)^2 - \frac{2m^3(m+1)(1+\beta_c)}{(m-1)^3(1+\beta)(m_c-\beta_c)m_c^2}, \quad (6.2.6)$$

where m and β are parameters for the hybrid. Substituting for K_1 in the coma and astigmatism coefficients in Table 6.4 gives

$$B_{2s} = \frac{\theta}{4f^2}\left[1 - \left(\frac{m}{m_c}\right)^2\left(\frac{m-\beta}{1+\beta}\right)\left(\frac{1+\beta_c}{m_c-\beta_c}\right)\right],$$

$$B_{1s} = -\frac{\theta^2}{2f}\left[\frac{m^2+\beta}{m(1+\beta)} + \frac{m}{2m_c^2}\left(\frac{m-\beta}{1+\beta}\right)^2\left(\frac{1+\beta_c}{m_c-\beta_c}\right)\right], \quad (6.2.7)$$

where f is the focal length of the hybrid telescope. It is evident from Eqs.

(6.2.7) that the aberrations are different from those of the aplanatic telescope and that coma is not zero. A good measure of the amounts of coma and astigmatism present is found by setting both β and β_c to zero, with the results

$$\text{ASC} = \frac{\theta}{16F^2}\left[1 - \left(\frac{m}{m_c}\right)^3\right], \quad \text{AAS} = \frac{\theta^2}{2F}\left[m + \frac{1}{2}\left(\frac{m}{m_c}\right)^3\right]. \quad (6.2.8)$$

For typical values of m/m_c, three or more, the coma of the hybrid telescope is much larger than that of the classical type of the same focal ratio and, although the astigmatism is also larger, the size of the usable field is set by coma.

Dropping the one in the expression for ASC in Eqs. (6.2.8) and ignoring the minus sign, we find $\text{ASC} = F\theta/16(m_c F_1)^3$. For a fixed ASC at the edge of the usable field, $F\theta$ is a constant for a given RC primary. Because $f\theta$ is the linear radius of this field at the hybrid focus, the larger the magnification of the hybrid secondary the smaller is the usable field in angular measure. Although the field size is relatively small, the observations made at a coudé or Nasmyth focus are most often made on or near the axis, where coma is not significant.

E. OTHER TWO-MIRROR TELESCOPES

In addition to the classical and aplanatic two-mirror telescopes, there are other less common types that deserve comment. Because each of these types has one or more serious drawbacks, our discussion of each is brief.

The Dall-Kirkham telescope is one in which the secondary is spherical and the primary is ellipsoidal, with the appropriate value of K_1 found from the relation in Table 6.4. With this value of K_1 it is straightforward to find the coma and, for the same normalized parameters, compare its value with that of a classical Cassegrain. For $\beta = 0$ the Dall-Kirkham has coma that is $(m^2+1)/2$ times larger than that of the classical Cassegrain, hence the field of good images for the Dall-Kirkham is smaller by the same factor. All other aberrations are negligible over this field.

Although the Dall-Kirkham is severely limited in its field coverage, the mirrors are relatively easy to build and test, as discussed in Section 6.V, and several telescopes of this type have been built. One other advantage of the Dall-Kirkham is that its on-axis image quality is relatively insensitive to misalignments between the mirrors, as compared to the classical and aplanatic types.

The remaining two-mirror telescopes considered are those which are variations of the aplanatic type, specifically those for which another aberration is corrected. Because the conic constants of the mirrors in an aplanat are chosen to give zero spherical aberration and coma, elimination of another aberration will put restrictions on the remaining normalized parameters. The available choices are easily found by setting each expression in Table 6.7 equal to zero in turn, with a specific combination of m and β now required. This combination, in turn, places restrictions on the remaining parameters.

The zero-distortion type is of little practical importance because distortion is quite small in two-mirror telescopes with small fields of view, and we will not discuss this type. The remaining choices are the zero-astigmatism type, or anastigmatic aplanat, and the flat-field aplanat.

For the anastigmatic aplanat the pertinent relations between the parameters are

$$\beta = -m(2m+1), \quad k = 1-2m, \quad mk = m(1-2m).$$

The condition for a real final focus requires magnification in the range $0 < m < 0.5$ when the primary is concave. For any m in this range, the secondary is also concave and the focal surface is located between the mirrors. This type of telescope, the so-called Couder, therefore suffers from the problem that the focal surface is relatively inaccessible. For a reasonable choice of m, say 0.25, it also has a relatively large secondary obscuration compared to the Ritchey-Chretien. One final thing to note is that the telescope focal length is one-half the distance between the primary and secondary mirrors. A diagram of a Couder design is shown in Fig. 6.3.

Another type of anastigmatic aplanat is found if the primary is convex and $k > 1$. From the relations above we find $m < 0$, hence there is a real final focus and the configuration is that of an inverse Cassegrain. For m in the range $-0.5 < m < 0$, the focal surface lies between the mirrors because $\beta > 0$. (Recall that βf_1 is the back focal distance, as defined in Section 2.V, with the focus outside the space between the mirrors when $\beta f_1 > 0$. For an inverse Cassegrain, $f_1 < 0$ and the focus is outside the mirrors when $\beta < 0$.)

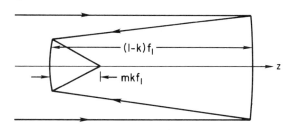

Fig. 6.3. Couder anastigmat with $m = 0.25$, $k = 0.5$.

II. Two-Mirror Telescopes

A sketch of this configuration with $\beta > 0$ shows that a blocking plate must be centered in the incident beam to prevent the focal surface from seeing the incident light directly. This configuration also has the problem that a significant fraction of the incident light is reflected back through the hole in the secondary.

For $m < -1$ some of the light reflected from the secondary passes outside the boundary of the primary, and if m is sufficiently negative a significant fraction reaches the focus. A feasible configuration of this type is one in which each mirror is a sphere. Substituting $\beta = -m(2m+1)$ into Eqs. (6.2.4) and (6.2.5), and setting $K_1 = K_2 = 0$, gives $m = -(1+\sqrt{5})/2$. The resulting configuration is the concentric Schwarzschild anastigmat, with the mirrors and curved focal surface having the same center, as shown in Fig. 6.4. The reader can verify that the fraction of light vignetted by the primary is 0.2 for this telescope.

The flat-field aplanat is defined by $\kappa_m = 0$, with the relations for selected parameters given by

$$\beta = \frac{m^2}{m^2 - 1}, \qquad mk = \frac{m(m^2 + m - 1)}{m^2 - 1}.$$

An analysis of these relations leads to two possible types: Cassegrain with concave secondary and focus between the mirrors, and inverse Cassegrain. Each of these types suffers from the same problems of image inaccessibility and relatively large vignetting as the corresponding anastigmat.

For more details on all of these variations of the aplanat, the reader should consult the article by Wetherell and Rimmer.

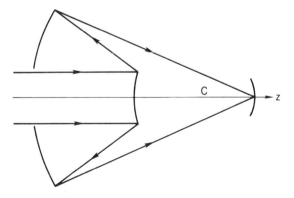

Fig. 6.4. Schwarzschild concentric anastigmat with C the center of curvature of the surfaces. Parameters are $m = -(1+\sqrt{5})/2$, $k = 2+\sqrt{5}$.

III. ALIGNMENT ERRORS IN TWO-MIRROR TELESCOPES

We now consider the consequences of an error in the position of the secondary mirror relative to the primary in a two-mirror telescope. This position error can be a decenter or tilt of the secondary, either of which is a misalignment between the mirrors. The error can also be an axial displacement of the secondary toward or away from the primary, in which case the error is called despace. In the discussion to follow, aberrations introduced by misalignment are treated separately from those resulting from despace.

A. TILT OR DECENTER MISALIGNMENT

In analyzing the effect of misalignnment, we begin by noting that the aperture stop of the telescope is the primary and the reference axis for the secondary is the axis through the vertex of the primary. In the discussion of misalignment, coma is the only aberration evaluated because astigmatism is negligible by comparison, as was the case for the Schmidt camera.

One possible layout of a misaligned secondary is shown in Fig. 6.5, where the secondary is decentered by an amount l in the y direction and tilted through an angle α about a line perpendicular to the plane of the diagram. In this particular case the displacement of the center of the stop from the axis of the secondary, its symmetry axis, is simply the sum of the separate displacements due to decenter and tilt. In the general case the displacements at the stop due to decenter and tilt are not colinear and must be combined by vector addition. We consider only the case shown in Fig. 6.5 because it is sufficient to establish the limits on misalignments.

From the geometry of Fig. 6.5 we see that

$$L' = -(l + \alpha W), \qquad \psi = -(\theta + \alpha), \tag{6.3.1}$$

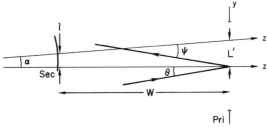

Fig. 6.5. Secondary (Sec) in two-mirror telescope decentered by l and tilted by angle α with respect to the axis of the primary (Pri). The relation between parameters is given in Eq. (6.3.1).

III. Alignment Errors in Two-Mirror Telescopes

where L' is the distance from the center of the stop to the axis of the secondary and ψ is the angle between the reflected chief ray and the secondary axis. Substitution of Eqs. (6.3.1) into the coma coefficients in Tables 5.6 and 5.9 gives

$$B_2 = B_2(\text{sec}) + \frac{1}{R_2^2}\left[\frac{l}{R_2}\left[K_2 - \left(\frac{m+1}{m-1}\right)\right] - \alpha\left(\frac{m+1}{m-1}\right)\right], \quad (6.3.2)$$

where $n = -1$ for the secondary. The factor $B_2(\text{sec})$ is the coefficient for the secondary taken from the relations preceding Eq. (5.6.9) and is the coma coefficient for a properly aligned secondary. In this section we are interested only in the effect of the other part of Eq. (6.3.2), and we denote this part by $B_2(\text{mis})$.

Taking Eq. (6.3.2), substituting into Eq. (5.6.8), and dividing by f to get the angular coma gives

$$\text{ATC} = 3k^3 B_2(\text{mis}) y_1^2$$

$$= \frac{3y_1^2 k^3}{R_2^2}\left[\frac{l}{R_2}\left[K_2 - \left(\frac{m+1}{m-1}\right)\right] - \alpha\left(\frac{m+1}{m-1}\right)\right]. \quad (6.3.3)$$

The separate contributions due to decenter and tilt can be written as

$$\text{ATC(dec)} = \frac{l}{f}\frac{3(m-1)^3}{32F^2}\left[K_2 - \left(\frac{m+1}{m-1}\right)\right],$$

$$\text{ATC(tilt)} = \alpha\frac{3(m-1)(1+\beta)}{16F^2}. \quad (6.3.4)$$

From the relations in Eqs. (6.3.4) we note that coma due to tilt is not dependent on the conic constant of the secondary, while coma due to decenter depends on the back focal distance β only insofar as it is part of K_2. If F is replaced by mF_1, we find that the coma introduced by misalignment is larger for smaller primary focal ratios, assuming other parameters are unchanged. The reader can also verify that for $|m| \gg 1$ the differences between the comas of the classical and aplanatic telescopes become insignificant for a given misalignment.

To illustrate the effect of misalignment, we take the set of telescopes whose parameters are given in Table 6.8 and evaluate the relations in Eqs. (6.3.4) for each type. The results are given in Table 6.10 for $l/f = 5E - 5$ and $\alpha = 5E - 4$ radians.

From the results in Table 6.10 we see that the secondary in a Gregorian telescope is more sensitive to tilt than is the secondary in a Cassegrain. There is little difference in the coma introduced by decenter, although the alplanatic types have slightly larger values. Recall, however, that the

Table 6.10
Angular Tangential Coma for Misaligned Secondary[a,b]

	CC	CG	RC	AG
ATC(dec)	1.16	1.16	1.26	1.21
ATC(tilt)	0.73	1.21	0.73	1.21

[a] $\alpha = 5E-4$ radians $= 1.72$ arc-min; $l/f = 5E-5$.
[b] Angular coma is given in units of arc-seconds. Parameters of telescopes are given in Table 6.8.

Gregorian is significantly longer than the Cassegrain and hence the required tolerances are more easily met with a Cassegrain.

Returning to Eq. (6.3.2) we see that for any two-mirror telescope there is a combination of tilt and decenter that introduces no coma due to misalignment. Setting $B_2(\text{mis}) = 0$ gives

$$\alpha = -\frac{l}{R_2}\left[1 - K_2\left(\frac{m-1}{m+1}\right)\right]. \tag{6.3.5}$$

This combination of tilt and decenter is equivalent to a rotation of the secondary around an axis that is perpendicular to the axis of the primary and intersects it. The intersection of these two axes is often called the neutral point, and its location on the primary mirror axis depends on the type of telescope. For a Dall-Kirkham with its spherical secondary, the neutral point is the center of curvature of the secondary. This position is the expected one because a sphere is invariant to a rotation about its center. For the classical and aplanatic telescopes we get

$$\alpha(\text{classical}) = -\frac{l}{s_2},$$

$$\alpha(\text{aplanatic}) = -\frac{l}{s_2}\left[1 + \frac{1}{(m-\beta)(m-1)}\right], \tag{6.3.6}$$

where s_2 is the distance from the vertex of the secondary to the focal point of the primary. Hence the neutral point of a classical telescope is at the focus of the primary, while for an aplanatic telescope it is located between the focus and the vertex of the secondary. The existence of a neutral point can be used to advantage if the secondary is deliberately displaced to bring a different source on to a fixed detector as, for example, is done with many infrared telescopes.

B. DESPACE ERROR

We now turn our attention to the aberrations that appear when the error in placement of the secondary is one of despace. If the secondary is not at its nominal design position, then spherical aberration and coma are introduced and images at all points in the image field are degraded. Although spherical aberration is larger than coma, results are given for both aberrations. Astigmatism is also introduced but its size is negligible by comparison.

The starting point for the calculation of spherical aberration resulting from despace is B_{3s} from Eq. (5.6.10). Substituting this relation into Eq. (5.6.11) and dividing by f gives the angular spherical aberration (ASA), which is zero at the nominal design position. If m is replaced in terms of k and ρ, the only variable parameter remaining is k. The position of the secondary relative to the focal point of the primary is given by $s_2 = -kf_1$, hence a change in k means an axial shift of the secondary.

Taking the derivative of ASA from Eq. (5.6.10) with respect to k, and resubstituting for k and ρ in terms of m, we find

$$\frac{d(\text{ASA})}{dk} = \frac{1}{16F^3} \left\{ m(m^2 - 1) - (m-1)^3 \left[K_2 + \left(\frac{m+1}{m-1}\right)^2 \right] \right\}.$$

If dk is the change in k starting from the correct position of the secondary, then $d(\text{ASA})$ is the angular spherical aberration resulting from the despace, or simply ASA.

It is now a simple matter to evaluate this relation for different types of telescopes and determine the sensitivity to despace. Using the relations for K_2 in Eqs. (6.2.3) and (6.2.5) for classical and aplanatic telescopes, respectively, the results are

$$\text{ASA(cl)} = \frac{m(m^2-1)}{16F^3} \frac{ds_2}{f_1},$$

$$\text{ASA(apl)} = \frac{m(m^2-1)}{16F^3} \left[1 + \frac{2}{(m-1)(m-\beta)} \right] \frac{ds_2}{f_1}. \quad (6.3.7)$$

Comparison of the relations in Eqs. (6.3.7) shows that aplanatic telescopes are somewhat more sensitive to despace error than are the classical type, though only by 10–15% for typical parameter values such as those in Table 6.8. Comparing ASA for the aplanatic telescopes in Table 6.11 shows that the Ritchey–Chretien is more sensitive by a few percent to error in secondary position.

A final thing to note about Eqs. (6.3.7) is that, to a good approximation, ASA is inversely proportional to the cube of primary mirror focal ratio. Hence a telescope with a "faster" primary is more sensitive to despace

Table 6.11

Angular Aberrations for Despaced Secondary[a,b]

	RC	AG
ASA	0.912	0.846
ATC	0.252	0.174

[a] $dk = ds_2/f_1 = 0.001$
[b] Aberrations are given in units of arcseconds. ATC is given at a field angle of 18 arc-min. Parameters of telescopes are given in Table 6.8.

error. A similar conclusion was noted above for secondary misalignments and therefore, in general, a telescope with a faster primary is more sensitive to alignment errors of any kind.

The calculation of coma introduced by a despaced secondary proceeds in a similar way. We start with the coma coefficients for the mirrors substituted into Eq. (5.6.9), express all variables in terms of k, and differentiate with respect to k. The result for the aplanatic telescope is

$$\text{ATC} = \frac{3\theta}{16F^2}\left[\frac{(2m^2-1)(m-\beta)+2m(m+1)}{1+\beta}\right]\frac{ds_2}{f_1}. \quad (6.3.8)$$

Corresponding results for other two-mirror telescopes are of little importance because the coma already present in the off-axis images is dominant over that introduced by despace.

A comparison of the relative sizes of ASA and ATC for aplanatic telescopes with despaced secondary is given in Table 6.11, with the parameters of the telescopes taken from Table 6.8.

IV. THREE-MIRROR TELESCOPES

With the addition of a third mirror to a reflecting system there are additional degrees of freedom to minimize or eliminate aberrations. It is possible, for example, to design a system free of third-order aberrations with a flat image surface. The general analysis of a three-mirror system in terms of aberration coefficients is considerably more complicated than that of a two-mirror system. Because of this complexity, we will only outline the procedure and apply it to one practical example, that of the Paul–Baker

IV. Three-Mirror Telescopes

telescope. For a discussion of the general approach the reader should consult the reference by Robb.

A. GENERAL FORMULATION

Setting up the general relations that describe a three-mirror system is a straightforward extension of results given in Chapter 5. The system aberration coefficients, referenced to the primary, are

$$B_{ks} = B_{k1} + B_{k2}\left(\frac{y_2}{y_1}\right)^{k+1} + B_{k3}\left(\frac{y_3}{y_1}\right)^{k+1}, \qquad (6.4.1)$$

where the subscripts 1, 2, and 3 refer to the primary, secondary, and tertiary mirrors, respectively. Note that Eq. (6.4.1) is simply an extension of Eq. (5.6.9).

The other relation of interest is that for the Petzval curvature, which, from Table 5.7, is given by

$$\kappa_p = 2\left(\frac{1}{R_1} - \frac{1}{R_2} + \frac{1}{R_3}\right). \qquad (6.4.2)$$

The general procedure is now one of selecting the mirror separations and radii of curvature, and adjusting the conic constants and Petzval curvature to eliminate the third-order aberrations. For the reasons noted above, we will not pursue this general approach but instead give one special case.

B. EXAMPLE: PAUL-BAKER TYPE

The starting point for the Paul-Baker telescope, hereafter denoted PB, is a pair of paraboloids, a concave primary and a convex secondary whose focal points coincide. This combination, shown in Fig. 6.6, is an afocal reducer that converts an input beam of diameter D into a collimated output beam of diameter kD, where $k = y_2/y_1 = R_2/R_1$. The pair of paraboloids is obviously free of spherical aberration because $B_3 = 0$ for each mirror.

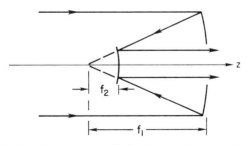

Fig. 6.6. Afocal beam reducer. Ratio of beam diameters $= k = f_2/f_1$.

The afocal reducer also has zero third-order coma and astigmatism, as the reader can easily verify by forming B_{1s} and B_{2s} with coefficients from Tables 5.4 and 5.6. The substitutions needed in the coefficients for the secondary are

$$\psi_2 = -\theta, \qquad m_2 = \infty, \qquad \frac{W_2}{R_2} = \frac{k-1}{2k}. \qquad (6.4.3)$$

We now add a concave spherical tertiary mirror whose center of curvature is at the vertex of the secondary mirror, as shown in Fig. 6.7. Note that the placement of the tertiary is similar to that of the mirror in a Schmidt camera. Because we have added a spherical mirror in collimated light the system now has spherical aberration and, to compensate, the paraboloidal secondary is replaced by a sphere, which introduces spherical aberration of opposite sign. If $R_3 = R_2$, these two contributions are equal in absolute magnitude and the system is again free of spherical aberration. It was first noted by Paul that this system is also free of third-order coma and astigmatism with a focal surface whose curvature $\kappa = 2/R_1$.

The Paul system can be generalized to give a system with zero Petzval curvature by choosing R_3 different from R_2. It was first shown by Baker that the third-order aberrations of this modified system are zero if the spherical secondary is replaced by an ellipsoid. An alternative solution for a flat-field three-mirror system, first noted by Angel, Wolff, and Epps, retains the spherical secondary but adds to it an aspheric figure to eliminate spherical aberration. For either type of solution it is necessary to include aspheric terms of higher order on the secondary to control higher-order aberrations. An example of a PB telescope with an $f/1$ primary and $f/2$ final focal ratio, and excellent image quality over a 1° diameter field, is discussed in the article by Angel *et al.*

Because of the excellent image quality achievable with the PB system, a more detailed look into its characteristics is in order. Substituting the spherical aberration coefficients for the secondary and tertiary mirrors from

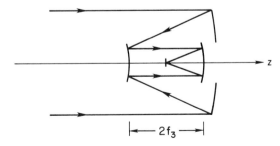

Fig. 6.7. Paul–Baker three-mirror telescope with focal length $f = f_3/k$.

Table 5.6 into Eq. (6.4.1) gives

$$B_{3s} = k^4 \left(\frac{1}{4R_3^3} - \frac{K_2+1}{4R_2^3} - \frac{b}{8} \right), \qquad (6.4.4)$$

where $k = y_2/y_1 = y_3/y_1$. The choice of parameters that makes $B_{3s} = 0$ includes

$$K_2 = 0, \quad b = \frac{2}{R_2^3}\left[\left(\frac{R_2}{R_3}\right)^3 - 1\right], \qquad (6.4.5a)$$

$$b = 0, \quad K_2 = -1 + \left(\frac{R_2}{R_3}\right)^3. \qquad (6.4.5b)$$

The first of these parameter combinations puts an aspheric figure on a convex spherical mirror; the other is an ellipsoidal mirror.

With either of these combinations, it is a straightforward exercise to verify that $B_{1s} = 0$, $B_{2s} = 0$, where the substitutions needed in the coefficients for the tertiary are

$$\psi_3 = \frac{\theta}{k}, \quad m_3 = 0, \quad \frac{W_3}{R_3} = 1 + \frac{R_2}{R_3}\left(\frac{1-k}{2}\right). \qquad (6.4.6)$$

The condition for zero Petzval curvature is found by first writing Eq. (6.4.2) in terms of k, with the result

$$\kappa_p = \frac{2}{R_1}\left[1 - \frac{1}{k}\left(1 - \frac{R_2}{R_3}\right)\right]. \qquad (6.4.7)$$

Setting $\kappa_p = 0$ gives $R_2/R_3 = 1 - k$.

The first large telescope of this type is the charge-coupled device (CCD)/transit instrument described by McGraw *et al.* and located at the Steward Observatory of the University of Arizona. This telescope has a 1.8-m $f/2.2$ primary with $k \approx 0.32$ and near-diffraction-limited images over a 1° field of view. Because of the baffles required to prevent light reflected directly from the primary or tertiary from reaching the detector, the vignetting by the central obscuration is approximately 22%. For further details, the reader should consult the article by McGraw *et al.*

V. TESTING OF LARGE MIRRORS

The aberration characteristics of the telescopes discussed above are derived assuming the mirrors are essentially perfect, thus the figure on the surface of each mirror is according to the prescription given by Eq. (5.1.1). Mirrors produced in the optical shop are not perfect, but if the actual figures agree with the prescriptions to within a specified tolerance the telescope performance is not noticeably degraded. In this section we discuss briefly some testing methods used to ensure that a mirror figure is within a given tolerance. This discussion is only a quick look at a large subject and the interested reader should consult the literature for details. A good overview of a large number of testing methods is given in the book by Malacara cited at the end of the chapter.

For our discussion we assume that a mirror is "perfect" if the root-mean-square (rms) deviation of the reflected wavefront compared to the ideal wavefront is $\leq \lambda/14$, where λ is the test wavelength. If this condition is satisfied, the mirror is "diffraction-limited." The definition of rms wavefront error and details on the origin of the stated condition are given in Section 10.II. The sensitivity of the test methods discussed here is considerably better than the diffraction-limited criterion.

One of the first testing methods applied to optical elements is the Foucault or knife-edge test used to detect the presence of transverse aberrations. For a concave mirror the test setup usually consists of an illuminated pinhole located slightly off-axis near the center of curvature of the mirror and a knife edge in the image plane that can be moved in this plane through the image. In this setup the pinhole and knife edge are the same distance from the mirror, hence $s = s' = R$ and $m = -1$.

For a perfect spherical mirror there is no spherical aberration or coma in this setup, as is evident from the relations for TSA and TSC in Table 5.4. Furthermore, transverse astigmatism, given by TAS in Table 5.4, is negligible if the pinhole and image are close together and θ is sufficiently small. For a perfect conic mirror of conic constant K, spherical aberration is present in this setup with $\text{TSA} = -y^3 K/R^2$.

Consider first the case of a concave spherical mirror. When the knife edge in this setup is moved through the image, an observer just behind the knife edge sees a uniformly illuminated mirror surface suddenly darken uniformly, if the mirror is perfect and the knife edge is exactly at the focus. If the mirror is irregular, rays reflected from the irregular zones will not pass through the focus and the observed pattern is one of bright and dark zones on the mirror. This pattern provides the optician with a visual picture of the irregularities on the mirror which must be removed. For a discussion of the patterns seen in the presence of different aberrations, the reader should consult the reference by Malacara.

If the intended concave mirror is a conic other than a sphere, each annular zone has a focus that is displaced toward or away from the mirror. This displacement Δ depends on the amount of spherical aberration and, relative to the paraxial focus, is given by $\Delta = -y^2 K/R$. The axial shift Δ given here satisfies the usual sign convention. When the knife edge is shifted axially by Δ and moved through the image of a conic mirror, only the

V. Testing of Large Mirrors   119

annular zone at the appropriate radius y will darken if the mirror is perfect. If the mirror is not perfect, the observed pattern will again give a visual picture of the irregularities.

A variation of the Foucault test suitable for the testing of a concave ellipsoidal mirror is one in which the illuminated pinhole is placed at one focus of the mirror and the knife edge is placed at the other focus. In this setup there is no spherical aberration and, because the object and image are on the mirror axis, no coma or astigmatism. Thus the observed pattern on a perfect mirror will show the same uniform darkening when the knife edge is moved through the image. Because of the simple setup in this case, the testing of the primary mirror of a Dall-Kirkham is easily done, as pointed out in Section II.E.

A second method now widely used for testing many kinds of optical elements, including large concave mirrors, uses a modified Michelson interferometer called the Twyman-Green interferometer. The basic arrangement of the Michelson interferometer is shown in Fig. 13.15. For testing a concave mirror, the conversion of a Michelson interferometer into a Twyman-Green interferometer involves replacing the movable mirror B in Fig. 13.15 with a positive lens, usually called a null lens, and the test mirror. The lens and mirror are placed so that their optical axes coincide, with the focal point of the lens coincident with the center of curvature of the mirror. The other change is to use a laser plus beam expander as the light source, with the pinhole in the beam expander placed at the focal point of the input collimator lens.

The beamsplitter divides the amplitudes of the incident plane wavefront, which, after reflection in the two arms of the interferometer, returns to the beamsplitter. A portion of the two reflected beams is recombined at the beamsplitter and directed toward the output lens and detector, the latter designated D in Fig. 13.15. If the eye is placed at D, light from the entire field is seen. This field will be uniformly illuminated if the null lens-mirror combination is perfect and the reflected beam in each arm retraces exactly the path of the light from the beamsplitter. If, instead, the plane mirror designated A in Fig. 13.15 is tilted, a set of straight fringes with equal spacing is seen. If the lens-mirror combination is not perfect and mirror A is not tilted, then a pattern similar to that seen with the knife-edge test on an imperfect mirror is observed. If mirror A is tilted, a set of distorted fringes is seen. Examples of fringe patterns seen with various types of aberrations in the test element are displayed in the book by Malacara.

It should be evident from this discussion that the test using the Twyman-Green interferometer is one of the null lens and mirror combination, not of the mirror by itself. It is necessary, therefore, to independently test the lens and verify that it has the required characteristics. If, for example, the test mirror is a sphere, then the null lens by itself must be "perfect." If, on the other hand, the test mirror is a conic other than a sphere, then the null lens must introduce an amount of spherical aberration that exactly compensates the spherical aberration of the mirror in the test setup. Such a null lens was required, for example, in the testing of the 2.4-m hyperboloidal primary of the Hubble Space Telescope with a Twyman–Green interferometer.

As noted at the start of this section, this is only an introduction to the area of optical testing. We have not considered other widely used methods, such as the Hartmann screen test. This latter method was used to test the 4-m mirrors made at the Kitt Peak National Observatory. We have also omitted any discussion of test methods for convex conic mirrors, such as Hindle-type tests. For details on these and other test methods, and further references to the literature, the reader should consult the reference by Malacara.

VI. CONCLUDING REMARKS

The discussion of the image characteristics in this chapter is based entirely on the geometric theory derived with the aid of Fermat's principle, without taking into account the limit set by diffraction. The characteristics of images in the diffraction limit where geometric aberrations are negligible are discussed in detail in Chapter 10.

We have devoted most of our discussion to two-mirror telescopes because nearly all large reflectors are of this type. It should be evident, however, from our discussion of the Paul–Baker design that families of three-mirror telescopes with excellent image characteristics can be found, given the additional free parameters with another mirror. Although several three-mirror designs have been published, they have common problems of image surface accessibility and larger vignetting of the incident beam, compared to two-mirror telescopes. With careful attention given to these problems, however, practical three-mirror designs with excellent image characteristics can be found.

REFERENCES

Angel, R., Woolf, N., and Epps, H. (1982). *Proc. SPIE Int. Soc. Opt. Eng.* **332**, 134.
Baker, J. (1969). *IEEE Trans. Aerosp. Electron. Syst.* **5**, 261.
Malacara, D., ed. (1978). "Optical Shop Testing." Wiley, New York.

McGraw, J., Stockman, H., Angel, R., and Epps, H. (1982). *Proc. SPIE Int. Soc. Opt. Eng.* **331**, 137.
Paul, M. (1935). *Rev. Opt.* **14**, 169.
Robb, P. (1978). *Appl. Opt.* **17**, 2677.
Wetherell, W., and Rimmer, M. (1972). *Appl. Opt.* **11**, 2817.

BIBLIOGRAPHY

See listings under Telescopes in the bibliography in Chapter 4.

Hewitt, A. ed. (1980). "Optical and Infrared Telescopes for the 1990s", Vols. 1 and 2. Kitt Peak National Observatory, Tucson, Arizona.

West, R., ed. (1971). "ESO/CERN Conference on Large Telescope Design." European Space Organization, Geneva.

Chapter 7 | Schmidt Telescopes and Cameras

Typical ground-based two-mirror telescopes without correctors have usable field diameters of a fraction of a degree. In Chapter 9 we show that larger fields are obtained with the addition of a corrector system to such telescopes, reaching about 1° at prime focus and up to 3° at Cassegrain focus. Still larger fields require a telescope of the Schmidt type, or one of the many members of the family of telescopes based on the principle of the Schmidt. This principle is basically one of using a corrector plate to compensate for the spherical aberration of the reflecting optics and locating the plate and aperture stop to give zero coma and astigmatism for the system, at least to third order.

In this chapter we consider in more detail the classical Schmidt system first introduced in Chapter 4, including solid and semisolid Schmidt systems in which all or part of the air between the optical surfaces is replaced by glass. We discuss derivatives of the Schmidt design, such as Schmidt-Cassegrain and Bouwers-Maksutov systems in Chapter 8.

The classical Schmidt is the choice for a wide-field telescope if an aperture of 1 m or more is required. The principal reasons are its relative simplicity, only two large optical elements, and the smaller chromatic aberration of the aspheric corrector compared to that of the corrector in other types. In smaller apertures the choices for a wide-field instrument are a folded Schmidt or one of the two-mirror types. Whether the intended use is as a spectrograph camera or a telescope for visual observation, the requirement of an accessible focal surface is of overriding importance in this case.

I. GENERAL SCHMIDT CONFIGURATION

The Schmidt camera in its usual configuration is a corrector plate located at the center of curvature of a spherical mirror, as shown in Fig. 4.9. This arrangement was discussed in Section 4.IV, where it was introduced to illustrate the application of Fermat's principle to cancel the on-axis aberration of the mirror in collimated light. The importance of locating the aperture stop at the center of curvature of the mirror to eliminate off-axis aberrations was also noted there.

In this section we extend these discussions and consider the Schmidt configuration in a more general way. This is done to show the range of possibilities for placement of the aperture stop and corrector.

Consider the system of spherical mirror, corrector plate, and aperture stop shown in Fig. 7.1, with the object surface at distance s to the left of the mirror. The corrector plate is located a distance d to the left of the mirror, and the aperture stop is distance g to the left of the corrector. The distances s, d, and g in Fig. 7.1 are negative according to the sign convention.

Defining $k = y_2/y_1$, the ratio of the beam height at the mirror to that at the corrector, we see from Fig. 7.1 that $k = s/(s-d)$. The aberration coefficients for the corrector and mirror are found in Table 5.5 and 5.6, respectively, with only the b terms taken for the corrector. Substituting these results into Eq. (5.6.7) to get the system coefficients gives

$$B_{3s} = -\frac{1}{8}\left[b - \frac{2n}{R^3}k^4\left(\frac{m+1}{m-1}\right)^2\right], \tag{7.1.1}$$

$$B_{2s} = \frac{\psi}{2}\left[bg - \frac{2n}{R^2}k^3\left(\frac{m+1}{m-1}\right)\left(1 - \frac{W}{R}\right)\right], \tag{7.1.2}$$

$$B_{1s} = -\frac{\psi^2}{2}\left[bg^2 - \frac{2n}{R}k^2\left(1 - \frac{W}{R}\right)^2\right], \tag{7.1.3}$$

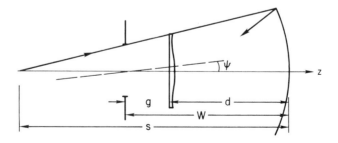

Fig. 7.1. Schmidt camera with stop displaced from corrector and object at distance s.

where $W = d + g$. From the system aberration coefficients and the requirement that each be zero, we can determine what freedom, if any, there is in their locations.

Setting Eq. (7.1.1) to zero, putting the result for b into Eqs. (7.1.2) and (7.1.3), and setting each equal to zero gives the condition

$$\frac{1}{R}\left(\frac{m+1}{m-1}\right)g = \frac{1}{k}\left(1 - \frac{W}{R}\right). \quad (7.1.4)$$

Using the relation between R and m in Table 5.2 and substituting for W and k in terms of s, d, and g, we find

$$g(R - s) = (s - d)(R - d - g).$$

Solving this equation for d gives two solutions: $d = R$ and $d = s - g$. The first of these solutions places the corrector at the center of curvature of the mirror, the same location as in the earlier discussions. The second solution gives $W = s$, hence the stop is at the object surface. This result is untenable and is discarded because it violates the condition that ψ is small, as is evident by putting $W = s$ into Eq. (5.5.2).

With $d = R$ we find $k = -(m-1)/(m+1)$, and therefore the aspheric factor is

$$b = \frac{2n}{R^3}\left(\frac{m-1}{m+1}\right)^2, \quad (7.1.5)$$

where $n = 1$ for the arrangement in Fig. 7.1. For collimated light $m = 0$ and $b = 2n/R^3$, the result noted in Section 5.V. For the configuration shown in Fig. 7.1, $m < 0$ and $|m| < 1$. Thus the factor in parentheses in Eq. (7.1.5) is larger than one, and b for noncollimated light is larger than for collimated light.

The upshot of this analysis is that for either collimated or noncollimated light the corrector plate must be located at the center of curvature of the mirror, but the stop location is arbitrary, provided W/s is not close to unity. Note that if an optical system precedes the Schmidt camera, the stop is the exit pupil of the preceding system.

This result is important because in some configurations using a Schmidt camera the stop or pupil is necessarily displaced from the corrector. An example of this is a camera in a spectrograph, where the pupil is usually at the prism or grating and different wavelengths leave the dispersing element in different directions. It is worth noting here that when the stop is displaced from the corrector, the corrector is larger and its chromatic effects are also larger. To minimize the chromatic effects, therefore, the pupil should be at the corrector or as close as can be arranged. We discuss the relation between the pupil location and the chromatic effect in a following section.

II. CHARACTERISTICS OF ASPHERIC PLATE

The aspheric plate is obviously the key to a properly configured Schmidt system and we now consider its aberration characteristics in some detail. In this section we consider the finite thickness of a real plate and its effect on the aberrations, and the effect of the radius term introduced in Section 4.IV to minimize the chromatic aberration of the plate. We also discuss chromatic aberration in more detail than in Section 4.IV and give relations for fifth-order spherical aberration of an aspheric plate and spherical mirror in collimated light.

The equation for an aspheric surface is given by an extension of Eq. (5.1.1), with $K = -1$, as follows:

$$z = \frac{\rho^2}{2R_c} + \frac{b\rho^4}{8(n'-n)} + \frac{b'\rho^6}{16(n'-n)} = \frac{\rho^2}{2R_c} + E\rho^4 + F\rho^6, \quad (7.2.1)$$

where the latter form in Eq. (7.2.1) is that usually used in ray-tracing programs. We include the terms in ρ^6 in anticipation of the section on fifth-order spherical aberration.

The difference between setting $K = -1$ and setting $K = 0$ is of no practical consequence for a refracting plate. If $K = 0$ the added terms, $\rho^4/8R_c^3$ and $\rho^6/16R_c^5$, are each several orders of magnitude smaller than the terms in b and b', respectively, for any practical plate.

A. CHROMATIC ABERRATION

One approach to finding the chromatic properties of an aspheric plate is given in Section 4.IV.A, where the analysis gives a relation for the minimum chromatic spherical aberration in Eq. (4.4.14). In this section we determine the chromatic properties in a more general way, including the effect of a displaced stop.

We first consider the Schmidt system shown in Fig. 4.9 with the aspheric figure on the surface facing the mirror. The profile of the aspheric surface can be written as

$$(n-1)z = \frac{1}{32f^3}(\rho^4 - a\rho_0^2\rho^2), \quad (7.2.2)$$

where $f = -R/2$ and a is an arbitrary parameter. Note that this relation is simply Eq. (4.4.10) written with $a/4$ replacing $3/8$ in the term containing ρ_0^2.

For a ray parallel to the z axis, the angle of deviation δ at height ρ at the aspheric surface is given by $\delta = i(n-1)$, as shown in Fig. 7.2. The angle

7. Schmidt Telescopes and Cameras

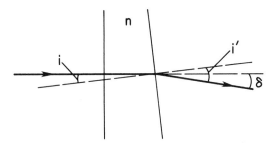

Fig. 7.2. Angle of deviation δ at wedge-shaped section of aspheric plate.

of incidence i at the aspheric surface is $i = -dz/d\rho$, where $dz/d\rho$ is the slope of the normal to the surface. Thus

$$\delta = -(n-1)\frac{dz}{d\rho} = \frac{1}{16f^3}(a\rho_0^2\rho - 2\rho^3). \qquad (7.2.3)$$

From Eq. (7.2.3) we see that $\delta = 0$ when $\rho = 0$ and $\rho = \rho_0\sqrt{a/2}$, where the latter value defines the radius of the neutral zone.

Inside the neutral zone the ray deviation is a maximum at the inflection zone, defined as that ρ for which $d\delta/d\rho = 0$, while outside the neutral zone δ is a maximum at the edge of the plate. The characteristics of the aspheric surface at these zones, given in terms of a, are found in Table 7.1.

From the entries in Table 7.1 it is evident that the deviations at the inflection zone and edge have opposite signs for $a < 2$. As a increases from zero, δ at the inflection zone increases while δ at the edge decreases. The net deviation across the plate is a minimum when the values of δ at these two radii are equal in magnitude. This is obtained with the choice $a = 1.5$ and the resulting magnitude of $\delta = \rho_0^3/32f^3$ at these radii. The neutral zone is then at $\rho = \rho_0\sqrt{3}/2 = 0.866\rho_0$.

Table 7.1

Deviations at Zones of Aspheric Plate

Zone	Radius	Deviation
Inflection	$\rho_0\left(\dfrac{a}{6}\right)^{1/2}$	$\dfrac{\rho_0^3}{4f^3}\left(\dfrac{a}{6}\right)^{3/2}$
Neutral	$\rho_0\left(\dfrac{a}{2}\right)^{1/2}$	0
Edge	ρ_0	$\dfrac{\rho_0^3}{16f^3}(a-2)$

II. Characteristics of Aspheric Plate

Differentiating Eq. (7.2.1) and setting it equal to zero, substituting for ρ at the neutral zone, and solving for R_c gives

$$R_c = -\frac{8(n'-n)}{3b\rho_0^2} = -\frac{1}{3E\rho_0^2}, \qquad (7.2.4)$$

where $n' = 1$ for the configuration in Fig. 4.9.

The sign of b depends only on the character of the plate. From Eq. (7.1.5) we see that $b < 0$ for a plate in a Schmidt camera because n and R are always of opposite sign. As shown in Section 4.IV.B and Fig. 4.11, a Schmidt plate has a "turned-up" edge. Conversely, $b > 0$ for a plate with a "turned-down" edge. The sign of E, on the other hand, depends on whether the aspheric is on the first or second surface of the corrector, and on the direction of light through the plate. Note also from Eq. (7.2.4) that E and R_c always have opposite signs in order to place the neutral zone at the desired radius.

The chromatic blur is obtained by finding the variation of δ with changing n. Using Eq. (7.2.3) gives

$$\frac{d\delta}{dn} = -\frac{dz}{d\rho} = \frac{\delta}{n-1}. \qquad (7.2.5)$$

Figure 7.3 shows two rays for different values of n leaving a point on the aspheric surface and intersecting the mirror a distance $R\,d\delta$ apart. The point on the aspheric surface can, in effect, be considered an object point at distance R that is reimaged at the corrector. Hence the blur at the focal surface for these two rays is $f\,d\delta$. Substituting the values of δ at the inflection zone and edge into Eq. (7.2.5) gives a blur diameter of $2f\,d\delta$, hence

$$\text{CSA} = \frac{f}{128F^3}\left(\frac{dn}{n-1}\right), \qquad (7.2.6)$$

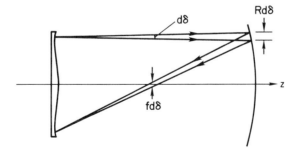

Fig. 7.3. Paths of rays of different wavelength through Schmidt camera. See Eq. (7.2.6).

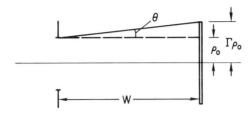

Fig. 7.4. Corrector size required to cover field when stop is displaced from plate.

where CSA is the chromatic spherical aberration and F is the focal ratio of the mirror. This result is, as expected, equivalent to that given in Eq. (4.4.14).

The results so far are appropriate for a corrector with aperture stop at the plate. When the stop or pupil is displaced from the plate, as shown in Fig. 7.4, the radius of the plate must be larger by a factor Γ to accept all of the light without vignetting. If ρ_0 is the radius of a collimated beam at the plate, then $\Gamma = 1 + W\theta/\rho_0$, and all of the results in Table 7.1 apply to the enlarged plate if ρ_0 is replaced by $\Gamma\rho_0$.

The chromatic effects are again minimized by choosing $a = 1.5$, hence the neutral zone is at $\rho = 0.866(\Gamma\rho_0)$. The deviations at the inflection zone and edge are now larger by the factor Γ^3, as is the blur diameter in Eq. (7.2.6). It is clear from this result that the placement of the stop or pupil at the corrector is the preferred choice to minimize chromatic effects. As a final item note that the relations for R_c in Eq. (7.2.4) apply to a plate of radius $\Gamma\rho_0$ with the substitution of $\Gamma\rho_0$ for ρ_0.

B. ABERRATION COEFFICIENTS

We now determine the effect of the plate thickness and radius R_c on the aberration coefficients of a real plate. Consider a plate of index n and thickness t whose first surface is plane and whose second surface has a radius of curvature R_c and an aspheric term b, as shown in Fig. 7.5, with

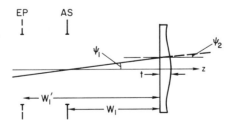

Fig. 7.5. Cross section of aspheric plate with stop AS and pupil EP. See Eq. (7.2.8).

II. Characteristics of Aspheric Plate

the pupil for the plate located a distance W_1 to the left of the first surface. The image of the pupil by the first surface is W_1' from this surface and W_2 from the second surface. Collimated light is incident on the plate and the beam heights at the two surfaces are equal.

The aberration coefficients for the first surface are zero when the light is collimated. The coefficients for the second surface from Table 5.5 are

$$B_{12} = -\frac{\psi_2^2}{2}\Gamma_2\left(1 - \frac{W_2}{R_c}\right)^2 - \frac{b}{2}(W_2\psi_2)^2,$$

$$B_{22} = -\frac{\psi_2}{2}\frac{\Gamma_2}{R_c}\left(1 - \frac{W_2}{R_c}\right) + \frac{b}{2}(W_2\psi_2), \quad (7.2.7)$$

$$B_{32} = -\frac{b}{8} - \frac{\Gamma_2}{8R_c^2}, \quad \Gamma_2 = -\frac{n^2(n-1)}{R_c},$$

where

$$\psi_2 = \frac{\psi_1}{n} = \frac{\theta}{n}, \quad W_2 = W_1' - t = nW_1 - t. \quad (7.2.8)$$

Substituting Eq. (7.2.8) into Eqs. (7.2.7), and assuming that $W_2 \ll R_c$ for all configurations using an aspheric plate, we find the following aberration coefficients for the corrector:

$$B_{1c} = \frac{\theta^2(n-1)}{2R_c} - \frac{b\theta^2}{2}\left(W_1 - \frac{t}{n}\right)^2,$$

$$B_{2c} = \frac{\theta n(n-1)}{2R_c^2} + \frac{b\theta}{2}\left(W_1 - \frac{t}{n}\right), \quad (7.2.9)$$

$$B_{3c} = -\frac{b}{8} + \frac{n^2(n-1)}{8R_c^3}.$$

Substituting Eq. (7.2.4) for one R_c in B_{3c} of Eqs. (7.2.9), we find that the right-hand term is of order $(\rho_0/R_c)^2$ smaller than the first term. For any practical plate $\rho_0 \ll R_c$ and the right-hand term is negligible.

For the remaining coefficients in Eqs. (7.2.9) we see that the contribution of the aspheric term in each is zero when $W_1 = t/n$. The remaining terms are simply those for a plano-convex lens in collimated light. If Eq. (7.2.4) is substituted into B_{1c} and B_{2c} and $W_1 - t/n$ is replaced by ε, the coefficient B_{1c} is dominated by the term in R_c when ε is small. For B_{2c}, on the other hand, the term in ε dominates when $|\varepsilon| > \rho_0^2/R_c$, hence the coma coefficient is sensitive to small changes in ε. This result can be used to minimize the effect of coma in a Schmidt system by adjusting the plate location.

Relations of a comparable form to those in Eqs. (7.2.9) are obtained for an aspheric plate with its figured surface facing the incident light, with the principal change one of substituting W_1 for $W_1 - t/n$. The comments in the previous paragraph on the dependence of B_{1c} and B_{2c} on small values W_1 hold without change.

If the incident light is coming from a source at a finite distance, then there is an additional contribution to each of the aberration coefficients from the plate thickness. These effects are easily derived with the aid of the geometry in Fig. 7.6 for a plane-parallel plate, with the coefficients for each surface taken from Table 5.1 with $b = 0$ and $R = \infty$. When these relations are substituted into Eq. (5.6.7), an exercise left to the reader, we find for a plate p

$$B_{1p} = \theta_1^2 (n^2 - 1) t / 2 n^3 s_1^2, \tag{7.2.10}$$

$$B_{2p} = -\theta_1 (n^2 - 1) t / 2 n^3 s_1^3, \tag{7.2.11}$$

$$B_{3p} = (n^2 - 1) t / 8 n^3 s_1^4. \tag{7.2.12}$$

The importance of these coefficients for an aspheric plate in noncollimated light depends on the specifics of a given configuration. In most configurations it turns out that their contributions are of little significance, with the details best left to computer ray-trace analysis.

Although the term in R_c in B_{3c} of Eqs. (7.2.9) is negligible, the value of R_c does affect the optimum choice of b required to zero the third-order spherical aberration of the system. With the addition of a radius term the corrector becomes, in effect, a weak positive lens with an aspheric figure. The effect of the lens part of the corrector is to convert the incident collimated light into a slightly converging beam. Thus the marginal rays intersect the mirror at a slightly smaller distance from the mirror vertex, as compared to the case where the corrector has no radius term. Omitting the details of the derivation, the spherical aberration coefficient for a Schmidt system in collimated light is given by

$$B_{3s} = -\frac{b}{8} + \frac{n}{4R^3} \left[1 - \frac{3}{2} \left(\frac{\rho_0}{R} \right)^2 \right], \tag{7.2.13}$$

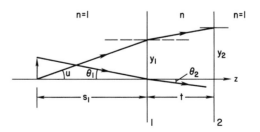

Fig. 7.6. Cross section of plane-parallel plate of index n in air.

II. Characteristics of Aspheric Plate

for the case where R_c is chosen according to Eq. (7.2.4). Setting B_{3s} to zero gives

$$b = \frac{2n}{R^3}\left[1 - \frac{3}{2}\left(\frac{\rho_0}{R}\right)^2\right], \tag{7.2.14}$$

where $\rho_0/R = 1/4F$. For typical values of F the result is a reduction of 1 or 2% in the magnitude of b needed to cancel the spherical aberration of the mirror.

With the exception of the correction given by Eq. (7.2.14), the effects of the plate thickness and radius of curvature on the aberration coefficients of an aspheric plate are usually small compared to those of terms containing b. Therefore the usual approach to the analysis of a system that includes one or more aspheric plates is to include only terms in b and let this description serve as the starting point for a ray-trace analysis.

C. FIFTH-ORDER SPHERICAL ABERRATION

The prescription of a Schmidt corrector plate usually includes higher-order aberration terms. Fifth-order spherical aberration is the most significant of these terms and we give here, without derivation, the necessary relations for a spherical mirror and aspheric corrector in collimated light.

The fifth-order spherical aberration coefficients, denoted by B_5, are obtained after a lengthy analysis paralleling that in Chapter 5, but with θ set equal to zero. The results of this analysis are given in Table 7.2, using Eq. (4.2.1) to get ASA5 for the mirror.

The calculation of the system aberration coefficient B_{5s} for the combined aspheric and spherical mirror is carried out using Eq. (5.6.7) with $k = 5$. Because the ray heights at the corrector and mirror are equal in a first approximation, B_{5s} is simply the sum of the coefficients in Table 7.2. Setting

Table 7.2

Fifth-Order Spherical Aberration Results[a]

Surface	B_5	ASA5
Aspheric	$-\dfrac{b'}{16}$	$-\dfrac{3b'}{8}\rho^5$
Spherical mirror	$\dfrac{3n}{8R^5}$	$\dfrac{9n}{4}\left(\dfrac{\rho}{R}\right)^5$

[a] Results valid for collimated light only. ASA5 = $6B_5\rho^5$.

132   7. Schmidt Telescopes and Cameras

the sum equal to zero gives $b' = 6n/R^5$; with this choice of b' the fifth-order spherical aberration of the system is zero.

III. SCHMIDT TELESCOPE EXAMPLE

We now apply the results above to an example of a 1-m Schmidt telescope with $F = 2.5$. The aspheric surface on the corrector plate faces the mirror; the plate material is SiO_2 and its thickness 10 mm at the vertex. The parameters R_c, E, and F in Eq. (7.2.1) are calculated at $\lambda = 548$ nm, at which wavelength the plate index is 1.46. Values of the telescope parameters are given in Table 7.3, with b given both for $R_c = \infty$ and according to Eq. (7.2.14). The depth of the corrector at the neutral zone, calculated from Eq. (7.2.1), is 0.1534 mm.

Results from a ray-trace analysis are given in Table 7.4, with all aberrations given in angular terms in units of arc-seconds. Various combinations of parameters from Table 7.3 are used to illustrate the effect of each of the parameters on the angular aberrations. Note that the on-axis angular aberrations of the mirror without corrector, given in arc-seconds, are ASA3 = 206.3 and ASA5 = 4.64, while the off-axis aberrations are zero in the third-order approximation.

Examination of the results in Table 7.4 clearly shows the improvement in the on-axis image quality when a fifth-order term is included in the aspheric and the third-order aspheric term is calculated from Eq. (7.2.14). We also see that there are small but nonzero off-axis aberrations which appear when the radius term is included on the corrector. These aberrations are a result of the terms in R_c in Eqs. (7.2.9). The presence of these off-axis

Rable 7.3

Parameters of 1-m Schmidt Telescope[a]

$R = -5000$ mm	$\rho_0 = 500$ mm
$b = 2/R^3 = -1.6E - 11$	$E = 4.34783E - 12$
$b' = 6/R^5 = -1.92E - 18$	$F = 2.6087E - 19$
$R_c = -1/3E\rho_0^2 = -306\,667$	
$b = \dfrac{2}{R^3}\left(1 - \dfrac{3}{32F^2}\right) = -1.576E - 11$	[from Eq. (7.2.14)]
$E = 4.28261E - 12$	

[a] Values of E and F are computed with $n = 1.46$.

III. Schmidt Telescope Example

Table 7.4

Ray-Trace Results for 1-m Schmidt Telescope[a]

System parameters							
$-b$	$-b'$	$-R_c$	ASA3	ASA5	ATC	AAS	
1.6E−11	0	∞	0.003	4.770	0.000	0.000	
1.6E−11	0	306 667	3.079	4.697	0.009	0.047	
1.6E−11	1.92E−18	306 667	3.079	0.034	0.009	0.047	
1.576E−11	1.92E−18	306 667	0.014	0.003	0.010	0.047	

[a] Ray traces at $\lambda = 548$ nm with $\theta = 1°$. Angular aberrations are given in arc-seconds.

aberrations limits the field size, and ray traces of the final system in Table 7.4 give an image blur diameter of about 1 arc-sec at a field angle of 3.5°.

Values for the angular chromatic spherical aberration, computed from Eq. (7.2.6), are shown in Table 7.5, where the indices are those of SiO_2. Because the index of refraction rises more steeply at shorter wavelengths, the chromatic blur increases rapidly for blue and ultraviolet wavelengths.

From the results in Table 7.5 it is evident that a single corrector does not give good images at all wavelengths over an extended spectral range. One alternative is to have several correctors, each designed to give good images over a selected range of wavelengths. Although this option is practical for a small telescope or camera, it is not considered practical for a Schmidt telescope of the 1-m class.

A different alternative, suggested by Bowen, is to design the corrector for a wavelength near the short end of the desired range and to use a flat glass plate of appropriate thickness to partially correct the chromatic aberration of the corrector at longer wavelengths. This plate, usually a filter to

Table 7.5

Image Diameters for 1-m Schmidt Telescope[a]

λ (nm)	n	ACSA
350	1.47689	3.78
400	1.47012	2.26
450	1.46577	1.24
548	1.46000	0.00
650	1.45650	0.79
700	1.45523	1.08

[a] Image diameters are given in arc-seconds. ACSA = angular chromatic spherical aberration.

remove shorter wavelengths, is placed in the converging beam close to the focal surface. For details on this approach the reader should consult the reference by Bowen. A final alternative is to use an achromatic corrector made of two different glasses, the subject of the next section.

The Schmidt telescope example in this section is intended primarily to illustrate the application of the theory to the design of a wide-field telescope. The focal surface is curved and further refinement of the design might include the addition of a field-flattener lens. Such a lens will introduce spherical aberration over the entire field and coma near the edge of the field, hence the parameters of the corrector will have to be adjusted to get an optimum system. The process of optimization is best carried out with a computer ray-trace program and will not be pursued here. For a theoretical discussion of the aberrations of a field-flattened Schmidt camera the reader should consult the reference by Linfoot.

A final point worth noting is the increasing importance of higher-order aberrations for smaller focal ratios. The importance of fifth-order spherical aberration is evident in our example, but in faster cameras it is necessary to consider the effects of still higher orders. In addition, fifth-order off-axis aberrations become important and attention must be given to their effects in the design of a fast camera.

IV. ACHROMATIC SCHMIDT TELESCOPE

The wavelength range over which a Schmidt telescope with a single-element corrector gives images of acceptable size is set by the dispersive characteristics of the corrector. This range can be extended by replacing the single-element plate with a two-element corrector, with each element a glass of different dispersive characteristics and plate parameters to make the combination achromatic. In this section we outline the procedure for making an achromatic corrector and apply the results to an example of a 1-m Schmidt telescope.

A cross section of a two-element corrector is shown in Fig. 7.7, with the plane surfaces of the elements in contact and the aspheric surfaces facing outward. The differential deviation for each element is given by Eq. (7.2.5), which can be written as

$$d\delta_1 = \delta_1/V_1, \qquad d\delta_2 = \delta_2/V_2, \qquad (7.4.1)$$

where

$$V_1 = \frac{\langle n_1 \rangle - 1}{n_1' - n_1}, \qquad V_2 = \frac{\langle n_2 \rangle - 1}{n_2' - n_2}. \qquad (7.4.2)$$

IV. Achromatic Schmidt Telescope

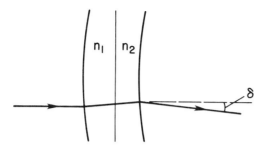

Fig. 7.7. Cross section of portion of achromatic corrector. The net deviation δ is the sum of the deviations of individual elements.

In Eqs. (7.4.2) V is the Abbe number and $\langle n \rangle$ is the mean of the indices in the denominator for each glass. The primed indices are taken at a shorter wavelength, by convention, hence V is positive.

The achromatic condition requires that $d\delta_1 = -d\delta_2$, hence a change in deviation with wavelength in one glass is compensated by a change of opposite sign in the other glass. Therefore

$$\delta_1/V_1 = -\delta_2/V_2. \qquad (7.4.3)$$

Because the Abbe numbers are positive, the deviations of the two elements are opposite in sign and one plate has a "turned-up" edge, as for a normal Schmidt plate, while the other has a "turned-down" edge, as shown in Fig. 7.7.

Assuming that the deviation at the plane interface is negligible, the net deviation of the achromatic plate is

$$\delta = \delta_1 + \delta_2 = \delta_1 \left(\frac{V_1 - V_2}{V_1} \right). \qquad (7.4.4)$$

Given that the achromatic plate is a replacement for a single plate, the deviation given by Eq. (7.4.4) must be the same as that in Eq. (7.2.3).

From Eqs. (7.4.3) and (7.4.4) we see that both δ_1 and δ_2 are larger than δ in magnitude, hence each aspheric surface has a larger local slope than on the single plate at the same ray height. If the achromatic plate is oriented as shown in Fig. 7.7, δ and δ_1 have opposite signs and from Eq. (7.4.4) we find $V_1 < V_2$. Thus the element with the turned-up edge is the one with the larger Abbe number, a result true for either orientation of the corrector.

It is evident from Eqs. (7.4.3) and (7.4.4) that the ratios δ_1/δ and δ_2/δ are independent of ray height ρ for a given set of glasses. Substituting Eq. (7.2.1) into Eq. (7.2.3) we see that each δ has the form

$$\delta_i = -(n_i - 1)(c_i\rho + 4E_i\rho^3 + 6F_i\rho^5), \qquad (7.4.5)$$

where c_i is the vertex curvature. If the ratio of one δ to another is independent of ρ, then it follows that the ratios of corresponding plate parameters must also have a common value. Substituting Eq. (7.4.5) into Eq. (7.4.4) and applying this condition gives

$$\frac{c_1}{c} = \frac{E_1}{E} = \frac{F_1}{F} = \left(\frac{\langle n \rangle - 1}{\langle n_1 \rangle - 1}\right)\left(\frac{V_1}{V_2 - V_1}\right), \quad (7.4.6)$$

where the unsubscripted parameters are those of the single-element corrector. Note the reversed order of the factors in the difference of the Abbe numbers between Eqs. (7.4.4) and (7.4.6), a consequence of the sign difference between δ and δ_1. Using Eq. (7.4.3) we find

$$\frac{c_2}{c_1} = \frac{E_2}{E_1} = \frac{F_2}{F_1} = \frac{n'_1 - n_1}{n'_2 - n_2}. \quad (7.4.7)$$

All of the relations needed to specify an achromatic plate are now in hand, and their application is straightforward once a suitable pair of glasses are chosen.

We choose two glasses from the Schott catalog, UBK7 and LLF2, the former a crown glass and the latter a light flint. Both glasses have good internal transmittance in the near ultraviolet, with values of 0.85 and 0.74 at $\lambda = 320$ nm for a 10-mm thickness of UBK7 and LLF2, respectively. The pair of chosen wavelengths at which to make the plate achromatic are 320 and 880 nm, with the indices at these wavelengths and Abbe numbers shown in Table 7.6.

Given the results in Table 7.6 and the discussion following Eq. (7.4.4), LLF2 and UBK7 are the glasses for elements 1 and 2, respectively, of the corrector shown in Fig. 7.7. The mean index $\langle n \rangle$ in Table 7.6 is approximately the index at $\lambda = 420$ nm for each glass. We use this wavelength to calculate the parameters of a single-element SiO_2 plate needed in Eq. (7.4.6).

The Schmidt telescope used in the following comparison is the same one used in the previous section, with Eqs. (7.4.6) and (7.4.7) used to calculate the parameters of the achromatic plate. The calculated parameters for both

Table 7.6

Indices and Abbe Numbers for UBK7 and LLF2[a]

	n (320 nm)	n (880 nm)	$n' - n$	$\langle n \rangle - 1$	V
UBK7	1.54634	1.50935	0.03699	0.52784	14.27
LLF2	1.58789	1.53081	0.05708	0.55935	9.799

[a] Indices of refraction taken from Schott catalog.

IV. Achromatic Schmidt Telescope

Table 7.7

Parameters of Single and Achromatic Correctors

	R_c	E	F
$SiO_2{}^a$	−312 067	4.2085E−12	2.5636E−19
LLF2	−170 120	7.7201E−12	4.7026E−19
UBK7	−110 240	1.1913E−11	7.2567E−19

[a] Parameters for SiO_2 plate are similar to those in Table 7.3, but computed with $n = 1.46810$.

the single and achromatic plate are found in Table 7.7. The sags at the neutral zone for the LLF2 and UBK7 elements are 0.2765 and 0.4268 mm, respectively.

Ray traces of a 1-m Schmidt telescope with an achromatic plate specified by the parameters in Table 7.7 show a well-corrected system at $\lambda 320$ and $\lambda 880$ nm, with the blur diameter on-axis set primarily by residual fifth-order spherical aberration. The blur diameters for on-axis images over the range 320 to 1000 nm are shown by the solid curve in Fig. 7.8. Although the correction is excellent at the ends of the range shown, the image diameters in the blue and near ultraviolet are larger than desired.

The corrector as specified provides the proper correction at the chosen wavelengths, but gives too large a correction over much of the range. This is easily remedied by making the aspherics on each surface slightly weaker.

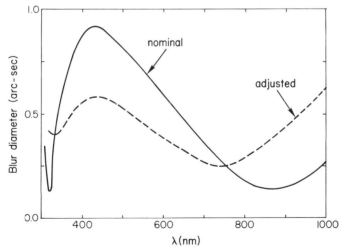

Fig. 7.8. Image diameters for $f/2.5$ achromatic Schmidt camera. Solid curve: parameters in Table 7.7; dashed curve: parameters adjusted as noted in text, Section 7.IV.

The dashed curve in Fig. 7.8 shows the image diameters when the values of E and F for the elements of the achromatic plate in Table 7.7 are reduced by 0.25%, with the values of R_c increased by the same amount. The overall improvement in on-axis image quality over much of the range is evident from a comparison of the two curves.

The quality of the off-axis images is acceptable for the modified corrector, provided it is moved about 30 mm away from the mirror. This shift reduces the coma to near-negligible levels and ray traces give symmetrical images of acceptable size over a field diameter of 6°.

As in the design of any Schmidt system, computer optimization is used to balance the various aberrations and find the best overall set of parameters. For a discussion of this process and the results found for an $f/3.5$ achromatic Schmidt, the reader should see the reference by Buchroeder.

In summary, the Schmidt telescope with an achromatic corrector has the advantage of an extended wavelength range over which good images are obtained. With the availability of several glasses that transmit well into the ultraviolet, the choice of an achromatic corrector over a standard one is a viable option.

V. SOLID AND SEMISOLID SCHMIDT CAMERAS

A common use of the standard Schmidt camera is as the camera in a spectrograph. In this application different wavelengths are in focus at different places on the focal surface, and it is no longer necessary that the camera be strictly achromatic. It is therefore possible to modify the standard air-Schmidt to achieve improvements that are otherwise not possible.

One such modified Schmidt is the so-called solid-Schmidt, one in which the space between the corrector and mirror is filled with glass, as shown in

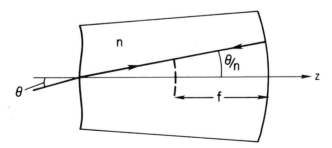

Fig. 7.9. Solid Schmidt camera of index n and effective focal length f/n. Aspheric figure, mirror on opposite ends of block.

Fig. 7.9. In this design the aspheric surface is on one end of the glass block and the mirror is on the other end. If the length of the block is equal to the radius of curvature of the mirror, then third-order off-axis aberrations are zero, just as for an air-Schmidt. The condition for zero spherical aberration and minimum chromatic aberration is given by Eq. (7.2.14), where n is now the index of the glass block. Compared to an air-Schmidt, the aspheric figure is n times stronger and the radius R_c is n times smaller.

From Fig. 7.9 we see that a chief ray entering the block at angle θ makes angle θ/n with the z axis inside the block. Because this ray is reflected back on itself, the height h of the corresponding image is $f\theta/n$ from the z axis. Thus the effective focal length f' of the solid-Schmidt is f/n, where f is the focal length of the equivalent air-Schmidt.

The reduction in focal length by a factor of n is significant for several reasons. First, the focal ratio is reduced by this factor and thus the "speed" of the camera is effectively larger by a factor of n^2. The term "speed" for a spectrograph is defined in Chapter 12; at this point it is sufficient to note that photographic exposure time to a given level is inversely proportional to the speed. Second, the off-axis aberrations present in an optimized air-Schmidt camera are smaller by a factor of n^2 in a solid-Schmidt. As a consequence, a solid camera will have comparable image quality at a field angle which is n times larger than that of an air-Schmidt of the same size. Alternatively, a solid-Schmidt will cover the same field as that of an air-Schmidt, where the former is n times shorter.

Given height $h = f\theta/n$, we find the variation of h with changing index is given by

$$dh = -\theta f \, dn/n^2. \tag{7.5.1}$$

If, for example, we take the values of n for SiO_2 from Table 7.5 at 400 and 700 nm, and assume $f = 500$ mm, then $dh = 61$ μm for a field angle of 1°. A lateral shift of this amount is not acceptable in direct photography because a point source would be imaged as a short spectrum, with its length proportional to the field angle. In a spectrograph camera, on the other hand, each image of the slit is quasi-monochromatic and the lateral shift is simply an offset without additional blurring.

The effect of index n on the aberrations is most easily seen from an example. Ignore for the moment the aspheric term on the surface of the solid-Schmidt and consider only the radius term. Because the stop is at the surface, the astigmatism coefficient is that given in Table 5.1. For $s = \infty$ we find

$$A_1 = -\frac{\theta^2}{2}\left(\frac{n-1}{n^2 R_c}\right). \tag{7.5.2}$$

The corrector for the air-Schmidt, in the absence of the aspheric term, is a plano-convex lens of thickness d. The astigmatism coefficient of the lens is the sum of the surface coefficients. With the convex surface facing the incident light, and the stop at this surface, we find that the astigmatism coefficient of the lens for $d \ll R_c$ is given by

$$B_1 = -\frac{\theta^2}{2}\left(\frac{n-1}{R_c}\right). \qquad (7.5.3)$$

In comparing Eqs. (7.5.2) and (7.5.3) it is important to note that Eq. (7.5.2) applies to the solid-Schmidt and R_c is n times smaller than in Eq. (7.5.3). Therefore A_1 for the solid-Schmidt is n times smaller than B_1 for the air-Schmidt. Substituting each of these coefficients into Eq. (5.6.6), we see that the transverse aberration for the solid-Schmidt is smaller by another factor of n. Hence the net reduction in the astigmatism due to the radius term on the corrector is smaller by a factor of n^2, as stated above. The same factor is found in a comparison of the coma coefficients.

The fabrication of the solid-Schmidt is obviously difficult because the curved focal surface lies in the center of the block. To avoid the complication of preparing this surface in a hole in the block, an alternative is the so-called semisolid or thick-mirror Schmidt. This camera is one in which glass fills the space between the focal surface and the mirror, with a conventional aspheric plate in front of the block, as shown in Fig. 7.10. Except for the curved focal surface, the face of the block toward the corrector is plane.

From Fig. 7.10 we see that the location of the corrector is such that the chief ray, after refraction at the surface of the block, appears to come from the center of curvature of the mirror. Because the refracted chief ray makes angle θ/n with the z axis, the distance from the axis to the image point is the same as that of the solid-Schmidt. Hence the focal length of the thick-mirror Schmidt is the same as that of the solid-Schmidt and all of the comments above also apply. The aspheric figure and radius R_c are also the same as those for the solid-Schmidt.

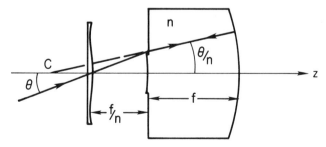

Fig. 7.10. Semisolid Schmidt camera with center of curvature at C. Focal length $= f/n$.

Ray traces of a solid-Schmidt and thick-mirror Schmidt, with a b' aspheric term added to control fifth-order spherical aberration, show very similar image characteristics. For $F = 2.5$, the focal ratio of the equivalent air-Schmidt, the image blur diameters are 1 arc-sec at a field angle of 5°. Compared with the $f/2.5$ design example in Section 7.III, the field is about n times larger, as expected.

REFERENCES

Bowen, I. (1960). See reference in Chapter 4.
Buchroeder, R. (1972). *Appl. Opt.* **11**, 2968.

BIBLIOGRAPHY

See listings under Schmidt cameras in the bibliography in Chapter 4.

Chapter 8 Catadioptric Telescopes and Cameras

In this chapter we discuss various derivatives of the Schmidt type of telescope, including Schmidt-Cassegrain, Baker-Schmidt, and Bouwers-Maksutov systems. Each of these is a type of catadioptric telescope in which a full-aperture refracting element provides the aberration correction needed to get good imagery over a wide field. Given this definition, the classical Schmidt telescope is also of this type.

The Schmidt-Cassegrain, as the name suggests, is a two-mirror system with an aspheric corrector in the collimated beam ahead of the primary mirror. Baker-Schmidt systems are a subclass of the Schmidt-Cassegrain with a flat focal surface, of which examples of two specific types are given. The Bouwers-Maksutov type is one in which the aspheric corrector is replaced by a meniscus lens with spherical surfaces. This type of corrector, in combination with one or two mirrors, is the basis for a wide variety of wide-field systems. The design parameters are given for selected examples of systems using a meniscus corrector.

I. SCHMIDT-CASSEGRAIN TELESCOPES

The Schmidt-Cassegrain telescope, hereafter designated SC, is a two-mirror telescope with a corrector plate in the collimated beam, as shown in Fig. 8.1. Compared to an all-reflective Cassegrain, the principal differences are the addition of an aspheric plate to compensate for the spherical

I. Schmidt–Cassegrain Telescopes

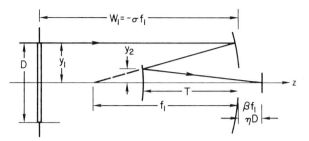

Fig. 8.1. Schematic of Schmidt–Cassegrain telescope with stop at corrector plate. Distance from stop to primary = σf_1.

aberration of the mirrors and the shift of the aperture stop from the primary to the corrector. With these changes there are additional free parameters available for the elimination of other aberrations, and a host of wide-field SC systems are possible.

In this section we outline the procedure by which the aberration characteristics of a general SC are found. Rather than exploring the features of the general SC, however, we choose to apply these results to a selected number of SC types to illustrate their basic features. The types considered include the flat-field anastigmat, the anastigmat with spherical mirrors, and the "short" SC with the corrector a distance f_1 from the primary. For further details on these and other types of SC systems, the reader should consult the work by Linfoot.

A. GENERAL PARAMETERS OF SCHMIDT–CASSEGRAIN TELESCOPES

The notation used in writing the aberration coefficients for each surface is the same as that used for two-mirror telescopes. The subscripts 1 and 2 refer to the primary and secondary mirrors, respectively, while the subscript c is used for the corrector. For a concave primary, the only type considered, the focal length f_1 is positive.

The relative locations of the mirrors and focal surface of the SC are described in terms of the normalized parameters in Table 6.2 used for two-mirror telescopes. An additional normalized parameter introduced for the SC is σ, the location of the aspheric plate relative to the primary in units of the primary focal length. According to our sign convention, the distance W_1 from the primary to the stop is negative and we therefore choose to define $\sigma = -W_1/f_1$ to make σ positive.

With the stop at the corrector, hence $W_c = 0$, the only nonzero aberration coefficient for the corrector is $B_{3c} = -b/8$. In writing this result we ignore the radius term added to minimize chromatic aberration.

8. Catadioptric Telescopes and Cameras

Table 8.1

Aberration Coefficients of SC Primary

$$B_{11} = \frac{\theta^2}{R_1}\left[\frac{\sigma^2}{4}K_1 + \left(1 - \frac{\sigma}{2}\right)^2\right]$$

$$B_{21} = \frac{\theta}{R_1^2}\left[1 - \frac{\sigma}{2}(K_1 + 1)\right]$$

$$B_{31} = \frac{K_1 + 1}{4R_1^3}$$

For the primary mirror, the stop is at a distance $W_1 = -\sigma f_1$, the chief ray angle ψ_1 is the field angle θ, and the magnification m is zero. Substituting these results into the equations in Table 5.6 gives the coefficients for the primary, in the form shown in Table 8.1. Note that $n = 1$ for the primary.

To find the aberration coefficients for the secondary, we first determine the location of the pupil for the secondary using the paraxial relations. As shown in Fig. 8.2, the primary images the stop at a distance $W_1' = f_1\sigma/(1-\sigma)$, where W_1' is negative when $\sigma > 1$. The location of the pupil relative to the secondary is $W_2 = W_1' + T$, where $T = (1 - k)f_1$ is the separation between the primary and secondary.

To find the chief ray angle ψ_2 for the secondary, we see from Fig. 8.2 that the chief ray is directed toward the center of pupil after reflection from the primary. Therefore $W_1\psi_1 = W_1'\psi_2$, where ψ_2 is the chief ray angle for the secondary. Substituting for ψ_1 and W_1' gives

$$\psi_2 = \theta(\sigma - 1), \qquad \frac{W_2}{f_1} = -\frac{1 + k(\sigma - 1)}{\sigma - 1}. \tag{8.1.1}$$

The resulting aberration coefficients for the secondary, taken from Table 5.6, are shown in Table 8.2. Note that $n = -1$ for the secondary and that R_2 has been replaced by ρR_1 in writing these relations.

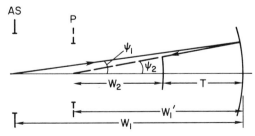

Fig. 8.2. Geometry of aperture stop AS, pupil P, and chief ray angles for Cassegrain telescope. See Eq. (8.1.1) and preceding discussion.

Table 8.2
Aberration Coefficients for SC Secondary

$$B_{12} = -\frac{\theta^2(\sigma-1)^2}{\rho R_1}\left[\left(\frac{W_2}{\rho R_1}\right)^2 \cdot K_2 + \left(1 - \frac{W_2}{\rho R_1}\right)^2\right]$$

$$B_{22} = \frac{\theta(\sigma-1)}{(\rho R_1)^2}\left[\left(\frac{W_2}{\rho R_1}\right)K_2 + \left(\frac{m+1}{m-1}\right)\left(1 - \frac{W_2}{\rho R_1}\right)\right]$$

$$B_{32} = -\frac{1}{4(\rho R_1)^3}\left[K_2 + \left(\frac{m+1}{m-1}\right)^2\right], \quad \frac{W_2}{R_1} = \frac{1+k(\sigma-1)}{2(\sigma-1)}$$

The system aberration coefficients are found by applying Eq. (5.6.7) to corresponding sets of surface coefficients. It is convenient to reference the system coefficients to the primary by choosing $y_t = y_1$ in Eq. (5.6.7). Before proceeding to apply these results to selected examples, we develop some additional useful relations between the normalized parameters.

From Section 2.V we find $\eta = \beta F_1 = \beta F/m$, where η is the back focal distance in units of the telescope diameter D. Using the relations in Table 6.2 we can write k in terms of ρ, η, and F, with the result

$$k^2 F - kF(2\rho+1) + \rho(F+\eta) = 0.$$

Solving this relation for k we get

$$k = \rho + \tfrac{1}{2} - \left(\rho^2 + \tfrac{1}{4} - \frac{\rho\eta}{F}\right)^{1/2}, \tag{8.1.2}$$

where the minus sign in front of the radical is chosen to ensure that $k < 1$. From Eq. (8.1.2) we see that a specification of ρ, η, and F sets the value of k, which in turn fixes the values of m and F_1. Note that k is independent of F when $\eta = 0$.

It is also important to determine the sizes of the primary and secondary required to cover a given field without vignetting. If D is the diameter of the aperture stop, then from the geometry in Fig. 8.3 we find that D_1, the

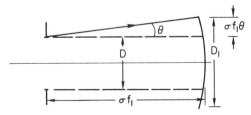

Fig. 8.3. Geometry showing primary mirror diameter needed to cover field of angular radius θ without vignetting. See Eq. (8.1.3).

diameter of the primary, is given by

$$D_1 = D(1 + 2\theta\sigma F_1), \tag{8.1.3}$$

where θ is the angular radius of the field.

To determine the size of the secondary, we use the geometry in Fig. 8.4. The diameter D_2, expressed in terms of the diameter of the corrector, is given by

$$D_2 = D[k + 2\theta F_1(k(\sigma - 1) + 1)]. \tag{8.1.4}$$

Note that Eqs. (8.1.3) and (8.1.4) with $\sigma = 0$ apply to the two-mirror telescopes discussed in Chapter 6. Note also that D_1 and D_2 can be expressed in terms of η and F with the substitution $F_1 = F(\rho - k)/\rho$. We now apply the results in this section to some specific types of SC telescopes.

B. FLAT-FIELD ANASTIGMATIC SCHMIDT-CASSEGRAIN

An anastigmatic optical system is one with zero astigmatism, coma, and spherical aberration. The condition for a flat field is zero Petzval curvature, hence $\rho = 1$, and the surface of best images is a plane. Setting $\rho = 1$ also fixes m in terms of k, with $m = 1/(1-k)$.

With these substitutions, and after some straightforward algebra, the system aberration coefficients take the form

$$B_{1s} = \frac{\theta^2}{4R_1}(A_2\sigma^2 - 2A_1\sigma + A_0), \tag{8.1.5}$$

$$B_{2s} = \frac{\theta}{2R_1^2}(A_1 - A_2\sigma), \tag{8.1.6}$$

$$B_{3s} = \frac{A_2}{4R_1^3} - \frac{b}{8}, \tag{8.1.7}$$

where

$$A_0 = 4 - k^2[(3-k)^2 + (1-k)^2 K_2], \tag{8.1.8}$$

$$A_1 = 2 - k^2[(2-k)(3-k) - k(1-k)K_2], \tag{8.1.9}$$

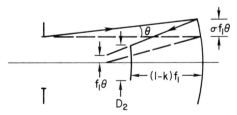

Fig. 8.4. Geometry of secondary mirror diameter needed to cover field of radius θ without vignetting at outer edge. See Eq. (8.1.4).

I. Schmidt–Cassegrain Telescopes

$$A_2 = 1 + K_1 - k^2(2-k)^2 - k^4 K_2. \qquad (8.1.10)$$

Applying the anastigmatic condition to Eqs. (8.1.5)-(8.1.7) gives

$$A_1 = \sigma A_2, \qquad A_0 = \sigma A_1, \qquad bR_1^3 = 2A_2. \qquad (8.1.11)$$

The first step in the procedure for solving the relations in Eqs. (8.1.11) to find the system parameters is to calculate k from Eq. (8.1.2) for selected η and F. Substituting this value of k into $A_0 = \sigma A_1$ gives a relation between σ and K_2, and one can be found after the other is specified. With σ and K_2 now known, K_1 is computed using $A_1 = \sigma A_2$. Finally, using the known values of the conic constants, b is calculated using the last relation in Eq. (8.1.11).

Carrying out the first step for $F = 3$ and selected nonnegative values of η to ensure an accessible focal plane, we get the results shown in Table 8.3.

Using the values of k from Table 8.3 we find the characteristics of two specific systems, first analyzed by Baker. The first is the so-called Baker A design with $\sigma = 1$; the second is the Baker B design, for which $K_2 = 0$. For the Baker A design, the solution of the relations in Eqs. (8.1.11) gives

$$K_1 = 1 + 2k, \qquad K_2 = -1 + \frac{2(1+k)}{k^2},$$

and for the Baker B

$$K_1 = \frac{k^2(1-k)^2}{4 - k^2(3-k)^2}, \qquad \sigma = \frac{4 - k^2(3-k)^2}{2 - k^2(2-k)(3-k)}.$$

Table 8.4 gives the parameters of the Baker A design, including the radius of the secondary needed to cover a field 0.1 radians in diameter. Table 8.5 gives the results for the Baker B design.

From the tabulated data in Tables 8.4 and 8.5 it is evident that there is a significant difference between the two designs. The mirrors in the Baker A design are strongly elliptical, whereas the primary in the other system differs only slightly from a sphere. The diameter of the secondary needed

Table 8.3

Parameters of $f/3$ Schmidt-Cassegrain[a]

η	k	m	F_1
0.00	0.3820	1.618	1.854
0.05	0.3894	1.638	1.832
0.10	0.3970	1.658	1.809
0.15	0.4046	1.679	1.786
0.20	0.4122	1.701	1.763

[a] Zero Petzval curvature, hence $\rho = 1$.

Table 8.4

Parameters of $f/3$ Baker A Design

η	K_1	K_2	A_2	D_2/D
0.00	1.7639	17.944	2.0000	0.567
0.05	1.7789	17.322	1.9870	0.573
0.10	1.7940	16.729	1.9735	0.578
0.15	1.8091	16.164	1.9595	0.583
0.20	1.8244	15.624	1.9450	0.589

to cover the given field is about 15% smaller for the A version, hence there is about 30% less vignetting in this design. This difference is a direct consequence of the difference in lengths between the two designs, about a factor of two.

The most significant difference between the two designs is in the size of A_2, which is approximately 3.2 times larger for the A version. As seen from Eqs. (8.1.11), this means that the aspheric term b is larger by this factor, as are the chromatic effects. This is most easily seen by noting that CSA in Eq. (7.2.6) is proportional to b, the aspheric parameter. By definition, CSA is proportional to the slope of the surface at the edge of a corrector configured for minimum chromatic aberration. Putting Eq. (7.2.4) into Eq. (7.2.1), setting $\rho = \rho_0$, and ignoring the term in b', we find

$$\frac{dz}{d\rho} = \frac{b\rho_0^3}{8(n'-n)}.$$

Hence larger b means larger CSA in direct proportion, and the chromatic aberration of the Baker A design is 3.2 times that of the B design.

It is also important to compare the chromatic properties of each SC system with those of a standard Schmidt of the same final focal ratio. The ratio of the chromatic aberrations is simply the ratio of the corresponding

Table 8.5

Parameters of $f/3$ Baker B Design

η	K_1	σ	A_2	D_2/D
0.00	0.01858	2.1708	0.6366	0.650
0.05	0.01906	2.1775	0.6256	0.657
0.10	0.01954	2.1843	0.6146	0.663
0.15	0.02003	2.1913	0.6034	0.669
0.20	0.02051	2.1985	0.5922	0.676

b terms, hence

$$\frac{\text{CSA(SC)}}{\text{CSA(SS)}} = \frac{2A_2}{R_1^3}\frac{R^3}{2} = m^3 A_2, \tag{8.1.12}$$

where SS denotes a standard Schmidt and m is the magnification of the Schmidt–Cassegrain. Combining m from Table 8.3 with the values of A_2 and $\eta = 0.1$, we find from Eq. (8.1.12) the ratios A:B:SS = 9.0:2.8:1. It is evident from these results that a single-element corrector in a Schmidt–Cassegrain as fast as $f/3$ has chromatic aberration that is relatively large compared with a standard Schmidt.

Ray traces of the Baker B system, with the addition of a b' aspheric parameter to control fifth-order spherical aberration, show acceptable images to a field radius of about 4° at the design wavelength. An acceptable image is defined as one for which the blur diameter is no larger than 10 μm for an overall focal length of 900 mm. This diameter corresponds to an angular blur of about 2.3 arc-sec for this focal length, with the image size determined primarily by residual astigmatism.

Ray traces of the Baker A design give acceptable on-axis images, as defined above, only with the addition of aspheric parameters of still higher order. This is not surprising given the large spherical aberration of the pair of highly elliptical mirrors. The field radius for acceptable images is about one-half that of the Baker B design. Thus the A version, in spite of its shorter length and smaller vignetting by the secondary, is inferior to the B version in field size and chromatic aberration.

Compared to a standard Schmidt, the Baker B design has the advantages of a flat accessible focal surface and a shorter length by about 40%. If these advantages more than outweigh the disadvantages of larger chromatic aberration and vignetting by the secondary of 40% or more, then this system or another of the flat-field anastigmats with conic constants near zero is a viable alternative to the Schmidt.

An analysis of the general solution of the relations in Eqs. (8.1.11) shows that the product σA_2 decreases slowly as K_2 decreases. As can be verified from Eq. (8.1.9), a change in K_2 from 10 to zero gives a decrease in σA_2 of roughly 30%. For the same decrease in K_2, the factor A_2 decreases by a bit over a factor of 2 while σ increases by about a factor of 1.6. Hence there is a trade-off between chromatic aberration and vignetting of the secondary, with a reasonable balance achieved when the conic constants of the mirrors are near zero.

It is worth noting that Baker also gave the results for other flat-field systems; the Baker C design has a spherical primary and the Baker D design is free of distortion. The parameters of the C design are little different from those of the B design given above, while those of the D design lie between those of the Baker A and B systems. For specifics on these other versions, see the reference by Linfoot.

C. SCHMIDT-CASSEGRAIN WITH SPHERICAL MIRRORS

An alternative to the flat-field anastigmatic SC is the family in which both mirrors are spherical and the focal surface is curved. The analysis of this type of SC proceeds in a way very similar to that in the last section. Putting $K_1 = K_2 = 0$ in the relations in Tables 8.1 and 8.2, the system aberration coefficients take the same form as Eqs. (8.1.5)-(8.1.7), where

$$\rho^3 A_0 = 4\rho^3 - k^2(2\rho + 1 - k)^2, \tag{8.1.13}$$

$$\rho^3 A_1 = 2\rho^3 - k^2(2\rho - k)(2\rho + 1 - k), \tag{8.1.14}$$

$$\rho^3 A_2 = \rho^3 - k^2(2\rho - k)^2. \tag{8.1.15}$$

If we require that the system be anastigmatic, then the relations in Eqs. (8.1.11) apply. The only solution from these relations that gives an accessible focal surface is $2\rho = 1 + k$. Substituting for ρ in Eqs. (8.1.13)-(8.1.15) gives $A_0 = 2A_1 = 4A_2$, hence $\sigma = 2$ by Eqs. (8.1.11). Writing ρ in terms of R_1 and R_2, we find the relation $R_2 - R_1 = (1 - k)f_1$. Hence the two mirrors have a common center of curvature with the vertex of the corrector at this common point. This is the so-called concentric Schmidt-Cassegrain. Because the mirrors are concentric, so also is the Petzval surface, the focal surface when the astigmatism is zero. As can be shown using Eq. (8.1.2), this surface is accessible when $k > 1/3$.

Another version of the SC with spherical mirrors is the aplanat, a system in which spherical aberration and coma are zero but astigmatism is not. In this case only the first and last of the relations in Eqs. (8.1.11) hold. Substituting $A_1 = \sigma A_2$ and Eqs. (8.1.13)-(8.1.15) into Eq. (8.1.5) we find

$$B_{1s} = -\frac{\theta^2}{4R_1} \frac{k^2(2\rho - k - 1)^2}{A_2 \rho^3}. \tag{8.1.16}$$

With astigmatism not equal to zero, we choose to set κ_m, the curvature of the median image surface, to zero. Following the procedure in Section 6.II we find B_{1s} referenced to the secondary from Eq. (8.1.16) and calculate κ_m from the relations in Table 5.7. The result is

$$\kappa_m = \frac{1}{\rho R_1} \left[2(1 - \rho) - \frac{k^2(2\rho - k - 1)^2}{A_2 \rho^3} \right]. \tag{8.1.17}$$

Setting $\kappa_m = 0$ and substituting into Eq. (8.1.16) gives

$$\text{AAS} = 2B_{1s}y = \frac{m\theta^2}{4F} \left(\frac{1 - \rho}{\rho} \right). \tag{8.1.18}$$

It is evident from Eq. (8.1.18) that it is necessary to have ρ near one to keep the astigmatism small. If, for example, we choose $\rho = 0.95$, $m = 1.7$, and $F = 3$, then AAS is approximately 1 arc-sec when $\theta = 1.5°$.

Table 8.6 gives the calculated results for two SC designs with spherical mirrors, where $\eta = 0.1$ and $F = 3$ for each. Note that the chromatic aberration, the main discriminant between these two designs, is significantly smaller for the aplanat. We also see that the CSAs for the aplanat and Baker B design are comparable, which is not surprising given mirrors that are similar.

Ray traces of the aplanat in Table 8.6, with the addition of a b' parameter to control fifth-order spherical aberration, show that astigmatism limits the field diameter to about 4°, as compared with roughly twice this value for the Baker B design. Thus the Baker B system has an edge over the aplanat, if its larger field is a requirement.

D. CONCLUDING COMMENTS

The SC designs discussed in the preceding sections are either anastigmats or aplanats with the stop located at the corrector plate. Because both spherical aberration and coma are zero in all of these designs, it follows from the discussion in Section 5.V that both coma and astigmatism are independent of the stop position. Thus all of the results given, except the chromatic aberrations, are valid for an arbitrary stop position. If the stop is displaced from the corrector, the chromatic effects increase by the factor Γ^3, as described in Section 7.II. Given the already large chromatic effects in the SC compared to those of the standard Schmidt, it is evident that an SC with a stop displaced from the corrector is of limited usefulness.

We have limited our discussion of the Schmidt–Cassegrain to designs that are possible alternatives to the standard Schmidt, that is, designs with wide field and fast focal ratios. If these conditions are changed to smaller field, on the order of 1° in diameter, and Cassegrain focal ratios ≈ 10, then a family of aplanatic SC designs is found with $\sigma \approx 1$. Although various

Table 8.6

Parameters of $f/3$ Schmidt-Cassegrains[a]

	σ	k	m	A_2	$m^3 A_2$
Concentric	2.000	0.348	2.069	0.6040	5.35
Aplanat	2.231	0.391	1.700	0.5943	2.92

[a] Each SC has spherical mirrors with $\eta = 0.1$, $F = 3$. For the aplanat $\rho = 0.95$.

combinations of K_1 and K_2 are possible, the usual choice is a spherical primary. With this choice the secondary is ellipsoidal and $m \approx 5$. Small telescopes of this type are available from several manufacturers and are popular choices among amateur astronomers. The characteristics of aplanatic SC telescopes of this type are found by applying the general theory above, an exercise left to the reader.

II. CAMERAS WITH MENISCUS CORRECTORS

We now turn our attention to another type of wide-field camera, one in which the aspheric corrector is replaced by a meniscus lens. The purpose of the meniscus is the same as that of the corrector, to compensate for the spherical aberration of the following mirror(s). The theory of the meniscus corrector was developed independently by Bouwers, Maksutov, and Baker in the 1940s, and their names are attached to various versions of meniscus cameras. In this section we consider a subset of the many types of meniscus cameras that have been described in the literature. The reader should consult the references at the end of the chapter, including the monograph by Maxwell, for details on these and other designs.

A. CONCENTRIC MENISCUS CORRECTOR

A type of meniscus lens is one in which the two surfaces of the lens are concentric with the surface of a spherical mirror, as shown in Fig. 8.5. If an aperture stop is placed at the common center of curvature, as in a standard Schmidt, then the system has no unique axis and all off-axis aberrations are zero. The Petzval surface is also concentric with the other surfaces and the image surface is curved, as in a standard Schmidt. The characteristics of the images are determined entirely by the spherical aberration and any chromatic aberration introduced by the meniscus.

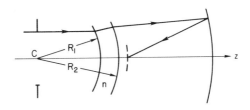

Fig. 8.5. Bouwers concentric camera with meniscus corrector. All surfaces are spherical with common center of curvature C.

II. Cameras with Meniscus Correctors

Complete analysis of the spherical aberration of the system shown in Fig. 8.5 involves the application of Eq. (5.6.7) with $k = 3$ together with the corresponding coefficients from Tables 5.5 and 5.6. The result of this exercise, with all surfaces concentric, is a cubic equation involving the thickness and location of the meniscus. Although the solution of this equation gives results in good agreement with those derived from ray traces, the form of the equation is quite complicated and gives little insight into the workings of the meniscus lens. It is more instructive to follow the approach by Bouwers and we choose to use his method.

The starting point in the Bouwers method is the assumption that the spherical aberration coefficient of the meniscus is that of a thin lens for which the source is at infinity. Although the derivation of this result is straightforward using the results in Chapter 5, we take the expression given by Bouwers and convert it into the desired coefficient. The result is

$$B_{31} = -\frac{1}{8f^3}\left[\left(\frac{n}{n-1}\right)^2 - \frac{f}{R_1}\left(\frac{2n+1}{n-1}\right) + \frac{f^2}{R_1^2}\left(\frac{n+2}{n}\right)\right], \quad (8.2.1)$$

where f is the focal length of the lens and R_1 is the radius of curvature of its first surface. For a concentric lens, as we show below, $f \gg R_1$ and to a good approximation

$$B_{31} = -\frac{1}{8fR_1^2}\left(\frac{n+2}{n}\right). \quad (8.2.2)$$

The condition for a concentric lens is $d = R_1 - R_2$, where $d > 0$ and the radii are negative according to the sign convention. Substituting Eq. (2.4.1) into Eq. (2.4.3) we find that the focal length of a concentric lens is given by

$$\frac{1}{f} = -\frac{d}{R_1 R_2}\left(\frac{n-1}{n}\right). \quad (8.2.3)$$

Practical values of d are 10 or more times smaller than R_2, hence f is typically 30 times or more larger than R_1. Thus we are justified in taking Eq. (8.2.2) for the spherical aberration coefficient of the lens.

Note that the concentric meniscus lens has a large negative focal length and is therefore a weak diverging lens. Thus the lens is thicker at the margin than at the center, the same as that of an aspheric corrector without an added radius term, and the signs of the spherical aberration coefficients of the lens and aspheric plate are the same. To find the system spherical aberration coefficient we simply add Eq. (8.2.2) to the aberration coefficient of a spherical mirror in collimated light from Table 5.2. The result, after substitution of Eq. (8.2.3), is

$$B_{3s} = \frac{(n-1)(n+2)}{8n^2}\frac{d}{R_1^3 R_2} + \frac{1}{4R^3}, \quad (8.2.4)$$

where R is the radius of curvature of the mirror. Note that by adding the coefficients to get Eq. (8.2.4), we have ignored the divergence of the beam from the lens and taken the same ray heights at the mirror and lens. This is acceptable in view of the other approximations made.

At this point we express d, R_1, and R_2 in terms of R as follows:

$$d = -\beta R, \qquad R_1 = \alpha R, \qquad R_2 = (\alpha + \beta)R,$$

where α and β are positive. Setting Eq. (8.2.4) equal to zero, substituting in terms of R, and solving for β, we find

$$\beta = \alpha^4 \left[\frac{(n-1)(n+2)}{2n^2} - \alpha^3 \right]^{-1}. \qquad (8.2.5)$$

By taking $n = 1.46$, values of β for a selected set of α values are found in Table 8.7. Note that the normalized thickness of the meniscus increases rapidly as the lens is placed farther from the stop.

The values in Table 8.7 serve as the starting point for ray-trace analysis of the meniscus camera, with β adjusted for each α to make third-order spherical aberration zero. The results found from this analysis for an $f/2.5$ system are shown in Table 8.8. Note that the values of β derived from ray trace analysis are significantly larger than those from Eq. (8.2.5). Note also that the camera focal lengths decrease and the back focal lengths increase as the lens thickness increases.

From Table 8.8 we see that there is significant fifth-order spherical aberration, which decreases for thicker lenses. This trend is not surprising because a thicker lens is less strongly curved, and in the limit of zero curvature there is no spherical aberration contribution from a glass plate in collimated light. The angular diameter of the image at the circle of least confusion for a given α is one-half of the corresponding value given in Table 8.8.

Table 8.7

Parameters for Concentric Meniscus Lens[a]

α	β
0.200	0.00438
0.225	0.00708
0.250	0.01092
0.275	0.01622
0.300	0.02339

[a] Values derived from Eq. (8.2.5) with $n = 1.46$.

II. Cameras with Meniscus Correctors

Table 8.8
Parameters for Concentric Meniscus Cameras[a]

α	β	f/R	BFD/R	ASA5
0.200	0.00539	0.490	0.510	64.3
0.225	0.00908	0.487	0.513	46.2
0.250	0.01478	0.483	0.517	34.6
0.275	0.02357	0.478	0.522	26.6
0.300	0.03751	0.472	0.528	20.5

[a] Results derived from ray traces with ASA3 = 0. ASA5 = fifth-order angular spherical aberration in units of arc-seconds; BFD = distance from mirror to focal surface.

It is clear from the results in Table 8.8 that the choice of zero third-order aberration is probably far from optimum. By changing β at a given α, third-order aberration of an amount equal in magnitude but opposite in sign to that of the fifth-order contribution for rays at the margin can be introduced; the result is an image diameter that is significantly smaller. With $\alpha = 0.25$, for example, this condition is met when $\beta = 0.01246$, a decrease of about 15% in the lens thickness. From ray traces the diameter of the best image is reduced nearly a factor of five, giving an angular diameter of 3.7 arc-sec. This corresponds to a linear diameter of 13 μm for a focal length of 750 mm. Hence a significant improvement is achieved by balancing the spherical aberration contributions.

The final item of interest for the meniscus camera is the chromatic aberration introduced by the lens. From Eq. (8.2.3) we find that the focal length of the lens changes with index according to the relation

$$df/f = -dn/n(n-1). \tag{8.2.6}$$

Because the rays incident on the mirror appear to come from the focal point of the lens, a shift of this point translates into a shift of the camera focal point. Denoting the camera focal length by f_c and applying Eq. (2.5.7) we find $df_c = -m^2 \, df$, where m is approximately $-f_c/f$, the magnification due to the mirror. Combining these results with Eqs. (8.2.3) and (8.2.6) we find

$$\frac{df_c}{f_c} = \frac{\beta}{2\alpha^2} \frac{dn}{n^2}, \tag{8.2.7}$$

where df_c is the axial shift of focus with changing index. For the balanced system given above, with $dn = 0.0056$, Eq. (8.2.7) gives $df_c/f_c = 0.00026$, a result in good agreement with that found from ray traces. The diameter of the image over the range of wavelengths spanned by this change of index is about 23 arc-sec, a significant increase over the monochromatic diameter of 3.7 arc-sec.

There are two methods for reducing the chromatic aberration of the meniscus. One, proposed by Bouwers, is to use an achromatic meniscus made of two different glasses cemented together. In this case the cemented interface cannot be made concentric with the outer surfaces, and the system is no longer strictly concentric. For further details the reader should consult the references by Bouwers and Maxwell.

A second method, first proposed by Maksutov, is to use an achromatic meniscus corrector made of a single glass with f invariant to a change in index. To achieve this condition, however, it is necessary to depart from the concentric lens surfaces. We examine briefly the characteristics of this type of corrector in the next section.

B. MAKSUTOV ACHROMATIC CORRECTOR

As noted in the previous paragraph, the achromatic corrector proposed by Maksutov is one in which the focal length is invariant to a change in index. This condition is easily derived by taking f for a thick lens and setting $df/dn = 0$. Applying this condition to Eq. (2.4.3) combined with Eq. (2.4.1) we find

$$d = (R_1 - R_2) \left(\frac{n^2}{n^2 - 1} \right). \tag{8.2.8}$$

Relative to a concentric lens, we see from Eq. (8.2.8) that this lens is roughly 2× thicker. Using Eq. (8.2.8) we can find the separation Δz between the centers of curvature of the surfaces of the meniscus, with the result

$$\Delta z = (R_1 - R_2) - d = -d/n^2, \tag{8.2.9}$$

where the minus sign indicates that the center of curvature of surface 2 is closer to the mirror than that of surface 1. It is evident from Eq. (8.2.9) that the surfaces of the meniscus are more nearly concentric for small d.

We are not going to discuss all of the details in the design of a Maksutov camera, but instead will illustrate the general characteristics with one example. For a mirror of radius of curvature R we take $d = -0.025R$ as a constant and vary R_1, R_2, and the lens-mirror separation until spherical aberration of the marginal rays is balanced. The value of R_2, of course, is tied to that of R_1 by Eq. (8.2.8). The axial position of the stop is then altered until coma is balanced, with the parameters of the final system shown in Table 8.9.

II. Cameras with Meniscus Correctors

Table 8.9
Parameters of a Maksutov Achromatic Camera[a]

	Distance from stop
Surface 1 of lens	$0.2667|R|$
Mirror	$0.8543|R|$
Image	$0.3455|R|$

[a] Index n used in Eq. (8.2.8) is 1.46. $d = 0.025|R|$, $R_1 = 0.2087R$, $R_2 = 0.2219R$. Radius of curvature of image surface $= 0.538R$.

Ray traces for an $f/2.5$ system with parameters given in Table 8.9 show an on-axis image whose angular diameter is about 5.5 arc-sec, and a slow increase in image size to 10 arc-sec at a field radius of 3°. This on-axis image diameter is about 20 μm for a camera whose focal length is 750 mm. The chromatic effects of the corrector are much smaller than those of the concentric meniscus, as expected, with df_c some 30 times smaller than that given by Eq. (8.2.7). For an SiO_2 corrector the on-axis image diameters on a fixed focal surface do not exceed 10 arc-sec over the wavelength range from 400 to 900 nm.

Although this system was not given a detailed optimization, it is evident that its image quality is satisfactory, provided the camera is not too large. Compared to the concentric meniscus camera, the Maksutov achromatic camera is clearly superior in its chromatic characteristics.

C. CONCLUDING COMMENTS

The discussion of meniscus lens cameras in the preceding sections is only an introduction to cameras based on this type of corrector. Among other types are those in which the meniscus is split, with part of it preceding the aperture stop. In addition, the meniscus on either side of the stop may itself be split into two separate pieces of glass. There are also so-called hybrid systems in which an aspheric plate located at the stop is used in conjunction with a meniscus corrector and others in which an aspheric surface is put on one of the surfaces of the meniscus.

A widely used hybrid system is the super-Schmidt or Baker-Nunn camera used for wide-field photography to record trails of meteors and artificial satellites. This type of camera has a double concentric meniscus, half on either side of the stop, with a doublet Schmidt plate at the stop. The light reflected from the mirror passes through the meniscus nearer the mirror a

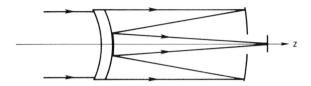

Fig. 8.6. Maksutov Cassegrain telescope with achromatic meniscus corrector. The secondary mirror shown is an aluminized circular area on the back face of the corrector.

second time before coming to the curved focal surface. Because the design is based on the concentric principle, an angular field diameter of about 50° with good imagery is achieved.

A meniscus lens can also be used in place of an aspheric corrector in a Cassegrain camera, as shown schematically in Fig. 8.6. The secondary mirror in this type of system can be a separate mirror or a centered reflecting area on the back surface of the meniscus lens.

For more details on these and other meniscus lens systems, the reader should consult the references at the end of the chapter; those by Bouwers and Maxwell are a good introduction.

III. SEMISOLID SCHMIDT–CASSEGRAIN CAMERAS

The same principles discussed in Section 7.V apply to the design of semisolid Schmidt–Cassegrain cameras for spectrographs. A schematic layout of a semisolid SC camera is shown in Fig. 8.7, with glass filling the

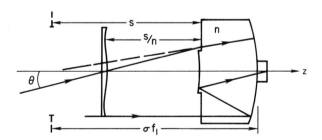

Fig. 8.7. Semisolid Schmidt–Cassegrain camera. The effective focal length is n times shorter than that of an equivalent air Schmidt–Cassegrain.

space between the primary and secondary mirrors. Note that the separation between the corrector and secondary in Fig. 8.7 is smaller by a factor of n than that in Fig. 8.1. This change in corrector location is analogous to that shown in Fig. 7.10 and is required to place the effective stop for the primary at a distance σf_1. As in the conversion of an air-Schmidt to a semisolid Schmidt, the aspheric figure is n times stronger and the radius R_c is n times smaller.

Ray traces of a semisolid Baker B design show that the field radius for acceptable images is about n times larger than that of the equivalent design with air between the mirrors, with the focal length of the former n times shorter. The larger field and shorter focal length follow from the arguments given in Section 7.V for the solid-Schmidt. The chromatic effects given in Eq. (7.5.1) also apply to the semisolid Schmidt–Cassegrain.

IV. CONCLUDING REMARKS, ALL-REFLECTING SYSTEMS

The discussion in this chapter and the preceding one is intended to show the basic characteristics of so-called wide-field cameras and telescopes. It is evident from this discussion that many designs are capable of good imagery over fields of several degrees, with chromatic effects generally setting the limit to the range of wavelengths that can be covered.

The principles used for these catadioptric designs also apply to all-reflecting wide-field systems. One obvious use of such a system is a space-based ultraviolet telescope. In these systems a reflecting aspheric corrector replaces the refracting plate and chromatic effects are absent. To separate the incident beam from the beam reflected from the corrector, the corrector axis is tilted by angle θ relative to the mirror axis. The angle between the incident and reflected chief ray at the corrector is then 2θ.

The main complication of a tilted corrector is that its surface figure must be modified so that a collimated beam from the center of the field "sees" a circular profile on the corrector. This is achieved by placing an elliptical figure on the corrector with ρ^2 in Eq. (7.2.1) given by $\rho^2 = x^2 \cos^2 \theta + y^2$, instead of $\rho^2 = x^2 + y^2$. The other change required is to replace z in Eq. (7.2.1) by $z \cos \theta$; this ensures that the OPD introduced by the tilted corrector is the same as that of an untilted plate. An aspheric plate with an elliptical figure is obviously more difficult to make than one with a circular figure, and no large systems of this type have been made. We do not discuss this type of system in detail here; the interested reader should consult the paper by Schroeder.

REFERENCES

Baker, J. (1940). *Proc. Am. Philos. Soc.* **82**, 339.
Bouwers, A. (1946). "Achievements in Optics," Chap. 1. Elsevier, Amsterdam.
Linfoot, E. (1955). "Recent Advances in Optics," Chap. 4. Oxford Univ. Press, London and New York.
Maksutov, D. (1944). *J. Opt. Soc. Am.* **34**, 270.
Maxwell, J. (1972). "Catadioptric Imaging Systems." Amer. Elsevier, New York.
Schroeder, D. (1978). *Appl. Opt.* **17**, 141.

Chapter 9 | Auxiliary Optics for Telescopes

Telescopes are often used for direct photography without extra optics in the light beam, but there are many observations for which auxiliary optics are required. Examples of some types of observations that require additional optics are spectroscopy and photometry. In the case of photometry this is often no more than a field lens to reimage the exit pupil of the telescope onto a detector. For spectroscopy the extra optics may be as simple as a prism or diffraction grating placed in the light beam, or it may be a separate spectrograph with many optical elements whose entrance aperture is at one of the telescope foci. The characteristics of spectrometers are discussed in Chapters 14 and 15.

Even for direct photography, it may be important to enhance the imaging characteristics of an existing telescope by adding optical elements to widen the field, flatten the image surface, or reimage at a different focal ratio. For a new telescope it is now customary to include such optics in the early design stages and often to design the telescope with its auxiliary optics as a system in itself.

The kinds of auxiliary optics discussed in this chapter include field lenses and field flatteners, corrector systems for both prime and Cassegrain focus, and focal reducers for Cassegrain telescopes. Attention is given to the aberration characteristics of these systems, with examples given for each type of system considered.

I. FIELD LENSES, FLATTENERS

A field lens is an element that is placed at or near an image plane in an optical system. One application for such a lens, as noted in Section 5.VII, is that of flattening a curved image surface. Before discussing this application, and others, we consider the aberrations introduced when an object lies close to an optical surface.

A. ABERRATIONS

The aberration coefficients for a general surface are given in Table 5.5. For an object close to the surface, take Γ from Table 5.5, replace s' using Eq. (2.2.1), and let $s \ll R$. The result obtained is

$$\Gamma = \frac{n}{s}\left(\frac{n^2}{n'^2} - 1\right). \tag{9.1.1}$$

Given the condition that s is small, the dominant terms in the coefficients in Table 5.5 are those containing the factor Γ. Taking only the dominant terms, putting the coefficients into Eqs. (5.5.9), and letting $s'/n' = s/n$ gives

$$\text{TSA} = -\frac{s}{2}\left(\frac{y}{s}\right)^3 \Lambda, \qquad \text{TSC} = \frac{s\psi}{2}\left(\frac{y}{s}\right)^2\left(1 - \frac{W}{R}\right)\Lambda,$$

$$\text{TAS} = -s\psi^2\left(\frac{y}{s}\right)\left(1 - \frac{W}{R}\right)^2 \Lambda, \qquad \Lambda = \frac{n^2}{n'^2} - 1. \tag{9.1.2}$$

Because y/s is finite for all s, each of the transverse aberrations in Eqs. (9.1.2) goes to zero as s approaches zero and the image is free of aberrations in this limit. This result is not surprising because $s' \to 0$ as $s \to 0$, and the image and object coincide when $s = 0$.

For a real lens placed at an image surface, s cannot be zero for both surfaces, but it is small enough for each surface so that its contributions are usually of little consequence.

B. FIELD-FLATTENED RITCHEY-CHRETIEN TELESCOPES

As an example, we consider a lens placed near the Cassegrain focus of a Ritchey-Chretien telescope, as shown in Fig. 9.1. The lens parameters are chosen so that the median astigmatic surface of the telescope-lens combination is flat, thus

$$\kappa_m(\text{RC}) + \kappa_m(\text{lens}) = 0. \tag{9.1.3}$$

I. Field Lenses, Flatteners

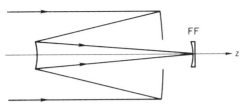

Fig. 9.1. Ritchey-Chretien telescope with field flattener lens FF. The lens parameters are given in Eq. (9.1.4).

The relation for the first term in Eq. (9.1.3) is given in Table 6.7. We find the second term by noting that a thin lens has negligible aberrations according to Eqs. (9.1.2), hence $\kappa_m = \kappa_p$.

The Petzval curvature of a lens is derived from the relation in Table 5.7, with the choice $R_2 = \infty$ so that the distance between the flat image and lens can be made as small as desired. Substituting the derived result into Eq. (9.1.3) gives

$$(n-1)/nR_l = \kappa_m(\text{RC}), \tag{9.1.4}$$

where R_l is the radius of curvature of the lens surface facing the secondary. Because κ_m for a Ritchey-Chretien is negative (the image surface is concave as seen from the secondary), the lens has $R_l < 0$ and is plano-concave in cross section, as shown in Fig. 9.1.

Using the parameters of the RC telescope in Table 6.8 and letting $R_1 = -6000$ gives $R_l = -248$ for $n = 1.46$, the index of SiO_2 at a wavelength of 548 nm. The aberrations of the telescope with and without the field-flattener lens are given in Table 9.1, with the results taken from the output of a computer ray-trace program.

Table 9.1

Aberrations for Field-Flattened Ritchey-Chretien Telescope[a,b]

	Without lens	With lens	
		$R_l = -248$	$R_l = -260$
ASA	0.000	0.002	0.002
ATC	0.000	0.039	0.037
AAS	1.025	0.913	0.924
$\kappa_m R_1$	7.625	−0.393	−0.003
f	12 000	12 101	12 096

[a] Lens: plano-concave shape, thickness at vertex = 6, back focal distance = 0.43. Telescope: $R_1 = -6000$, other parameters in Table 6.8.

[b] Aberrations are given at a field angle of 18 arc-min in units of arc-seconds.

Note that the field lens does change the system aberrations, but only slightly. Because the lens has nonzero astigmatism, the assumption that $\kappa_m = \kappa_p$ for the lens is not quite true, and the image surface curvature is not zero. However, this assumption is a good first approximation, and a change of R_l to -260 gives a flat image surface.

C. FIELD-FLATTENED SCHMIDT CAMERA

As a second example we consider a field-flattener lens placed near the curved focal surface of a Schmidt camera. In Section 5.VII we derived the condition for a flat Petzval surface for the combination of a spherical mirror and thin lens, with (5.7.14) giving the required condition for a plano-convex lens. The spherical aberration of the mirror alone is canceled by an aspheric plate with b chosen according to Eq. (7.1.5) with $m = 0$.

Choosing $R = -1000$ and $R_c = \infty$ gives $b = -2.\mathrm{E}-9$ and $R_1 = -157.5$ for $n = 1.46$. The aberrations of the camera with and without the flattener lens are shown in Table 9.2, with $F = 2.5$ for the camera without the lens. Note that the lens flattens the field but introduces significant aberrations, especially coma and spherical aberration.

The coma due to the field-flattener lens can be largely removed by reducing the corrector-mirror separation by about 2%, but this displacement does not, of course, affect the spherical aberration. For this example it turns out that higher-order aberrations are not negligible. The fifth-order spherical aberration compensates in part for the third-order effect and the result from

Table 9.2

Aberrations for Field-Flattened Schmidt Camera[a,b]

		With lens	
	Without lens	$W/R = 1.0$	$W/R = 0.979$
ASA	0.002	2.807	2.807
ATC	0.000	2.247	0.010
AAS	0.000	0.208	0.217
$\kappa_p R$	2.000	-0.0004	-0.0004
f	500.0	494.2	494.2

[a] Lens: plano-convex shape, $R_1 = -157.5$, thickness at vertex = 5, back focal distance = -0.53. Mirror: $R = -1000$. Corrector: $b = -2.\mathrm{E}-9$, $R_c = \infty$, thickness at vertex = 10.

[b] Aberrations are given at a field angle of 1° in units of arc-seconds.

ray tracing is an on-axis blur diameter of about 1.4 arc-sec. This blur can be reduced to a negligible value by adding an aspheric term of higher order to the corrector, as noted in Section 7.II.

Comparing the effects of the flattener lens in these two examples, it is evident that the aberrations it introduces are significantly larger for the Schmidt camera. For spherical aberration this is entirely a consequence of the different focal ratios, $F = 2.5$ for the camera versus $F = 10$ for the RC telescope, where $y/s = 1/2F$ in Eqs. (9.1.2). For astigmatism and coma the different pupil position for the lens is also a factor. For the RC telescope the pupil location is given by Eq. (2.5.4), from which we find $W/R_1 \approx 15$. For the Schmidt camera the mirror images the aperture stop back on itself and $W/R_1 \approx 3$. Substitution of these results into the relations in Eqs. (9.1.2) accounts for the relative sizes of the aberrations introduced by the field lens.

In two other applications of a field lens its primary purpose is to reimage the exit pupil of the telescope. For a photometer an aperture at the telescope focus passes the light of a single star and a field lens at the aperture images the telescope exit pupil on the photosensitive surface of a detector. If the star should wander in the aperture because of atmospheric effects, the effect is not seen by the detector because the reimaged exit pupil does not wander on its surface. Such a lens is often called a Fabry lens. When the instrument on a telescope is a spectrograph a lens is often placed at the entrance aperture so that the lens, in combination with the spectrometer collimator, reimages the exit pupil onto the grating or prism that follows the collimator in the spectrometer.

II. PRIME FOCUS CORRECTORS

A large Ritchey–Chretien telescope is generally equipped with interchangeable secondaries to provide a range of focal ratios, as noted in Section 6.II. The focal ratio at the Cassegrain focus is usually the smallest, typically 7 or 8, which for a 4-m telescope gives an image scale of about 7 arc-sec/mm. With this scale the typical blur diameter of a star image is often not well matched to the size of the detector resolution element, be it photographic grain or individual pixels in a solid-state detector. A better match is usually achieved if the focal ratio is smaller, which in a Cassegrain configuration means a larger secondary and more obscuration. An alternative approach to getting a smaller focal ratio is to use the primary mirror only, which, in combination with a corrector system, can provide a usable field at a focal ratio of 2 or 3.

A. ASPHERIC PLATES

The simplest prime focus corrector system is a single aspheric plate in the converging beam near the image surface, as shown in Fig. 9.2. The aperture stop is the primary mirror and the plate is distance g from the focus, hence $W = f - g$ for the plate. The aberration coefficients for the primary are taken from Table 5.2, with $m = 0$, and the coefficients for the corrector are taken from Table 5.5, using only the terms in b. Substituting these results into Eq. (5.6.7) and choosing $y_\iota = y_1$, we get

$$B_{1s} = -\frac{\theta^2}{2f}\left[1 + \frac{b(f-g)^2 g^2}{f}\right],$$

$$B_{2s} = \frac{\theta}{4f^2}\left[1 - \frac{2b(f-g)g^3}{f}\right], \quad (9.2.1)$$

$$B_{3s} = -\frac{1}{8f^3}\left[\frac{K+1}{4} + \frac{bg^4}{f}\right],$$

where $\psi = -\theta$ and $y_2/y_1 = g/f$ have been substituted. For a given K there are two free parameters, b and g, in Eqs. (9.2.1) and two of the coefficients can be set to zero. The dominant aberrations at small field angle, for any primary other than a paraboloid, are spherical aberration and coma. Setting these coefficients to zero gives

$$b = -\frac{(K+1)f}{4g^4}, \quad \frac{g}{f} = \left(\frac{K+1}{K-1}\right). \quad (9.2.2)$$

It is evident from Eq. (9.2.2) that the location of the plate is set by the conic constant of the primary and this, in turn, sets the value of b. Because g and f are each positive, so also is b for $K < -1$. Note also that the condition $g > 0$ means that correction of both spherical aberration and coma with a single plate is not possible for an ellipsoidal primary.

Taking b from Eq. (9.2.2) and substituting into B_{1s} we find

$$\text{AAS} = 2B_{1s}y_1 = \frac{\theta^2}{2F}\left(\frac{K}{K+1}\right). \quad (9.2.3)$$

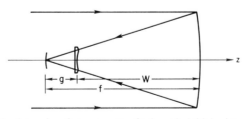

Fig. 9.2. Aspheric plate prime-focus corrector for hyperboloidal primary at distance g from the focal surface.

II. Prime Focus Correctors

Comparing the astigmatism given by Eq. (9.2.3) with that of the primary only, we see that, depending on the value of K, the radius of the usable field is limited to a few arc-minutes. From Eq. (9.2.3) we see that the larger the value of K for the hyperboloid in absolute terms, the smaller is the astigmatism at a given field angle and the larger is the usable field. A larger difference also means a greater distance between the plate and the focus and a larger plate size, as seen from Eq. (9.2.2). As pointed out by Gascoigne, these conclusions also hold for more complex corrector systems.

The final parameter of interest for this system is the curvature of the median image surface. Following the procedure in Section 6.II we find

$$k^2 B_{1s}(\text{cor}) = B_{1s}(\text{pri}), \qquad \theta = k\theta',$$

where $k = g/f$. Using these relations we get

$$\kappa_m = \frac{1}{f}\left(\frac{K-1}{K+1}\right) = \frac{1}{g}, \qquad (9.2.4)$$

hence the focal surface of best images is strongly curved and is concave as seen from the primary.

An example of these results applied to an $f/3$ Ritchey–Chretien primary is shown in Table 9.3. The conic constant chosen is that for an RC telescope with $m = 2.5$ and $\beta = 0.25$ at the Cassegrain focus. The plate has a diameter-to-thickness ratio of 25 and its radius of curvature is chosen to give minimum chromatic effect. The radius of the plate, ρ_0, is chosen to cover a field radius of about $0.12°$.

From the results in Table 9.3 we see that the coma and spherical aberration of the primary have been largely, but not entirely, eliminated. The size of

Table 9.3

Characteristics of a Prime Focus Corrector[a,b]

	Without plate	With plate	
		Theory	Actual[c]
ASA	21.221	0.000	0.850
ATC	7.500	0.000	0.057
AAS	0.105	0.696	0.707
$\kappa_m f$	1.000	12.25	12.20

[a] Primary: diameter = 4.0 m, $f = 12.0$ m, $K = -1.17778$. Plate: $b = 5.792\text{E}-10$, $R = -52\,800$ mm, $\rho_0 = 200$ mm, $n = 1.46$, thickness = 16 mm, $g = 0.08163f$, $W = 11\,000$ mm.

[b] Aberrations are given at a field angle of $0.1°$ in units of arc-seconds.

[c] Values in "actual" column are from ray-trace program.

the residuals depends on plate thickness and orientation. Although ASA for the example in Table 9.3 can be reduced to zero either by moving the plate closer to the primary or by adjusting the value of b, ray-trace results show that spherical aberration and coma of higher order are not negligible. These aberrations are reduced to negligible levels by including a fifth-order aspheric coefficient b' and adjusting the aspheric parameters and plate position, details that are omitted here.

This example illustrates a general procedure in the design of any system in which one or more of the elements is an aspheric plate. The procedure is one of taking only the aspheric terms in the aberration coefficients to get a first-order design and using computer analysis to refine the design. In this way one reduces the effort required in the theoretical analysis leading to the original design and uses the computer to help in arriving at the final design.

Although the single-plate corrector makes the prime focus of a Ritchey-Chretien primary usable, the surface of best images is curved enough so that a field-flattener lens is also needed. The sag of the image surface at a field angle of 6 arc-min is about 220 μm. This sag, in combination with blur already present in the off-axis image, gives an unacceptably large blur on a flat detector in focus on the on-axis image.

Before going on to other prime-focus correctors, it is worth noting that a single plate will not improve the images of a paraboloidal primary. Putting a corrector in the beam will, for example, introduce spherical aberration and astigmatism of unacceptable amounts if b is chosen to eliminate coma. Verification of this statement using Eq. (9.2.1) is left as an exercise for the reader.

Getting a larger and flatter image field at prime focus requires more complex correctors, of which many are discussed in the literature. Here we consider briefly a few of these, but without the detailed consideration given to the single-plate corrector. One kind of system that has been explored in detail is a set of corrector plates in series in the converging beam, as shown in Fig. 9.3 for three plates. Taking the aberration coefficients for the mirror from Table 5.2 and those for the correctors from Table 5.5, it is a straightforward step to the system coefficients. The results are

$$B_{1s} = -\frac{\theta^2}{2}\left(\frac{1}{f} + \sum b_i W_i^2 k_i^2\right),$$

$$B_{2s} = \frac{\theta}{2}\left(\frac{1}{2f^2} - \sum b_i W_i k_i^3\right), \quad (9.2.5)$$

$$B_{3s} = -\frac{1}{8}\left(\frac{K+1}{4f^3} + \sum b_i k_i^4\right),$$

II. Prime Focus Correctors

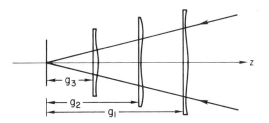

Fig. 9.3. Schematic of three-plate prime focus corrector.

where the ith corrector is a distance W_i from the primary, g_i from the image surface, and $k_i = g_i/f$.

For two plates there are four free parameters, b_1, g_1, b_2, g_2, and each of the coefficients in Eqs. (9.2.5) can be made zero for a hyperboloidal primary. In this case the signs of the aspheric coefficients are opposite, with $b > 0$ for the plate nearer the primary. For the same primary mirror parameters as in Table 9.3, ray-trace analysis of a two-plate corrector shows that the field of acceptable images is about two times larger in diameter than that of a single-plate system. The curvature of the median image surface is about 10 times smaller for the two-plate system, though its curvature is significant over the larger field. As noted by Gascoigne, a paraboloid with a two-plate corrector has image blurs and surface characteristics comparable to those of a hyperboloid with a single-plate corrector.

The design for a three-plate corrector, first proposed by Meinel, has been described in the literature, and the reader should consult the references at the end of the chapter for details. The field is larger than that achieved with the two-plate corrector by about a factor of two but, as noted by Wilson, the complete corrector set is not easy to manufacture because of the several large aspheric surfaces required. Wilson also points out that the optical performance of the three-plate corrector system is no better than that of three-lens systems with spherical surfaces, a type we consider briefly in the next section.

B. WYNNE TRIPLETS

An alternative approach to prime focus correctors is to use lenses with spherical surfaces, such as the Wynne corrector for a hyperboloidal primary shown in Fig. 9.4. Designs of this type give fields of good images up to 50 arc-min in diameter for an $f/2.7$ mirror and a somewhat larger field for a slower primary. The major advantages of this type, compared to the multi-plate type, are ease of fabrication, flatter fields, and more compact size. For the corrector shown in Fig. 9.4 the length L is approximately $0.06f$, less

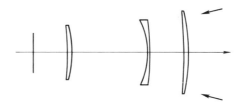

Fig. 9.4. Wynne triplet corrector for prime focus of hyperboloidal primary. See the article by Wynne for characteristics of the lens elements for an $f/3.25$ primary.

than that of aspheric plate systems, hence the diameters of the separate lenses are less than those of the aspheric plates.

Wynne has also shown that a three-lens corrector with a paraboloidal primary gives fields of comparable size and image quality. A schematic of this type of corrector, given by Wynne for an $f/3.25$ primary, is shown in Fig. 9.5. The correctors shown in Figs. 9.4 and 9.5 are drawn to the same scale for ease of comparison. Although the general forms of the corresponding lenses in the two correctors are similar, there are obvious differences in shape and spacing. The interested reader should consult the papers by Wynne for further details on these types of correctors.

III. CASSEGRAIN FOCUS CORRECTORS

Of all of the two-mirror telescopes discussed in Chapter 6, the Ritchey–Chretien type has the largest field at the Cassegrain focus. To third order, the only significant aberrations are astigmatism and field curvature. It was first shown by Gascoigne that the placement of an aspheric plate in the Cassegrain beam removes the astigmatism without introducing a significant amount of coma and spherical aberration. This plate also reduces the field curvature because, as noted in Section 6.II, the median image surface is

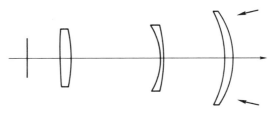

Fig. 9.5. Wynne triplet corrector for prime focus of paraboloidal primary. See the article by Wynne for characteristics of the lens elements for an $f/3.25$ primary.

III. Cassegrain Focus Correctors

more strongly curved than the Petzval surface. In this section we discuss the characteristics of this type of corrector for the Cassegrain focus.

A. ASPHERIC PLATE

A diagram of an RC telescope with aspheric plate in the Cassegrain beam is shown in Fig. 9.6, with the plate located a distance g from the focus and W from the telescope exit pupil. The aberration coefficients of the system, referenced to the primary, are

$$B_{3s} = -\frac{b}{8}\left(\frac{g}{f}\right)^4, \quad B_{2s} = \frac{b}{2}\left(\frac{g}{f}\right)^3 (W\psi),$$

$$B_{1s} = -\frac{\theta^2}{2f}\left[\frac{m(2m+1)+\beta}{2m(1+\beta)}\right] - \frac{b}{2}\left(\frac{g}{f}\right)^2 (W\psi)^2, \quad (9.3.1)$$

where the astigmatism coefficient for the telescope is taken from Table 6.4, with Eq. (6.2.4) substituted for (K_1+1).

To evaluate the coefficients we first express W and ψ in terms of the telescope parameters and plate location. The location of the telescope pupil is given by Eq. (2.5.4), and $W = g - f_1\delta$, where W is negative for the plate. The relation between ψ and θ, given by Eq. (2.5.5), is $\psi = \theta(m/\delta)$. In terms of the telescope parameters we get

$$W\psi = -f\theta\left[1 - \frac{g}{f}\frac{m^2+\beta}{m(1+\beta)}\right]. \quad (9.3.2)$$

A good first approximation to zero astigmatism is obtained by assuming that $g/f \ll 1$, substituting $W\psi = -f\theta$ into Eqs. (9.3.1), and setting $B_{1s}=0$. The relation found is

$$bfg^2 = -\frac{m(2m+1)+\beta}{2m(1+\beta)} = -\Lambda. \quad (9.3.3)$$

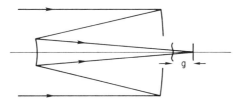

Fig. 9.6. Aspheric corrector for Cassegrain focus of Ritchey–Chretien telescope. The aspheric figure is similar to that of a Schmidt plate.

This is one relation between b and g, with the other relation found by requiring that the coma at a given field angle does not exceed a specified amount. Substituting Eq. (9.3.3) into B_{2s} and B_{3s} we find the following angular aberrations:

$$\text{ATC} = \frac{3\theta\Lambda}{8F^2}\left(\frac{g}{f}\right), \qquad \text{ASA} = \frac{\Lambda}{16F^3}\left(\frac{g}{f}\right)^2. \tag{9.3.4}$$

As an example we take the parameters for the RC telescope in Table 6.8 and find $\Lambda = 3.625$. Assuming that $\text{ATC} = 0.25$ arc-sec at $\theta = 0.3°$, we find from Eqs. (9.3.4) that $g/f = 0.01703$ and $\text{ASA} = 0.014$ arc-sec. Putting this value of g/f into $W\psi$ in Eq. (9.3.2), and substituting $W\psi$ into Eqs. (9.3.1), we get better values for the off-axis aberrations. The results in arc-seconds are $\text{ATC} = 0.236$ and $\text{AAS} = 0.056$, and all of the aberrations are clearly at a tolerable level.

The one remaining calculation is that of finding b using Eq. (9.3.3). Because $b < 0$, the plate has the shape of a Schmidt plate. With a radius added to the plate to minimize chromatic aberration, a ray-trace analysis of the plate in this example shows that the image diameters are 1 arc-sec or less over a field diameter of about 1.4°. Because the plate in this example is in an $f/10$ beam, higher-order aberrations are negligible. The computed image surface curvature is found to be $\kappa_m R_1 = 4.19$, a value about 5% larger than the normalized Petzval curvature in Table 6.9.

B. MODIFIED RITCHEY–CHRETIEN TELESCOPE

The effectiveness of an aspheric plate in the Cassegrain beam suggests that still larger fields are possible if the telescope plus plate is designed as a system. In this case the conic constants of the primary and secondary are also adjustable parameters, and all of the aberrations can be made zero. Examples of such designs are the 1.0- and 2.5-m telescopes described by Bowen and Vaughan and located at Las Campanas Observatory in Chile. The design of the 1.0-m telescope has the additional feature of a flat Petzval field; thus the need for bending a photographic plate to match the median image surface is avoided. Well-corrected fields over 2° in diameter are achieved with these designs.

The first step in the procedure for designing a flat-field Ritchey–Chretien to cover a wide field is to specify zero Petzval curvature for the telescope. Thus $\rho = R_2/R_1 = 1$, as is evident from Eq. (5.7.15). This condition, in turn, requires that

$$m^2 - 1 = m(1 + \beta). \tag{9.3.5}$$

Substitution of Eq. (9.3.5) into Eqs. (6.2.4) and (6.2.5) gives the conic constants of the mirrors, and all of the telescope parameters are now specified. If, for example, we choose $\beta = 0.2$, then $m = 1.7762$, $K_1 = -1.4912$, $K_2 = -26.905$, and $k = 0.4338$. From these results we see that the mirrors, especially the secondary, are strongly hyperbolic and that the obscuration of the secondary is larger than that of the typical RC telescope.

The next step in the design is simply one of substituting m, β, and a choice of g into Eq. (9.3.3) and finding a first-order solution for b. The telescope parameters and the values of b and g are then the starting point for computer optimization of the complete system, telescope plus corrector plate.

IV. CASSEGRAIN FOCAL REDUCERS

A focal reducer is an optical system whose function is to change the focal ratio of a telescope. It is most often used at the Cassegrain focus to reduce the focal ratio so that a given field can be placed on a detector of smaller area. In this section we discuss the general characteristics of focal reducers used at the Cassegrain focus. With the exception of a Schmidt camera example for a Ritchey-Chretien telescope, we omit the details of specific designs.

A. GENERAL CONFIGURATION

A schematic of a Cassegrain focal reducer is shown in Fig. 9.7. Its components include a field lens at the Cassegrain focus to image the exit pupil of the telescope onto the aperture stop of the focal reducer, a collimator to render the light parallel, and a camera. Other optical elements, such as a grating or filter, can be put in the space between the collimator and camera. Because such elements are located in a collimated beam, they introduce no additional aberrations.

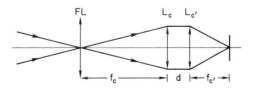

Fig. 9.7. Schematic of focal reducer, where c and c' denote collimator and camera, respectively, and FL is the field lens at the telescope focus.

The diameters of the focal reducer components depend on the field to be covered. If the angular radius of the field on the sky is θ, then the diameter of the field lens is $2f\theta$, where f is the telescope focal length. The diagram in Fig. 9.8 shows the chief ray from the center of the telescope exit pupil for an object at the edge of the field. Assuming the aperture stop of the focal reducer is at the collimator lens, the angle α at which this chief ray enters the collimator is given by

$$\alpha f_c = \psi f_p \delta = f\theta, \qquad (9.4.1)$$

where f_c is the focal length of the collimator, f_p is the focal length of the primary, and Eq. (2.5.5) is substituted to eliminate ψ. We also see from Fig. 9.7 that D_c, the diameter of the collimator lens, is f_c/F. Therefore

$$\alpha/\theta = f/f_c = D/D_c. \qquad (9.4.2)$$

For a real lens pair the stop is often located in the space between the lenses, but the distance d in Fig. 9.7 is usually small compared to f_c and, to a good approximation, the stop is effectively at the collimator. If d is small, the diameters of the collimator and camera lens are nearly equal, and Eq. (9.4.2) can be used to find the diameter of either.

For a given D and θ, we see from Eq. (9.4.2) that a smaller D_c implies a larger α. A larger value of α, in turn, generally means that the design of the lenses in the focal reducer is more difficult. We also see that the size of the focal reducer scales directly with the size of the telescope for a given α and θ.

The focal length of the telescope–focal reducer combination is f of the telescope times the magnification of the focal reducer. The reader can verify that the magnification of the focal reducer is the ratio of the camera to collimator focal lengths or focal ratios, hence f of the combination is the diameter D of the telescope times the focal ratio of the camera.

As an example, let $\theta = 0.5$, $\alpha = 20$, in degrees, and $D = 1$ m. Substituting these values into Eq. (9.4.2) gives $D_c = 25$ mm, and a well-designed commercial camera lens is appropriate for the camera lens of the focal reducer. If

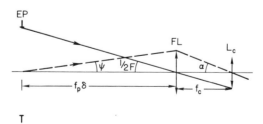

Fig. 9.8. Schematic of focal reducer in relation to telescope exit pupil EP. See text, Section 9.IV, for definitions of symbols.

$F = 10$ for the telescope, then $f_c = 250$ mm is the focal length of the collimator, and the constraints on its design are relatively modest compared to those on the camera lens. With the same telescope magnification and angles, but $D = 4$ m, the dimensions of the lenses are $4 \times$ larger and their design and mounting is a more difficult problem.

B. TYPES OF FOCAL REDUCERS

A variety of focal reducer types have been analyzed by a number of investigators, and a good summary of these is given by Wilson. His paper includes examples of multilens systems that convert an $f/8$ telescope beam to $f/3$ with good image quality of a field $0.9°$ in diameter. The main difficulty with lens systems, as noted by Wilson, is the chromatic aberration over an extended wavelength range.

Catadioptric systems have the advantage of smaller chromatic aberration but the disadvantage of obstruction due to either the detector or one of the mirrors. Wilson describes briefly some catadioptric systems that have been proposed for focal reducers, such as the standard Schmidt, Bouwers-Maksutov, and Schmidt-Cassegrain cameras, where each is used with a field lens to reimage the telescope exit pupil. The reference by Wilson should be consulted for further details and references.

Meinel *et al.* have described a four-mirror focal reducer for the Nasmyth focus of a large telescope with good imagery over a field radius of 8 arc-min. Their paper should be consulted for details.

C. EXAMPLE: SCHMIDT CAMERA

To illustrate the approach to the design of a focal reducer, we consider a Schmidt camera modified for the required conditions. The basic configuration adopted, field lens plus camera, is shown in Fig. 9.9. The field lens images the telescope exit pupil on the aspheric plate with the chief ray shown entering the camera at angle ψ, where $\psi/\theta = f/d$ for a telescope of focal length f with field angle θ. We assume a Ritchey-Chretien telescope

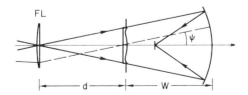

Fig. 9.9. Schematic of Schmidt focal reducer. See text, Section 9.IV, for discussion.

and adjust the camera parameters to eliminate the astigmatism present at the RC focal surface.

The RC telescope is free of coma and spherical aberration, while the aspheric plate has no coma and astigmatism when it is at a pupil. Thus the coma of the system is that of the mirror only, with the coma coefficient given in Table 5.6. Setting this coefficient to zero gives

$$\frac{W}{R} = \left[1 - K\left(\frac{m-1}{m+1}\right)\right]^{-1}, \quad (9.4.3)$$

where K is the conic constant of the mirror. As we show shortly, K must be chosen nonzero to eliminate the telescope astigmatism.

The system spherical aberration is that of the aspheric plate and mirror, while the astigmatism is that of the telescope and focal reducer mirror. Taking the appropriate coefficients for the corrector and mirror from Tables 5.5 and 5.6 and the telescope astigmatism from Table 6.4, we get

$$B_{1s} = B_1(\text{RC}) + \frac{n\psi^2}{R}\left(\frac{y_4}{y_1}\right)^2 \left[\frac{W^2}{R^2}K + \left(1 - \frac{W}{R}\right)^2\right], \quad (9.4.4)$$

$$B_{3s} = -\frac{b}{8}\left(\frac{y_3}{y_4}\right)^4 + \frac{n}{4R^3}\left[K + \left(\frac{m+1}{m-1}\right)^2\right], \quad (9.4.5)$$

where the subscripts 1, 3, and 4 refer to the telescope primary, aspheric plate, and camera mirror, respectively, and $n = 1$ for the mirror. In writing these system coefficients we assume the field lens and aspheric plate thickness and radius do not contribute to the aberrations.

The terms in Eqs. (9.4.4) and (9.4.5) are simplified by noting that $y_3/y_4 = d/s$, $y_4/y_1 = s/f$ and $\psi = \theta(f/d)$, where s is the distance from the field lens to the camera mirror.

The procedure is now one of setting B_{1s} to zero, substituting Eq. (9.4.3) to eliminate W/R, and solving for K in terms of $B_1(\text{RC})$. Letting $B_1(\text{RC}) = -\theta^2\Gamma/2f$, where Γ is the quantity in brackets in AAS in Table 6.7, the result is

$$K = \frac{\Gamma R}{2f}\left(\frac{m+1}{m-1}\right)^2 \left(1 - \frac{\Gamma R}{2f}\right)^{-1}. \quad (9.4.6)$$

The value of K from Eq. (9.4.6) is substituted into Eq. (9.4.3) to find W/R, which, in turn, is used to find the ratio d/s. In terms of the camera parameters we find

$$\frac{d}{s} = 1 - \frac{W}{s} = 1 - \frac{W}{R}\left(\frac{2m}{m-1}\right). \quad (9.4.7)$$

Values derived from Eqs. (9.4.6) and (9.4.7) are substituted into Eq. (9.4.5), which is solved for b after setting B_{3s} to zero.

IV. Cassegrain Focal Reducers

Table 9.4

Parameters of 1.5-m Ritchey-Chretien Telescope

Overall	$m = 2.667$,	$k = 0.3273$,	$\beta = 0.2$
	$f = 12.0$ m,	$F = 8$	
Primary	$R_1 = -9000$ mm,	$K_1 = -1.1368$	
Secondary	$R_2 = -4712.7$ mm,	$K_2 = -6.5524$	

All of the relations needed to specify the Schmidt focal reducer, hereafter denoted by SFR, are now in hand. For the telescope we take the design parameters of the 1.5-m $f/8$ telescope shown in Table 9.4. For the SFR we assume a final focal ratio of 2.67, hence $m = -1/3$ for the camera, with $s = -2000$ mm, $R = -1000$ mm.

With these parameters we find $\Gamma = 2.67$, $K = -0.025$, $W = 1.053R$, $d/s = 0.4737$, and $b = -8.937E-9$ for an SiO_2 corrector at $\lambda = 500$ nm. A listing of all the SFR parameters, including the field lens, is shown in Table 9.5.

Ray traces of the system with the nominal SFR parameters given in Table 9.5 show an image diameter of about 1 arc-sec at a field angle of 0.5° and wavelength of 500 nm. With a 2-mm change in the corrector location and a 10% increase in the focal length of the field lens, the image diameter is reduced to approximately 0.25 arc-sec at the same field angle and wavelength. Over the range from 320 to 1000 nm, the image diameter is 0.5 arc-sec or smaller at the edge of the field, hence the broadband image quality is satisfactory.

Although this type of camera would appear to be an obvious choice for a focal reducer, it has several disadvantages. One problem is its curved focal surface, but a field-flattener lens can be added to remove this curvature. A much more serious problem is the location of the focal surface inside

Table 9.5

Parameters of Schmidt Focal Reducer

Overall	$m = -0.333$, $\quad s = -2000$ mm
Mirror	$R = -1000$ mm, $\quad K = -0.025$
	$W = 1.053R$ (nominal)
	$\quad = 1.055R$ (optimized)
Corrector	$b = -8.937E-9$, $\quad E = 2.416E-9$
	$R_c = -35\,320$ mm, $\quad$ thickness $= 10$ mm
Field lens	plano-convex, thickness $= 18$ mm,
	$R_2 = -380$ mm (nominal)
	$\quad = -420$ mm (optimized)
Lens and corrector material	SiO_2

the camera. It is not possible to locate large detectors such as image tubes or cooled solid-state arrays at an internal focus without vignetting most of the light before it reaches the mirror. Getting an external focus requires a folded Schmidt camera with a tilted plane mirror between the corrector and sphere, as shown in Fig. 15.3, and the detector behind a hole in the plane mirror. The size of the hole and the position of the detector determine the amount of vignetting, and for a large field this amount is significant.

REFERENCES

Bowen, I., and Vaughan, A. (1973). *Appl. Opt.* **12**, 1430.
Gascoigne, S. (1973). *Appl. Opt.* **12**, 1419.
Meinel, A. (1953). *Astrophys. J.* **118**, 335.
Meinel, A., Meinel, M., and Wang, Y. (1985). *Appl. Opt.* **24**, 2751.
Wilson, R. (1971). "ESO/CERN Conference on Large Telescope Design," p. 131. European Space Organization, Geneva.
Wynne, C. (1972). "Progress in Optics," Vol. 10, Chap. 4. North-Holland, Amsterdam.
Wynne, C. (1974). *Mon. Not. R. Astron. Soc.* **167**, 189.

BIBLIOGRAPHY

Courtes, G. (1972). *Vistas Astron.* **14**, 81.

Chapter 10 | Diffraction Theory and Aberrations

The discussion of telescopes and their aberrations in previous chapters is entirely from the point of view of geometric optics. This approach is one in which the ray is a well-defined entity, with the wavelength in the geometric optics limit effectively zero. The paths of rays through an optical system are governed by Fermat's principle and aberrations occur when rays do not pass through the paraxial image point. An aberration-free image in the geometric optics limit is, according to Fermat's principle, a true point image. It was pointed out in Section 3.VI, however, that the wave nature of light sets the image size for an otherwise perfect or diffraction-limited optical system, with the analysis there intended only to give an estimate of the size of the diffraction image.

In this chapter and the next we discuss the character of the perfect image from the point of view of diffraction theory. Because no optical system is strictly perfect, we also consider the effect of the aberrations of a nearly perfect optical system on the diffraction image. Our analysis proceeds along two lines. In this chapter the starting point is Huygens' principle and the superposition of waves from points on a wavefront. In the following chapter the analysis is in terms of transfer functions, with application to the expected imaging capability of the Hubble Space Telescope.

I. THE PERFECT IMAGE

The characteristics of the image of a point source object formed by a perfect optical system are completely described by the *point spread function* (PSF) and quantities derivable from the PSF. One of the quantities derived from the PSF is the *encircled energy* fraction (or EE), the fraction of the total energy in the image within a circle of a given radius. The intensity in units of flux per unit area at a point on the image is directly proportional to the PSF, while the average intensity over a centered portion of an image depends on both the PSF and EE. Each of these items is discussed in this section.

A. POINT SPREAD FUNCTION

The intensity at point P of an image is the absolute square of the time-averaged amplitude of the electromagnetic wave at P, while the PSF at P is the intensity normalized to unity at the point where the intensity is a maximum. Our approach, therefore, is to develop the expression for the wave amplitude at a point P, an expression usually called the diffraction integral.

The analysis of the diffraction integral to follow parallels that given in most optics texts, such as Hecht and Zajac or Born and Wolf, with the notation that of the latter authors. We limit the discussion to the case of a circular aperture with a central obscuration, from which results for a clear circular aperture easily follow.

Consider the exit pupil of an optical system with radius a and central obscuration of radius εa, as shown in Fig. 10.1, with a spherical wavefront W of radius of curvature R emerging from the pupil. In Fig. 10.1 an arbitrary point Q on the wavefront has coordinates (ξ, η, ζ) and is a distance $a\rho$ from the z axis at angle θ with the η axis. An arbitrary field point P is a

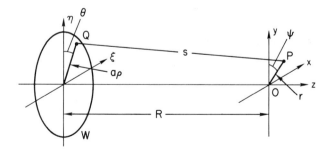

Fig. 10.1. Coordinate frames at exit pupil (η, ξ, ζ) and image surface (x, y, z) of optical system. W is a spherical wavefront of radius R centered at O. See Eq. (10.1.1).

I. The Perfect Image

distance r from the z axis at angle ψ with y axis. Relative to the origin at O, the coordinates for points Q and P are

$$\xi = a\rho \sin\theta, \quad x = r\sin\psi,$$
$$\eta = a\rho \cos\theta, \quad y = r\cos\psi,$$
$$\zeta = -\sqrt{(R^2 - a^2\rho^2)}. \quad (10.1.1)$$

According to Huygens' principle, the amplitude U at point P is the sum of all amplitude contributions dU from each area dS on the wavefront. For all systems of interest we assume the distances z, r, and a are small compared to R. When these conditions are satisfied, the sum of the contributions from each dS is a simple scalar sum given by

$$U(P) = A \int\int_W \exp[ik(s-R)]\, dS, \quad (10.1.2)$$

where $k = 2\pi/\lambda$, s is the distance from Q to P, and A is a constant proportional to the amplitude at Q.

As noted above, the center of curvature of the wavefront is at point O. For this particular point P we have $s = R$ for all points Q on the wavefront, hence all waves are in phase at O. Therefore the argument in Eq. (10.1.2) is zero and the integral gives the area of the exit pupil. Because all waves are in phase at O, the amplitude $U(O)$ is a maximum. For any other P the path and phase differences are $(s - R)$ and $k(s - R)$, respectively, and the amplitude at P is less than that at O.

Expressing s in terms of the coordinates of Q and P, we get

$$s^2 = (\xi - x)^2 + (\eta - y)^2 + (\zeta - z)^2$$
$$= R^2 - 2(x\xi + y\eta + z\zeta), \quad (10.1.3)$$

where all squared terms in x, y, and z are negligible. With s nearly equal to R, given our assumption about distances above, we have $s^2 - R^2$ nearly equal to $2R(s - R)$.

Substituting this relation and Eqs. (10.1.1) into Eq. (10.1.3) gives

$$s - R = -\frac{a\rho r}{R}\cdot\cos(\theta - \psi) + z\left[1 - \frac{1}{2}\left(\frac{a\rho}{R}\right)^2\right], \quad (10.1.4)$$

where ζ is transformed by the binomial expansion. Following Born and Wolf, we define dimensionless variables u and v in the form

$$u = \frac{2\pi}{\lambda}\left(\frac{a}{R}\right)^2 z, \quad v = \frac{2\pi}{\lambda}\left(\frac{a}{R}\right) r. \quad (10.1.5)$$

Substituting Eqs. (10.1.5) into Eq. (10.1.4) gives

$$k(s - R) = -v\rho\cos(\theta - \psi) - \frac{u\rho^2}{2} + u\left(\frac{R}{a}\right)^2. \quad (10.1.6)$$

The introduction of these dimensionless variables is made for convenience in relations involving aberrations to follow in subsequent sections.

At this point we set $u = 0$ and evaluate $U(P)$ in the paraxial focal plane. Noting that the area element $dS = \rho\, d\rho\, d\theta$, the amplitude in the paraxial focal plane is given by substituting Eq. (10.1.6) with $u = 0$ into Eq. (10.1.2). The result is

$$U(P) = A \int_0^{2\pi} \int_\varepsilon^1 \exp[-iv\rho \cos(\theta - \psi)]\rho\, d\rho\, d\theta. \qquad (10.1.7)$$

To carry out the integration over θ we note that $U(P)$ is independent of ψ because the system is symmetric about the z axis. Therefore we can choose any convenient value for ψ; we choose $\psi = \pi$.

Carrying out the integration over θ involves substituting the integral representation of J_0, the Bessel function of order zero, with the result

$$U(P) = 2\pi A \int_\varepsilon^1 J_0(v\rho)\rho\, d\rho = \frac{2\pi A}{v^2} \int_\varepsilon^1 d[v\rho J_1(v\rho)]. \qquad (10.1.8)$$

The second step in Eq. (10.1.8) follows after substituting one of the recurrence relations for Bessel functions found in tables of mathematical functions (see, e.g., "Tables of Integrals and Other Mathematical Data" by Dwight). Integration of Eq. (10.1.8) gives

$$U(P) = \pi A \left[\frac{2J_1(v)}{v} - \varepsilon^2 \frac{2J_1(\varepsilon v)}{\varepsilon v} \right], \qquad (10.1.9)$$

where J_1 is a Bessel function of order one. The ratio $2J_1(w)/w$ approaches one as w approaches zero, hence $U(0) = \pi A(1 - \varepsilon^2)$.

Using this result, we write the PSF at point P as

$$i(P) = \frac{I(P)}{I(0)} = \frac{1}{(1 - \varepsilon^2)^2} \left[\frac{2J_1(v)}{v} - \varepsilon^2 \frac{2J_1(\varepsilon v)}{\varepsilon v} \right]^2, \qquad (10.1.10)$$

where $I(P) = |U(P)|^2$ and $i(P) = \text{PSF}$. It is convenient to represent the intensity in this form to facilitate comparison of intensity profiles for apertures with different central obscurations. As examples, PSFs are shown in Fig. 10.2 for $\varepsilon = 0$, a clear aperture, and $\varepsilon = 0.33$, the obscuration of the Hubble Space Telescope. Note that the ordinate in Fig. 10.2 is v/π.

The PSF given by Eq. (10.1.10) and shown in Fig. 10.2 is in the paraxial focal plane and characterizes the so-called Airy pattern. The intensity is a maximum at the paraxial image point at $v = 0$, and the pattern is a central bright disk, the Airy disk, surrounded by concentric bright and dark rings. For a clear aperture the peak intensity of the bright rings decreases as v increases; for an obstructed aperture the intensity of the bright rings decreases in a cyclic manner, depending on the specific value of ε.

I. The Perfect Image

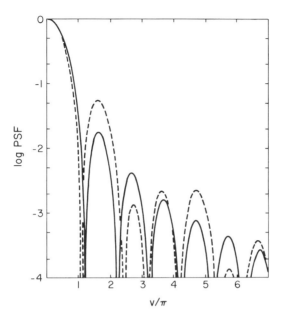

Fig. 10.2. Point spread function of perfect image for obscuration ratios $\varepsilon = 0$ (solid line) and 0.33 (dashed line).

The radii of the dark rings in the Airy pattern are found by setting $i(P) = 0$. Radii for the first three dark rings are given in Table 10.1 for several values of ε. Note that the radius of the first dark ring, which encloses the Airy disk, decreases as ε increases, while the radius of the second dark ring is a maximum near $\varepsilon = 0.3$ and decreases for larger obscurations.

One important descriptor of the Airy pattern is the radius of the first ring. Substituting v_1 from Table 10.1 with $\varepsilon = 0$ into Eqs. (10.1.5) gives

$$r_1 = 1.22\lambda F, \qquad \alpha_1 = r_1/f = 1.22\lambda/D, \qquad (10.1.11)$$

where r_1 and α_1 are the linear and angular radii, respectively, of the Airy

Table 10.1
Radii of Dark Rings in Airy Pattern

ε	v_1	v_2	v_3
0.00	1.220π	2.233π	3.238π
0.10	1.205π	2.269π	3.182π
0.20	1.167π	2.357π	3.087π
0.33	1.098π	2.424π	3.137π
0.40	1.058π	2.388π	3.300π

disk, f is the system focal length, D is the diameter of the entrance pupil, and F is the focal ratio. Substituting f for R and D for $2a$ assumes the point source object is effectively at infinity. For other values of ε in Table 10.1, the factor 1.22 is replaced by the appropriate numerical factor.

For a distant point source, the variable v is related to the system parameters and a dimensionless radius w by the relations

$$v = w\pi = \pi r/\lambda F = \pi D\alpha/\lambda, \tag{10.1.12}$$

where r and α are the linear and angular radii, respectively. The radii w for the first and second dark rings and the radius at which $i(P) = 0.5$ are shown in Fig. 10.3 for a range of ε.

The description of the PSF in the wings of the Airy pattern beyond a few bright rings is derived by using the following asymptotic relation for the function J_n:

$$J_n(\beta) = \left(\frac{2}{\pi\beta}\right)^{1/2} \cos\left(\beta - \frac{n\pi}{2} - \frac{\pi}{4}\right). \tag{10.1.13}$$

This approximation is good to 1% or better for $\beta > 15$. Choosing $n = 1$ and $\beta = v$ or εv, as appropriate, Eq. (10.1.10) becomes

$$i(P) = \frac{1}{\pi(1-\varepsilon^2)^2}\left(\frac{2}{v}\right)^3 \left[\cos\left(v - \frac{3\pi}{4}\right) - \sqrt{\varepsilon}\cos\left(\varepsilon v - \frac{3\pi}{4}\right)\right]^2.$$

Squaring and expanding the factor in brackets, we find a locally smoothed PSF by setting each of the $\cos^2$ terms to one-half and the cross term to zero,

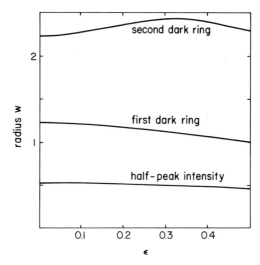

Fig. 10.3. Dimensionless radii at first two dark rings and one-half of peak intensity as a function of obscuration ratio. Linear radius = $w\lambda F$; angular radius = $w\lambda/D$.

I. The Perfect Image

with the result

$$\langle i(P)\rangle = \frac{4(1+\varepsilon)}{\pi(1-\varepsilon^2)^2} \frac{1}{v^3} = \frac{4(1+\varepsilon)}{\pi^4(1-\varepsilon^2)^2}\left(\frac{\lambda}{D}\right)^3 \frac{1}{\alpha^3}, \qquad (10.1.14)$$

where α is the field angle in radians. The quantity $\langle i(P)\rangle$ is a good measure of the average intensity over one or two Airy rings in the range where the asymptotic relation for J_1 is a good approximation. For $\varepsilon = 0.33$, for example, Eq. (10.1.14) is valid beyond the tenth bright ring.

From Eq. (10.1.14) we see that the average intensity in the wings of the Airy pattern is larger for larger values of ε. It is apparent that the effect of the central obscuration is to transfer some of the energy from the disk and nearest bright rings into the wings. A quantitative measure for the fraction of the energy in the wings of the Airy pattern is developed in the following section.

B. ENCIRCLED ENERGY

The encircled energy EE is defined as the fraction of the total energy E enclosed within a circle of radius r centered on the PSF peak. In terms of the variable v we have

$$\mathrm{EE}(v_0) = \frac{1}{E} \int_0^{2\pi} \int_0^{v_0} i(v) v \, dv \, d\psi, \qquad (10.1.15)$$

where v_0 is a dimensionless radius, and $i(v)$ is given by Eq. (10.1.10).
We first evaluate Eq. (10.1.15) for the clear aperture, and find

$$\begin{aligned}
\mathrm{EE}(v_0) &= \frac{8\pi}{E}\int_0^{v_0}\left[\frac{J_1(v)}{v}\right]^2 v\, dv \\
&= -\frac{4\pi}{E}\int_0^{v_0} d[J_0^2(v)+J_1^2(v)] \\
&= 1 - J_0^2(v_0) - J_1^2(v_0).
\end{aligned} \qquad (10.1.16)$$

The intermediate step in Eq. (10.1.16) follows after substitution of a recurrence relation. When v_0 is taken at a dark ring, $J_1(v_0)$ is zero and the fraction of the energy outside the dark ring is given by $J_0^2(v_0)$.

Following the same procedure for the obstructed aperture, Eq. (10.1.15) becomes

$$\mathrm{EE}(v_0) = \frac{1}{1-\varepsilon^2}\left[1 - J_0^2(v_0) - J_1^2(v_0) + \varepsilon^2(1 - J_0^2(\varepsilon v_0) - J_1^2(\varepsilon v_0)) \right.$$
$$\left. - 2\varepsilon \int_0^{v_0} J_1(\varepsilon v)\frac{2J_1(v)}{v}\, dv \right]. \qquad (10.1.17)$$

10. Diffraction Theory and Aberrations

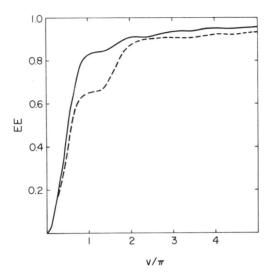

Fig. 10.4. Encircled energy fraction for obscuration ratios $\varepsilon = 0$ (solid line) and 0.33 (dashed line), for perfect image.

Results derived from these relations for EE are shown in Fig. 10.4 for $\varepsilon = 0$ and $\varepsilon = 0.33$. Encircled energy values within each of the first three dark rings are given in Table 10.2, with data for the first two dark rings plotted in Fig. 10.5. Also shown in Fig. 10.5 are EEs within the radii at which the intensity is one-half of the peak.

Examination of the results in Table 10.2 and Fig. 10.5 shows that there is a significant transfer of energy from the Airy disk to the first bright ring with increasing ε. We also see that EE in the disk and first bright ring decreases very slowly as ε increases from zero to 0.35. From Fig. 10.5 it is

Table 10.2

Encircled Energy within Airy Dark Rings[a]

ε	EE_1	EE_2	EE_3
0.00	0.838	0.910	0.938
0.10	0.818	0.906	0.925
0.20	0.764	0.900	0.908
0.33	0.654	0.898	0.904
0.40	0.584	0.885	0.903

[a] The subscript on EE is the number of the dark ring starting at the innermost ring.

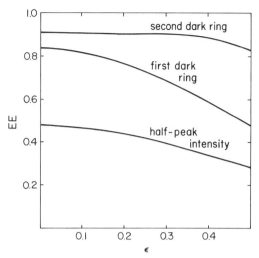

Fig. 10.5. Encircled energy fraction within first and second dark rings and within one-half of peak intensity. Results are given as a function of obscuration ratio.

also evident that noticeable energy transfer to the second bright ring begins when ε is approximately equal to 0.4.

Returning to Eq. (10.1.17), we note that the value of the integral in this relation is equal to $\varepsilon \pm \delta$ for $v_0 \gg 1$, with $\delta \ll \varepsilon$. As an example, with $\varepsilon = 0.33$ we find $\delta < 0.01$ for $v_0 > 15$. Therefore a good approximation to EE for large v_0 is found by substituting ε for the integral in Eq. (10.1.17) and substituting Eq. (10.1.13) for J_0 and J_1. Combining the terms involving the Bessel functions, we find

$$J_0^2(\beta) + J_1^2(\beta) = 2/\pi\beta,$$

where β is either v_0 or εv_0. Therefore

$$\begin{aligned}\text{EE}(v_0 \gg 1) &= 1 - \frac{2}{\pi(1-\varepsilon)v_0} = 1 - \frac{2\lambda}{\pi^2(1-\varepsilon)D\alpha}, \\ \text{OE}(v_0 \gg 1) &= 1 - \text{EE}(v_0 \gg 1),\end{aligned} \quad (10.1.18)$$

where OE is the fraction of the energy outside radius v_0. Examination of Eq. (10.1.18) shows that the larger ε, the larger is the fraction of the energy outside a given large radius.

C. INTENSITY

The PSF defined in Section 10.I.A is a dimensionless measure of the intensity of the Airy pattern. It is also necessary to give the intensity in physical units. In this section we determine both the average intensity over the Airy disk and the intensity at the center of the Airy pattern.

The average intensity over the Airy disk is the flux in the disk divided by its area. Denoting the flux passing through the entrance pupil by Fl, the average intensity $\langle I(\text{disk})\rangle$ is given by

$$\langle I(\text{disk})\rangle = \frac{\eta \text{Fl}}{\pi \gamma^2 (1.22\lambda F)^2}, \qquad (10.1.19)$$

where η is the fraction of the flux in the Airy disk and γ is a numerical factor that is one for $\varepsilon = 0$ and less than one for other values of ε. The value of η also depends on ε, as noted in the discussion of EE in the previous section.

The intensity at the center of the Airy disk, given in Born and Wolf, is

$$I(0) = \frac{\pi^2 D^2}{16\lambda^2 F^2}(1-\varepsilon^2)B = \frac{\pi \text{Fl}}{4\lambda^2 F^2}, \qquad (10.1.20)$$

where B is the flux per unit area over a given wavelength passband at the entrance pupil of diameter D.

Dividing Eq. (10.1.19) by Eq. (10.1.20), we get

$$\frac{\langle I(\text{disk})\rangle}{I(0)} = \frac{4\eta}{(1.22\pi\gamma)^2} = 0.272\frac{\eta}{\gamma^2}. \qquad (10.1.21)$$

Taking $\varepsilon = 0$ and $\varepsilon = 0.33$, we use the results in Table 10.1 and find $\gamma = 1$ and $\gamma = 0.9$, respectively, and from Table 10.2 we get $\eta = 0.838$ and $\eta = 0.654$, respectively. Substituting these values into Eq. (10.1.21) gives $\langle I(\text{disk})\rangle/I(0) = 0.229$ and 0.222 for the apertures with $\varepsilon = 0$ and 0.33, respectively. We see that this ratio is only 3% smaller for the obstructed aperture, a consequence of the fact that a smaller Airy disk nearly compensates for the smaller encircled energy.

It is instructive to compute the peak intensity for a specific case. Consider a perfect Hubble Space Telescope with $D = 2.4$ m, $F = 24$, and $\varepsilon = 0.33$. Taking the canonical value for the photon flux as 1000 photons/(sec cm² Å) for a zero-magnitude star at a wavelength of 550 nm, we get Fl = 4.03E7 photons/(sec Å) = 3.22E − 5 ergs/(sec Å), for the flux per angstrom in the image of the HST with unit transmittance. Substituting into Eq. (10.1.20) with $\lambda = 0.55$ μm, we find

$$I(0) = 1.82\text{E}5 \text{ photons}/(\text{sec Å } \mu\text{m}^2)$$

$$= 2.91\text{E}-7 \text{ ergs}/(\text{sec Å } \mu\text{m}^2),$$

for a zero-magnitude star. Assuming a passband of 1000 Å centered at $\lambda = 550$ nm and a star of apparent magnitude 25, we find $I(0) = 0.0182$

photons/(sec μm^2) and $\langle I(\text{disk})\rangle = 0.004$ photons/(sec μm^2). The area of the Airy disk at 550 nm is 660 μm^2, hence the total flux over the disk is 2.67 photons/sec for this example. Assuming a reflectance of 0.9 for the primary and secondary HST mirrors, a more accurate flux value is 2.15 photons/sec for a twenty-fifth magnitude star at the $f/24$ focus.

II. THE NEAR-PERFECT IMAGE

An image is perfect if the wavefront emerging from the exit pupil is spherical; if there are any deviations of the wavefront from a sphere the result is a less-than-perfect image. These wavefront deviations may be due to the presence of geometric aberrations of the type discussed in Chapter 5, but they may also arise from random variations in optical surface quality as a result of the polishing process. Each of these wavefront deviations is characterized by a different scale at the exit pupil. Geometric aberrations vary slowly across the aperture and are specified in functional form; random variations occur on a much shorter scale and are usually treated with statistical models.

Wavefront errors may also arise if the shape or orientation of the wavefront changes with time, and such time-dependent errors may be regular or random. Examples of regular time-dependent errors include linear image drifts and sinusoidal oscillations of an image. These types of errors are also best treated with a statistical approach, and an introduction to this approach is given in the next chapter.

In this section we consider geometric aberrations and their effects on image quality. Our discussion is only an introduction to a large subject matter, and the interested reader should consult some of the references listed at the end of the chapter for more extensive discussions.

A. DIFFRACTION INTEGRAL WITH ABERRATIONS

A cross section of a wavefront with aberrations and the reference sphere are shown in Fig. 5.3, where Δ is the geometric path difference between the wavefront and the reference sphere. In the notation of Fig. 10.1, the center of curvature 0 of the reference sphere is the location of the Gaussian image for a perfect system. The coordinate systems used to locate points on the wavefront and near the image are given in Eq. (10.1.1).

To include aberrations in the diffraction integral given in Eq. (10.1.2), we substitute $(s - R + \Phi)$ for $(s - R)$, where Φ is the path difference between the aberrated wavefront and the reference sphere. If we consider only third-order aberrations, then from Eq. (5.5.1) we get

$$\Phi = B_0 y + B_1 y^2 + B_1' x^2 + B_2 y(x^2 + y^2) + B_3(x^2 + y^2)^2.$$

We choose, as is conventional, to describe the astigmatism at the sagittal image, hence $B_1' = 0$. To make the notation in Φ consistent with that used in this chapter, we replace x and y by ξ and η, respectively, using Eq. (10.1.1). The result is

$$\Phi = B_0 a\rho \cos\theta + B_1 a^2 \rho^2 \cos^2\theta + B_2 a^3 \rho^3 \cos\theta + B_3 a^4 \rho^4$$
$$= \lambda(A_1 \rho \cos\theta + A_2 \rho^2 \cos^2\theta + A_3 \rho^3 \cos\theta + A_4 \rho^4), \quad (10.2.1)$$

where the A coefficients include the radius of the exit pupil and are dimensionless. Note that the dimensions of Φ are included in the wavelength λ in Eq. (10.2.1). The factors in Eq. (10.2.1) represent, in turn, distortion, astigmatism, coma, and spherical aberration, with each corresponding A coefficient giving the amount of aberration in units of waves.

Note that the subscripts on the A coefficients are changed from those in the previous line. This is done to bring our notation more in line with that commonly used. The usual convention is to use a triple subscript, but for our purposes a single subscript equal to the power of ρ is sufficient.

The diffraction integral for a circular exit pupil, including aberrations, is found by substituting Eqs. (10.1.6) and (10.2.1) into Eq. (10.1.7), with the result

$$U(P) = C \int\int_W \exp\left[i\left(k\Phi - v\rho\cos(\theta - \psi) - u\rho^2/2\right)\right]\rho\, d\rho\, d\theta,$$
(10.2.2)

where the limits of integration are those given in Eq. (10.1.7). Note that the term $u(R/a)^2$ given in Eq. (10.1.6) is not included in Eq. (10.2.2); this term does not depend on the variables of integration and is removed from the integral.

A complete analysis of Eq. (10.2.2) is beyond the scope of our treatment. For such an analysis the interested reader should consult the references by Born and Wolf, Mahajan, and Wetherell given at the end of this chapter. We do present selected results after discussing the effect of aberrations on peak intensity.

B. PEAK INTENSITY AND AVERAGE WAVEFRONT ERROR

Before discussing specific aberrations, it is important to show the relation between the peak intensity and the average wavefront error. We take point

II. The Near-Perfect Image

P at the center of the reference sphere, hence $u = v = 0$, and assume the aberrations are small. Given that $i(P) = |U(P)|^2$ normalized to unity for a perfect image, we find

$$i(0) = \frac{1}{\pi^2(1-\varepsilon^2)^2} \left| \int_0^{2\pi} \int_\varepsilon^1 \exp(ik\Phi) \rho \, d\rho \, d\theta \right|^2$$

$$= C^2 \left| \int_0^{2\pi} \int_\varepsilon^1 \left[1 + ik\Phi + (ik\Phi)^2/2 + \cdots \right] \rho \, d\rho \, d\theta \right|^2, \quad (10.2.3)$$

where $C = 1/\pi(1-\varepsilon^2)$.

We now define $\langle \Phi^n \rangle$ as the average of the nth power of Φ, where

$$\langle \Phi^n \rangle = C \int_0^{2\pi} \int_\varepsilon^1 \Phi^n \rho \, d\rho \, d\theta, \quad (10.2.4)$$

and C is as given for Eq. (10.2.3). Neglecting all factors in $k\Phi$ higher than second power in Eq. (10.2.3), we can write the approximate intensity at the center of the reference sphere as

$$i'(0) = |1 + ik\langle\Phi\rangle - k^2\langle\Phi^2\rangle/2|^2$$

$$= 1 - k^2[\langle\Phi^2\rangle - \langle\Phi\rangle^2] = 1 - k^2\omega^2, \quad (10.2.5)$$

where ω, the rms wavefront error, is the square root of the quantity in brackets. The notation i' is used to indicate that this is an approximation. The rms wavefront error is a useful parameter for characterizing a high-quality optical system because its value can be calculated once the type and magnitude of aberrations are known. We also see from Eq. (10.2.5) that the normalized intensity at the location of the nominal focal point is independent of the type of aberration, with the decrease from unity proportional to ω^2 in this approximation. In the presence of aberrations, the normalized intensity $i'(0)$ is often used as one measure of image quality. This intensity, by convention, is called the *Strehl intensity* or *Strehl ratio*. A common convention is to consider a system as diffraction-limited if the Strehl ratio is greater than or equal to 0.8. Given this convention we find that $\omega = 0.0712\lambda = \lambda/14$.

The Strehl ratio given by Eq. (10.2.5) is an approximation to the normalized peak intensity valid for small ω. It was shown by Mahajan that a better approximation for the Strehl ratio S is given by

$$S = i'(0) = \exp(-k^2\omega^2). \quad (10.2.6)$$

A comparison of $i(0)$ calculated directly from Eq. (10.2.3) with S from Eq. (10.2.6) shows that the latter agrees with the former with an error of less than 10% for S greater than 0.3. This limit corresponds to an approximate value of $\omega = \lambda/5.7$.

C. CLASSICAL AND ORTHOGONAL ABERRATIONS

The classical third-order aberrations are those given in Eq. (10.2.1). Substituting each in turn into Eq. (10.2.4), it is a simple matter to calculate the rms wavefront error for each, with the results given in Table 10.3. These expressions for ω are appropriate for specific image locations: at the nominal paraxial or Gaussian focus for spherical aberration, coma, and distortion, and at the sagittal image for astigmatism.

Numerical calculations of $i(P)$ using Eq. (10.2.2) including focus shift show that the Strehl ratio is not a maximum at these image locations. In the presence of spherical aberration only, for example, the "best" image is not at the paraxial focus but between the paraxial and marginal foci, as a glance at Fig. 4.5 shows. Numerical calculations with $\varepsilon = 0$ show that $i(0)$ is largest at a point halfway between these two foci and the corresponding rms wavefront error is a minimum at this point. The dependence of ω and peak intensity on focus shift for $\varepsilon = 0$ is shown in Fig. 10.6; in this case the value of ω at the paraxial focus is 4 times larger than at the point where the peak intensity is a maximum.

Table 10.3

Classical Aberrations and RMS Wavefront Errors[a]

Aberration	RMS wavefront error
Spherical aberration: $A_4\rho^4$	$\dfrac{|A_4|}{3\sqrt{5}}(4 - \varepsilon^2 - 6\varepsilon^4 - \varepsilon^6 + 4\varepsilon^8)^{1/2}$
Coma: $A_3\rho^3 \cos\theta$	$\dfrac{|A_3|}{\sqrt{8}}(1 + \varepsilon^2 + \varepsilon^4 + \varepsilon^6)^{1/2}$
Astigmatism: $A_2\rho^2 \cos^2\theta$	$\dfrac{|A_2|}{4}(1 + \varepsilon^4)^{1/2}$
Distortion: $A_1\rho \cos\theta$	$\dfrac{|A_1|}{2}(1 + \varepsilon^2)^{1/2}$
Defocus: $A_0\rho^2$	$\dfrac{|A_0|}{2\sqrt{3}}(1 - \varepsilon^2)$

[a] Each expression is given in units of wavelength. For linear measure, multiply by wavelength.

II. The Near-Perfect Image

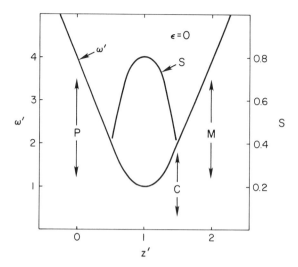

Fig. 10.6. Normalized rms wavefront error ω' and Strehl ratio S for image with spherical aberration as a function of image surface location. S is given for $\omega' = 1$, $\omega = 0.075\lambda$ at $z' = 1$. The normalized focus shift is z'. P Paraxial focus; M marginal focus; C, circle of least confusion.

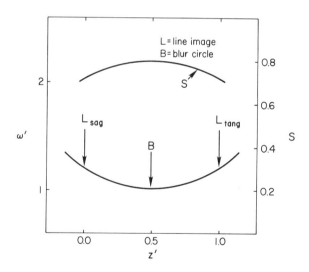

Fig. 10.7. Normalized rms wavefront error ω' and Strehl ratio S for astigmatic image as function of image surface location, with $z' = 0, 1$ at line images. S is given for $\omega' = 1$, $\omega = 0.075\lambda$ at blur circle B. The normalized focus shift is z'.

Following a similar procedure, it turns out that a system with astigmatism has a minimum ω and maximum $i(0)$ at a point halfway between the sagittal and tangential line images for any value of ε. The dependence of ω and peak intensity on focus shift is shown in Fig. 10.7 for an astigmatic image; in this case the value of ω at either line image is about $1.2 \times$ larger than at the midway point.

For both spherical aberration and astigmatism the point of maximum $i(0)$ is on the axis defined by $v = 0$. For coma and distortion $i(0)$ is a maximum for a point displaced transversely from the paraxial image point and v is not zero. The point at which the peak intensity is a maximum is called the *diffraction focus*. Table 10.4 gives the shifts from the Gaussian to the diffraction focus for each of the classical aberrations.

As seen from Table 10.4, the location of the diffraction focus depends on the type and magnitude of aberration present. Because of this dependence, it is appropriate to restructure the classical aberration terms and include explicitly the required image shift to place the diffraction focus at $u = v = 0$. These modified terms are called orthogonal aberrations, and the polynominals in ρ and θ are called *Zernike polynominals*. A list of the third-order orthogonal aberration terms and expressions for the rms wavefront error at the diffraction focus derived with the aid of Eq. (10.2.4) are shown in Table 10.5.

Examination of the spherical aberration and astigmatism terms shows the addition to each of a term in ρ^2, a focus shift term. The added term in the coma is proportional to $\rho \cos \theta$, which is effectively a tilt. The constant terms included in the spherical aberration and defocus make the average wavefront error $\langle \Phi \rangle = 0$ for these aberrations. For a more detailed discussion of the properties of the orthogonal aberrations, including derivations, consult the references by Born and Wolf, and Mahajan.

Table 10.4

Coordinate Shifts to Diffraction Focus

Aberration	x	y	z
Spherical	0	0	$8A_4\lambda(1+\varepsilon^2)F^2$
Coma	$\dfrac{4A_3\lambda F}{3}\dfrac{1+\varepsilon^2+\varepsilon^4}{1+\varepsilon^2}$	0	0
Astigmatism	0	0	$4A_2\lambda F^2$
Distortion	$2A_1\lambda F$	0	0
Defocus	0	0	$8A_0\lambda F^2$

II. The Near-Perfect Image

Table 10.5

Orthogonal Aberrations and RMS Wavefront Errors[a]

Aberration	RMS wavefront error
$A_4[\rho^4 - (1+\varepsilon^2)\rho^2 + \frac{1}{6}(1+4\varepsilon^2+\varepsilon^4)]$	$\dfrac{|A_4|}{6\sqrt{5}}(1-\varepsilon^2)^2$
$A_3\left[\rho^3 - \dfrac{2(1+\varepsilon^2+\varepsilon^4)}{3(1+\varepsilon^2)}\rho\right]\cos\theta$	$\dfrac{|A_3|}{6\sqrt{2}}\dfrac{(1-\varepsilon^2)(1+4\varepsilon^2+\varepsilon^4)^{1/2}}{(1+\varepsilon^2)^{1/2}}$
$\dfrac{A_2\rho^2}{2}(2\cos^2\theta-1)$	$\dfrac{|A_2|}{2\sqrt{6}}(1+\varepsilon^2+\varepsilon^4)^{1/2}$
$\dfrac{A_0}{2}[2\rho^2-(1+\varepsilon^2)]$	$\dfrac{|A_0|}{2\sqrt{3}}(1-\varepsilon^2)$

[a] Each expression is given in units of wavelength. For linear measure, multiply by wavelength. Distortion, not included here, is the same as in Table 10.3.

As illustrations of the effects of aberrations on the PSF, we take two examples: a perfect image subject to defocus and an image with spherical aberration at the diffraction focus. The results were obtained by numerical integration of Eq. (10.2.2) and apply to an unobstructed aperture.

Figure 10.8 shows image profiles for the disk and first two bright rings of an image with different amounts of defocus. Note that the ring structure, clearly visible for $A_0 = 0.25$ or $\omega = \lambda/14$, is essentially absent when $A_0 > 0.5$.

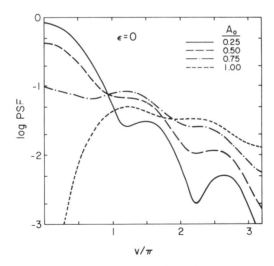

Fig. 10.8. Point spread function of perfect image with defocus. The aperture is unobstructed; shift from diffraction focus $= 8A_0\lambda F^2$.

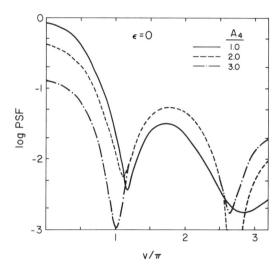

Fig. 10.9. Point spread function of image with spherical aberration at diffraction focus. The aperture is unobstructed.

The effect of defocus is clearly one of transferring energy from the disk to the nearby rings and filling in the dark rings. Although not shown in Fig. 10.8, the intensity $i(0) = 0$ when $A_0 = 1$. In general, the peak intensity is zero when $|A_0| = 1/(1 - \varepsilon^2)$.

Figure 10.9 shows image profiles for an image at the diffraction focus with different amounts of spherical aberration. In this case the separate rings remain well defined, but the energy within them grows at the expense of the disk.

The rms wavefront errors and encircled energy in the Airy disk for the profiles in Figs. 10.8 and 10.9 are given in Table 10.6. The results for EE were obtained by numerical integration of Eq. (10.1.15). It is evident from

Table 10.6

EE and ω for Images with Aberrations

	Figure 10.8			Figure 10.9	
A_0	EE	ω	A_4	EE	ω
0.25	0.733	0.072	1.0	0.668	0.075
0.50	0.490	0.144	2.0	0.324	0.149
0.75	0.248	0.217	3.0	0.094	0.224
1.00	0.105	0.289	4.0	0.068	0.298

these entries that the encircled energy drops dramatically with increasing rms wavefront error.

D. EXAMPLE: SPHERE IN COLLIMATED LIGHT

To illustrate the relation between the geometric and diffraction coefficients in Eq. (10.2.1) we evaluate the spherical aberration for an unobstructed spherical mirror in collimated light and compare the results with those given in Section 4.II. Taking the value of B_3 from Table 5.6 and equating the coefficients in Eq. (10.2.1) gives

$$B_3 a^4 = \lambda A_4 = a^4/4R^3 = a/256 F^3, \quad (10.2.7)$$

where $F = R/4a$. The angular, transverse, and longitudinal spherical aberrations are

$$\text{ASA} = 4 B_3 a^3 = 1/64 F^3 = 4\lambda A_4/a, \quad (10.2.8)$$

$$\text{TSA} = f\,\text{ASA} = a/32 F^2 = 8\lambda A_4 F, \quad (10.2.9)$$

$$\text{LSA} = 2F\,\text{TSA} = a/16 F = 16\lambda A_4 F^2, \quad (10.2.10)$$

where LSA is the distance between the paraxial and marginal foci. Note that LSA is 2× the shift from the paraxial focus to the diffraction focus for spherical aberration given in Table 10.4 for $\varepsilon = 0$.

We now assume that the system is diffraction-limited according to the criterion given in Section II.B, hence $\omega = \lambda/14$. With this value of the rms wavefront error we get $A_4 = 0.24$, using the first entry in Table 10.3 with $\varepsilon = 0$. Putting this value of A_4 into Eq. (10.2.1) and setting $\rho = 1$ gives $\Phi = \lambda/4$ for the marginal rays. This corresponds to the result given in Section 4.II and is often called Rayleigh's quarter-wavelength criterion for the amount of aberration that is tolerable in an imaging system.

Substituting $A_4 = 1/4$ into Eq. (10.2.8) gives $\text{ASA} = \lambda/a$. In Section 5.4 we noted that the diameter at the circle of least confusion is one-half of the blur radius at the paraxial focus. Therefore the angular blur diameter at the circle of least confusion is $\lambda/2a$ or λ/D, and the geometric blur size is comparable to the diffraction blur. It is evident that a larger value of A_4 gives images for which the geometric blur size is a good measure of the image quality.

REFERENCES

Born, M., and Wolf, E. (1980). "Principles of Optics," 6th ed., Chaps. 8 and 9. Pergamon, Oxford.

Dwight, H. (1961). "Tables of Integrals and Other Mathematical Data," 4th ed. Macmillan, New York.
Mahajan, V. (1983). *Appl. Opt.* **22**, 3035.
Wetherell, W. (1980). "Applied Optics and Optical Engineering," Vol. 8, Chap. 6. Academic Press, New York.

BIBLIOGRAPHY

For an introduction to the principles of diffraction see the intermediate-level texts in optics listed in the bibliography in Chapter 2.

For a discussion of aberrations for systems with annular pupils
Mahajan, V. (1981). *J. Opt. Soc. Am.* **71**, 75.
Mahajan, V. (1982). *J. Opt. Soc. Am.* **72**, 1258.

The perfect point spread function
Stoltzmann, D. (1983). "Applied Optics and Optical Engineering," Vol. 9, Chap. 4. Academic Press, New York.

Chapter 11 | Transfer Functions; Hubble Space Telescope

The results in the preceding chapter provide a complete description of the characteristics of a perfect or near-perfect image of a distant point object. The response of an optical system to a set of point objects or, more generally, an arbitrary intensity distribution was not considered in that analysis. Clearly, this response depends on factors in addition to the PSF, such as blurring due to image motion or detector pixel size. Factors such as these are most easily included by using the theory of transfer functions to describe the system response and image characteristics.

I. TRANSFER FUNCTIONS AND IMAGE CHARACTERISTICS

This approach to image analysis makes use of a complex function called the *optical transfer function* (OTF), with the real part of the OTF called the *modulation transfer function* (MTF). One advantage of this approach is that each independent component of a complete system, from the atmosphere to the detector, has its own OTF, and the system OTF is the product of the separate OTFs. This separation also applies to different types of wavefront error, with separate OTFs for geometric aberrations, random wavefront errors, and blurring due to image motion. The response of the system to an incident wavefront is determined by the system OTF comprising all these factors.

In this section, following a discussion of basic concepts, we draw on results derived from the theory of transfer functions and show how they are used to determine image characteristics. For derivations and discussion of the theory, the reader should consult references given at the end of the chapter.

A. DEFINITION OF THE TRANSFER FUNCTION

The concept of the transfer function is most easily seen by assuming a specific object intensity distribution. Consider a set of equally spaced line sources whose intensity in a direction perpendicular to the lines varies sinusoidally, as shown in Fig. 11.1a. Two parameters that describe this source are the spacing between the lines and the contrast. We let p_0 denote the spacing, or spatial period, where $\nu_0 = 1/p_0$ is the spatial frequency in cycles per unit length. The contrast C_o of the object, in the notation of Fig. 11.1a, is defined as

$$C_o = \frac{I_{\max} - I_{\min}}{I_{\max} + I_{\min}}, \qquad (11.1.1)$$

where C_o is assumed independent of ν_0.

Assuming an optical system of constant magnification, the image of this object is also a sinusoidal intensity distribution, as shown in Fig. 11.1b. Because each object point is imaged as a blur given by the PSF (or line spread function in one dimension), the image intensity is the superposition of all the individual spread functions. This addition of intensities assumes the illumination is incoherent.

We let p and ν denote the spatial period and frequency, respectively, at the image surface. The contrast C_i in the image is defined according to Eq. (11.1.1), with maximum and minimum intensities substituted. For a system with magnification m, we have $p = p_0 m$ and $\nu = \nu_0/m$. If the object distance is infinite, the spatial period and frequency of the object become angular period and frequency, with corresponding angular units. The image can also be described in angular terms in this case.

The *modulation transfer function* T is a measure of the change in contrast between the object and image, defined as

$$T(\nu) = \frac{C_i}{C_o} = \frac{\text{contrast in image at } \nu}{\text{contrast in object}}. \qquad (11.1.2)$$

Given that each object point is imaged as a blur described by the point or line spread function, we expect $T(\nu) < 1$ for all spatial frequencies. We also expect to find that $T(\nu) \to 1$ as $\nu \to 0$ and $T(\nu) \to 0$ as ν approaches the

I. Transfer Functions and Image Characteristics

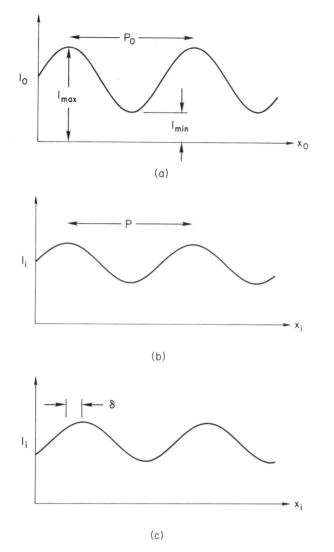

Fig. 11.1. (a) Object sine wave intensity; (b) image intensity profile, unshifted; and (c) image intensity profile, shifted.

resolution limit set by the width of the PSF. The spatial frequency at which contrast in a perfect image goes to zero is called the *cutoff frequency* ν_c. All information at frequencies higher than this frequency is lost.

To determine the cutoff frequency we apply the Rayleigh criterion of resolution to diffraction-limited images whose profiles are given by Eq.

(10.1.10). This criterion states that two images of equal intensity are just resolved when the peak of one coincides with the first minimum of the other, the first dark ring for circular images. From Eq. (10.1.11) we find that the angular separation of the peaks is approximately λ/D. Superposing two profiles like that in Fig. 10.2 at this separation gives an intensity midway between the peaks of approximately 0.8 that of either peak, and the peaks are said to be "just resolved."

For an object at infinity, the angle λ/D corresponds to a linear separation of $f\lambda/D$ at the image surface, or a spatial frequency of $1/\lambda F$. Though not a rigorous derivation, this is the cutoff frequency ν_c in linear units, with a corresponding cutoff frequency in angular units of D/λ.

It is convenient to define the *normalized spatial frequency* ν_n as

$$\nu_n = \nu/\nu_c = \nu_0/\nu_{0c}, \qquad (11.1.3)$$

where the range of this parameter is zero to one. Comparisons of different optical systems, or the same system at different wavelengths, are most often made in normalized units.

In addition to reduced contrast, the intensity pattern may also be shifted laterally on the image surface, as shown in Fig. 11.1c. This shift occurs if asymmetric aberrations, such as coma, are present. If the linear shift on the image surface is δ, the *phase transfer function* Φ is defined as

$$\Phi = 2\pi\delta/p. \qquad (11.1.4)$$

A combination of Eqs. (11.1.2) and (11.1.4) leads to the definition of the complex *optical transfer function* $Y(\nu)$ as

$$Y(\nu) = T(\nu)\exp[i\Phi(\nu)], \qquad (11.1.5)$$

where each independent component of a system has its own $Y(\nu)$. The two mirrors of a Cassegrain telescope, for example, are considered a single component because the image quality is determined by the mirror combination.

Given these definitions it is possible, in principle, to determine the response of a system to any object intensity distribution. From the theory of Fourier analysis, one finds that any such distribution can be synthesized by some combination of sinusoidal functions of different frequencies. The transformation of each harmonic component of the object into the corresponding harmonic part of the image is determined by Y at that frequency.

The discussion in this section is intended as an introduction to the basic characteristics of the transfer function. We now turn our attention to the relation between the transfer function and image characteristics for the important case where the PSF is symmetric about the system axis.

B. POINT SPREAD FUNCTION AND ENCIRCLED ENERGY BASED ON THE TRANSFER FUNCTION

The relations between image characteristics and the transfer function are derived using the theory of Fourier transforms. Given a PSF computed by the methods described in Section 10.II, the OTF is defined as the Fourier transform of the PSF. Because the PSF and OTF are a Fourier transform pair, the former can be calculated if the latter is known. For our purposes, we consider only the case where the phase transfer function Φ is zero and the OTF reduces to the MTF. This limitation rules out the treatment of asymmetric aberrations such as coma.

An alternative method of finding the OTF is by calculating the autocorrelation of the pupil function. The pupil function for a perfect system is the transmittance, usually constant, within the boundaries of the exit pupil and is zero outside. For a system with aberrations, the pupil function is complex and includes the aberrations. The autocorrelation integral is essentially one that gives the area of overlap between two pupil functions, with one shifted relative to the other by an amount proportional to the spatial frequency. The reader should consult the references by Born and Wolf and Wetherell for discussions of this approach to calculating the OTF.

In rectangular coordinates the MTF is given by

$$T(\nu_x, \nu_y) = A \iint_{-\infty}^{\infty} i(x, y) \exp[-2\pi i(\nu_x x + \nu_y y)] \, dx \, dy, \quad (11.1.6)$$

where $\nu_x = \nu \sin \gamma$, $\nu_y = \nu \cos \gamma$, x and y are given in Eq. (10.1.1), and A is a normalization factor chosen to give $T = 1$ at $\nu = 0$.

The relation corresponding to Eq. (11.1.6) in polar coordinates is

$$T(\nu, \gamma) = A \int_0^{2\pi} \int_0^{\infty} i(r, \psi) \exp[-2\pi i \nu r \cos(\psi - \gamma)] r \, dr \, d\psi, \quad (11.1.7)$$

where γ can be assigned any convenient value for the special but important case where the PSF is symmetric about the system axis. Letting $\gamma = \pi$, the integration over ψ in Eq. (11.1.7) is one of substituting the integral form of J_0, as done with Eq. (10.1.7).

Given a symmetric PSF and a circular aperture with a central obscuration, the methods of Fourier transforms give the following relations between the PSF, EE, and MTF:

$$T(\nu) = \frac{\pi^2(1-\varepsilon^2)D^2}{2\lambda^2} \int_0^{\infty} i(\alpha) J_0(2\pi\nu\alpha) \alpha \, d\alpha, \quad (11.1.8)$$

$$\mathrm{PSF}(\alpha) = \frac{8\lambda^2}{(1-\varepsilon^2)D^2} \int_0^{\nu_c} T(\nu) J_0(2\pi\nu\alpha) \nu\, d\nu, \qquad (11.1.9)$$

$$\mathrm{EE}(\alpha) = 2\pi\alpha \int_0^{\nu_c} T(\nu) J_1(2\pi\nu\alpha)\, d\nu, \qquad (11.1.10)$$

where ν is the frequency in angular units, α is the angular radius of the image, and J_0 and J_1 are Bessel functions. Because ν is given in angular units, the cutoff frequency ν_c in these units is D/λ. Equations (11.1.8)–(11.1.10) can be written in linear units by substituting $f\nu$(linear) for ν(angular) and r for α, where $r = f\alpha$. The factors outside the integrals in Eqs. (11.1.8)–(11.1.10) are normalization factors, with $T(0) = 1$, $\mathrm{PSF}(0) = 1$ for a perfect image, and $\mathrm{EE}(\infty) = 1$.

For ease of calculation and comparison of results for different systems or wavelengths, it is useful to rewrite these relations in terms of the normalized frequency ν_n. The results are

$$T(\nu_n) = \frac{\pi^2(1-\varepsilon^2)}{2} \int_0^\infty i(w) J_0(2\pi\nu_n w) w\, dw, \qquad (11.1.11)$$

$$\mathrm{PSF}(w) = \frac{8}{(1-\varepsilon^2)} \int_0^1 T(\nu_n) J_0(2\pi\nu_n w) \nu_n\, d\nu_n, \qquad (11.1.12)$$

$$\mathrm{EE}(w) = 2\pi w \int_0^1 T(\nu_n) J_1(2\pi\nu_n w)\, d\nu_n, \qquad (11.1.13)$$

where $w = \alpha D/\lambda = \alpha\nu_c$. Comparing the argument of each Bessel function with Eq. (10.1.12), we see that $2\pi w = 2v$, the dimensionless parameter used in Section 10.II.

For calculations of PSF and EE, all that is needed is the MTF. The general expression for the MTF of a perfect circular pupil with a central obscuration, taken from Appendix B of the reference by Wetherell, is given in slightly modified form in Table 11.1. For a clear circular aperture the factors B and C are zero. Substituting $2A/\pi$ in Eq. (11.1.12), it is a simple calculation to verify that $\mathrm{PSF}(0) = 1$ for a clear aperture, as required by normalization.

Modulation transfer functions for selected values of ε are shown in Fig. 11.2. The main effect of a larger central obscuration is a decrease in the MTF in the middle of the frequency range. This is expected because the effect of the obscuration on the PSF is to put more energy into the first bright ring of the Airy pattern, and the contrast in the image of an extended object is reduced because of the larger fraction of energy in this ring. For

I. Transfer Functions and Image Characteristics

Table 11.1
Modulation Transfer Function for Perfect Lens with Central Obscuration

$$T(\nu_n) = \frac{2}{\pi} \frac{(A+B+C)}{(1-\varepsilon^2)}$$

$A = [\cos^{-1}\nu_n - \nu_n(1-\nu_n^2)^{1/2}],$ $\qquad 0 \leq \nu_n \leq 1$

$B = \varepsilon^2 \left\{ \cos^{-1}\left(\frac{\nu_n}{\varepsilon}\right) - \left(\frac{\nu_n}{\varepsilon}\right)\left[1 - \left(\frac{\nu_n}{\varepsilon}\right)^2\right]^{1/2} \right\},$ $\qquad 0 \leq \nu_n \leq \varepsilon$

$ = 0,$ $\qquad \nu_n > \varepsilon$

$C = -\pi\varepsilon^2,$ $\qquad 0 \leq \nu_n \leq (1-\varepsilon)/2$

$ = -\pi\varepsilon^2 + \left\{ \varepsilon \sin\phi + \frac{\phi}{2}(1+\varepsilon^2) - (1-\varepsilon^2)\tan^{-1}\left[\frac{(1+\varepsilon)}{(1-\varepsilon)}\tan\frac{\phi}{2}\right] \right\},$ $\qquad (1-\varepsilon) \leq 2\nu_n \leq (1+\varepsilon)$

$ = 0$ $\qquad 2\nu_n > (1+\varepsilon)$

$\phi = \cos^{-1}\left(\frac{1+\varepsilon^2 - 4\nu_n^2}{2\varepsilon}\right)$

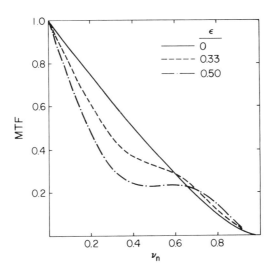

Fig. 11.2. Normalized modulation transfer function for several obscuration ratios calculated from relations in Table 11.1.

spatial frequencies near the cutoff frequency, on the other hand, the MTF is slightly larger when the pupil has an obscuration. This is also expected because the full width at half-maximum (FWHM) of the Airy disk is smaller for larger ε, and the "sharper" peak implies a smaller resolution limit according to the Rayleigh criterion.

When $T(\nu_n)$ from Table 11.1 is substituted into Eqs. (11.1.12) and (11.1.13) and the equations are integrated numerically, results like those shown in Fig. 10.2 and 10.4 are obtained. For a perfect image it is obviously easier to use Eqs. (10.1.10) and (10.1.16) to find the PSF and EE, but in the presence of aberrations it is usually easier to use the MTF approach.

The calculation of MTFs in the presence of symmetrical aberrations is done by either evaluating the autocorrelation integral, with the aberrations included in the pupil function, or integrating Eq. (11.1.11) with $i(w)$ for the aberrated image from Eq. (10.2.2). MTF curves for defocus and spherical aberration derived by the latter method are shown in Figs. 11.3 and 11.4, respectively.

C. MODULATION TRANSFER FUNCTIONS FOR OTHER WAVEFRONT ERRORS

In addition to wavefront errors due to classical aberrations, often called figure errors, we noted above that random errors on a finer scale due to the polishing process may also be present. Another source of image degradation is motion of an image due to effects from outside the optical system. Each of these nonfigure error contributions can be modeled with a factor in the MTF that is a statistical average of the effect. In this section we give an overview of some of these MTF models and their effects on the PSF and EE.

The usual way to include additional MTF factors is to write the system MTF as a product of independent factors in the form

$$T = T_d T_f T_r T_p, \qquad (11.1.14)$$

where T is the system MTF, T_d is the MTF for a perfect system, as given in Table 11.1, and the remaining factors are *degradation functions*. The subscripts f, r, and p denote, in turn, contributions due to figure, random, and pointing errors. This breakdown of the MTF into separate parts assumes that these contributions are independent of one another.

In considering fine-scale random errors on a wavefront, we assume that all figure errors have been subtracted from the wavefront map at the exit pupil. Figure error is usually taken to be components with spatial frequencies

I. Transfer Functions and Image Characteristics

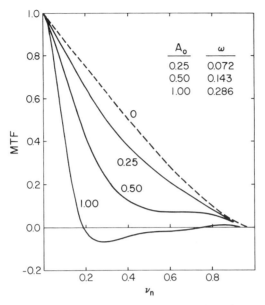

Fig. 11.3. Modulation transfer functions for perfect system ($\varepsilon = 0$) with defocus calculated from Eq. (11.1.11). The rms wavefront error ω is given in units of waves.

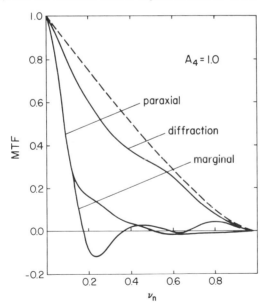

Fig. 11.4. Modulation transfer functions for unobstructed system ($\varepsilon = 0$) with spherical aberration calculated from Eq. (11.1.11). Curves are given for paraxial, diffraction, and marginal foci.

less than 5 cycles/radius. We also assume that the remaining wavefront errors of larger spatial frequency are distributed in a random fashion over the residual wavefront. The choice of the upper limit to the spatial frequency depends on the size of the spatial period selected. For the 2.4-m primary mirror of the Hubble Space Telescope (HST), a spatial period of 1 mm corresponds to a spatial frequency of 1200 cycles/radius. Random error in this middle range of spatial frequencies is often termed ripple.

A statistical analysis of this type of error has been made by O'Neill. The result of this analysis is an MTF degradation factor in the midfrequency range of the form

$$T_m = \exp\{-k^2\omega^2[1 - c(\nu_n)]\}, \qquad (11.1.15)$$

where $k = 2\pi/\lambda$, ω is the rms random wavefront error, and $c(\nu_n)$ is the normalized autocorrelation function of the residual pupil function. The characteristics of $c(\nu_n)$ are such that $c(0) = 1$ and $c(\nu_n) \to 0$ for a large shift of the residual wavefront in the autocorrelation integral. If the function $c(\nu_n)$ is modeled as a Gaussian of the form $c(\nu_n) = \exp(-4\nu_n^2/l^2)$, as given by Wetherell, the degradation function T_m has the form shown in Fig. 11.5. The parameter l is called the correlation length and is a measure of the structure on the wavefront. To a rough approximation, the spatial period of the dominant structure is $1/l$ cycles/diameter. For discussion of other forms of $c(\nu_n)$ and comparison with measured results, the reader should consult the reference by Wetherell.

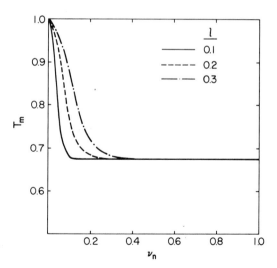

Fig. 11.5. Midfrequency degradation factor T_m with Gaussian correlation factor calculated from Eq. (11.1.15). The rms ripple error = 0.1 waves; l is the normalized correlation length.

Wavefront errors with high spatial frequencies, those larger than ones associated with ripple, are ascribed to microstructure on an optical surface and are sometimes called microripple. The degradation function for high-frequency microripple follows directly from Eq. (11.1.15) if we let $l \to 0$ in the autocorrelation function. In this limit

$$T_h = \exp(-k^2\omega^2) \qquad (11.1.16)$$

at all spatial frequencies except $\nu_n = 0$, where ω is the rms wavefront error due to microripple. The product of Eqs. (11.1.15) and (11.1.16) is the degradation function T_r in Eq. (11.1.14).

Degradation of an image due to random motion has been discussed by several authors, including Mahajan and Wetherell. The starting assumption of this analysis is an image motion that is rotationally symmetric and described by the unnormalized probability function

$$P(r) = \exp(-\alpha^2/2\sigma'^2), \qquad (11.1.17)$$

where σ' is the standard deviation and α is radius of the excursion of the image from the mean position, both in angular units. If we normalize σ' by multiplying by the cutoff frequency D/λ, then the pointing degradation function, as shown by Mahajan, is

$$T_p(\nu_n) = \exp(-2\pi^2\sigma^2\nu_n^2). \qquad (11.1.18)$$

Because T_p decreases as ν_n increases, it is evident that the effect of this degradation function is to depress the MTF more at higher spatial frequencies. Figure 11.6 shows the pointing degradation function for several values of σ. For an otherwise perfect system, the product of curves in Figs. 11.2 and 11.6 gives the system MTF with random pointing error.

We now give results for PSF and EE calculated from Eqs. (11.1.12) and (11.1.13) with different degradation factors multiplying T_d, the diffraction MTF. All of the results given are for $\varepsilon = 0.33$, the obscuration ratio of the HST, with zero figure error. A more complete discussion of the expected image characteristics of HST, with all factors taken together, follows in the next section.

Figures 11.7 and 11.8 show PSF and EE for a pupil wavefront with random error of the type described by Eq. (11.1.15), for three values of ω. The approximate correlation length assumed for these calculations is 0.04 cycles/diameter. Relative to the PSF for a perfect system, given in Fig. 10.2, the effect of this error is to depress the disk and inner ring and raise the outer rings. The Strehl intensity is given by Eq. (10.2.6). The transfer of energy outward from the center of the Airy pattern is clearly shown in Fig. 11.8. Taking values of EE at the right side of Fig. 11.8, we see that nearly five times as much energy is outside the fourth bright ring when $\omega = 0.1\lambda$, compared to a perfect image.

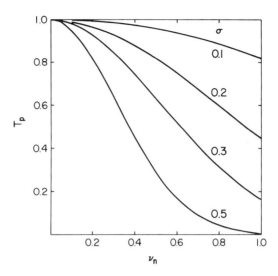

Fig. 11.6. Pointing degradation factor T_p for several normalized rms pointing errors calculated from Eq. (11.1.18).

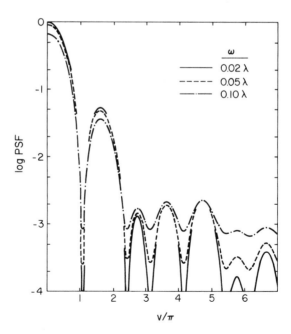

Fig. 11.7. Point spread function for obstructed system ($\varepsilon = 0.33$) with rms midfrequency error ω and $l = 0.04$. Results are calculated from Eq. (11.1.12).

I. Transfer Functions and Image Characteristics   211

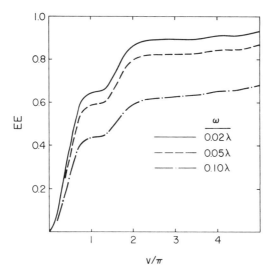

Fig. 11.8. Encircled energy fraction for obstructed system ($\varepsilon = 0.33$) with rms midfrequency error ω and $l = 0.04$. Results are calculated from Eq. (11.1.13).

When high-frequency microripple is present, as described by Eq. (11.1.16), the effect is to depress the PSF by the factor T_h at all image radii. This occurs because T_h is independent of spatial frequency, hence it can be taken out of the integral in Eq. (11.1.12). The effect of microripple on EE is similar, for the same reason. In theory, therefore, the energy scattered by microripple error disappears; in practice, the energy is scattered at angles large compared to the Airy disk diameter.

Figures 11.9 and 11.10 show PSF and EE for a perfect system with pointing error described by Eq. (11.1.18), for three values of σ. The effect of increased pointing error is clearly one of reducing the Strehl intensity, smoothing the PSF pattern, and distributing a given fraction of the encircled energy over a larger area. For the values of σ shown, the redistribution of energy takes place largely between the disk and the first bright ring. With specific reference to HST, the curve with $\sigma = 0.1$ corresponds to an rms pointing error on the sky of 0.005 arc-sec, at $\lambda = 580$ nm. Because σ is inversely proportional to λ, the curve with $\sigma = 0.3$ corresponds to the same pointing error at $\lambda = 190$ nm. As expected, a given pointing error on the sky has a greater effect on a "sharper" image.

With these examples we complete our introduction to the calculation of image characteristics by the MTF approach. It should be clear from the results shown that this is a powerful technique, especially when errors other than figure errors are present. In Chapter 16 this approach is used to examine

11. Transfer Function; Hubble Space Telescope

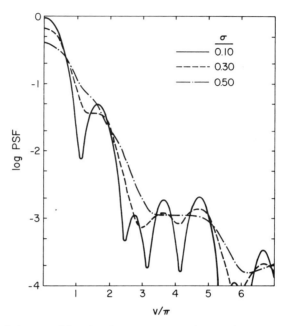

Figure 11.9. Point spread function for perfect system ($\varepsilon = 0.33$) with pointing error. σ is the normalized rms Gaussian error. Results are calculated from Eq. (11.1.12).

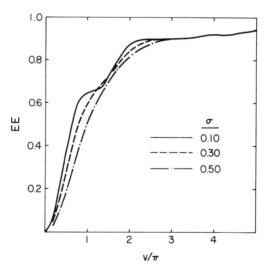

Fig. 11.10. Encircled energy fraction for perfect obstructed system ($\varepsilon = 0.33$) with pointing error. σ is the normalized rms Gaussian error. Results are calculated from Eq. (11.1.13).

the effects of other factors that influence system performance, such as detector pixel size and the atmosphere for ground-based telescopes.

II. THE HUBBLE SPACE TELESCOPE

The HST will be the first large astronomical observatory in space with a resolution capability an order of magnitude better than is possible with ground-based telescopes in the visible and near ultraviolet. This unique facility will enable astronomers to make observations not possible from the ground and obtain data needed to answer many fundamental astronomical questions. Given this promise, a brief description of HST and the expected image characteristics is in order.

A. BASIC CONFIGURATION

The HST is a 2.4-m Cassegrain telescope of the Ritchey–Chretien type, with the nominal parameters given in Table 11.2. The performance goals set by the National Aeronautics and Space Administration at the start of the project include spectral coverage from 115 nm to the far infrared, with diffraction-limited performance at visible wavelengths. Analysis of the completed system shows that HST is expected to meet or exceed the stated goal of $\lambda/20$ rms wavefront error at $\lambda = 633$ nm on the axis of the $f/24$ focal surface.

The complete observatory includes the following complement of instruments: wide-field/planetary camera (WFPC), faint object camera (FOC), faint object spectrograph (FOS), high-resolution spectrograph (HRS), and high-speed photometer (HSP). The fine guidance system (FGS) of the telescope will also be used for astrometric observations. For details of these instruments and their observing modes, the reader should consult the references at the end of the chapter, especially the "Instrument Handbook" distributed by the Space Telescope Science Institute.

Table 11.2

Nominal Parameters of Hubble Space Telescope

Primary	$D = 2400$ mm, $R_1 = -11\,040$ mm, $f/2.3$, $K_1 = -1.0022985$
Secondary	$R_2 = -1358$ mm, $K_2 = -1.496$
Overall	$m = 10.435$, $\beta = 0.2717$, $k = 0.1112$, scale = 3.58 arc-sec/mm = 279 μm/arc-sec

B. ON-AXIS IMAGE CHARACTERISTICS

In this section we describe the expected on-axis image characteristics at the $f/24$ focal surface. All of the results presented assume the mirrors are clean with no scattering due to dust. Examples of the expected limiting magnitude with selected instruments, given these image characteristics, are left to Chapter 16.

Analysis of HST performance proceeds along the lines described in the previous section. Each independent component is described by an MTF degradation function, and the product of these functions and the diffraction MTF is used as the basis of calculations of the image characteristics. Contributors to T_f in Eq. (11.1.14) include the aberrations of the mirrors, misalignments of the mirrors, thermal changes in orbit, ground-to-orbit changes, and errors of the optical system used to measure the wavefront in orbit. The errors that remain after figure errors are removed from the wavefront map are used to calculate T_m in the form given in Eq. (11.1.15). Surface errors derived from measurements on small parts of the mirrors are modeled as high-frequency errors in the form given in Eq. (11.1.16). The product of these functions, to which the figure error is the largest contributor, gives the system degradation function in the absence of pointing error. This combination leads to an overall rms system wavefront error of approximately $\lambda/21$ at a wavelength of 633 nm.

The product of the system degradation function with the diffraction MTF and the pointing degradation factor given in Eq. (11.1.18) gives the rotationally symmetric system MTF used in Eqs. (11.1.12) and (11.1.13). All of the following results are derived from calculations using these relations, with a nominal rms pointing error of 0.007 arc-sec assigned to image motion.

Figures 11.11 and 11.12 show PSFs at a number of wavelengths, with EE for each of these wavelengths shown in Figs. 11.13 and 11.14. It is evident from these curves that the PSFs show progressive degradation at shorter wavelengths. The ring structure in the Airy pattern, clearly seen in the visible and infrared wavelengths, is absent at the shortest wavelengths. This is a result of both pointing error and nonfigure contributors to the degradation function. We also see that the level of the PSF decreases in the ultraviolet, a consequence of the mid- and high-frequency components in the degradation function.

The Strehl ratio S and the FWHM of the image peak are shown in Fig. 11.15. The most notable feature of the FWHM curve is the limiting core diameter of about 0.023 arc-sec at the shortest wavelengths. Figure 11.16 shows the peak intensity $I(0)$ as a function of wavelength, normalized to unity at $\lambda = 633$ nm, assuming equal flux at each wavelength. The intensity at the peak is given by Eq. (10.1.20) with S included for a degraded image.

II. The Hubble Space Telescope

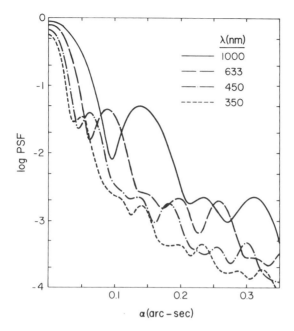

Fig. 11.11. Predicted PSFs for Hubble Space Telescope. The rms wavefront error is $\lambda/21$ at 633 nm; the rms pointing error is 0.007 arc-sec.

Therefore

$$\frac{I_\lambda(0)}{I_{633}(0)} = \frac{S_\lambda}{S_{633}} \left(\frac{633}{\lambda}\right)^2, \qquad (11.2.1)$$

where λ is in nanometers and S is given in Fig. 11.15. Also shown in Fig. 11.16 is the average intensity over an area enclosing 60% of the total energy. These results are derived using Eq. (10.1.19) with $\eta = 0.6$ and image radii taken from Figs. 11.13 and 11.14. The curves in Fig. 11.16 would show a λ^{-2} dependence for a perfect image; the actual curves show a peak in the ultraviolet.

Extension of PSF calculations to larger image radii than those shown in Figs. 11.11 and 11.12 shows that the average intensity far from the Airy disk falls off as α^{-3}, as for a perfect image. However, the intensity level is higher than that of a perfect image by an amount that depends on the wavelength. Comparing the average PSF at a radius of 1 arc-sec calculated from Eq. (11.1.12) with that given by Eq. (10.1.14), we get the results shown in Fig. 11.17. The increasing spread between the curves in Fig. 11.17 at shorter wavelengths is largely a consequence of the mid- and high-frequency factors in the degradation function. As noted in Section 11.I, the effect of

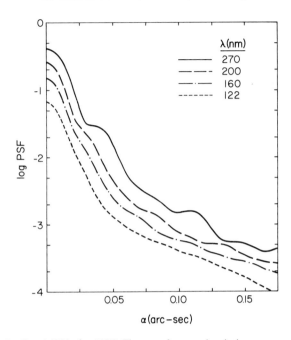

Fig. 11.12. Predicted PSFs for HST. The wavefront and pointing errors are given in the caption of Fig. 11.11.

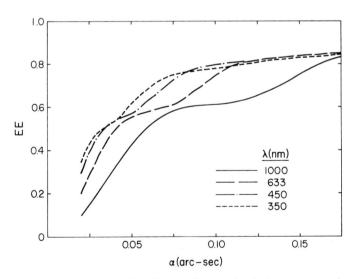

Fig. 11.13. Predicted EEs for HST. The wavefront and pointing errors are given in the caption of Fig. 11.11.

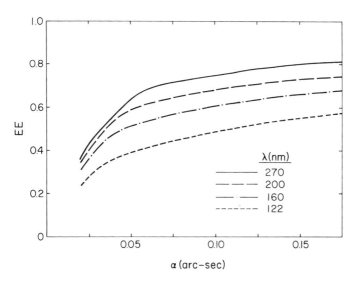

Fig. 11.14. Predicted EEs for HST. The wavefront and pointing errors are given in the caption of Fig. 11.11.

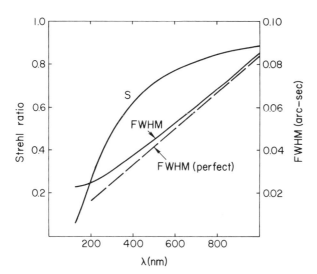

Fig. 11.15. Predicted Strehl ratio and FWHM for HST. The wavefront and pointing errors are given in the caption of Fig. 11.11.

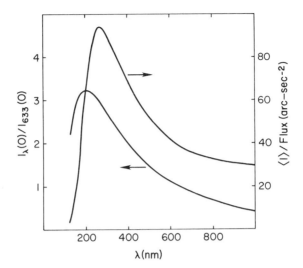

Fig. 11.16. Peak intensity for HST normalized to unity at 633 nm and average intensity per unit flux on area enclosing 60% of the encircled energy.

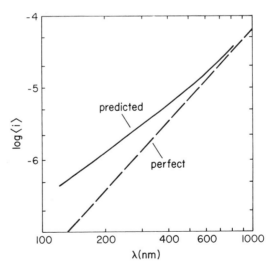

Fig. 11.17. Predicted average PSF for HST at 1 arc-sec from image peak. The wavefront and pointing errors are given in the caption of Fig. 11.11.

these factors is to transfer energy from the inner region of the Airy pattern to the wings.

C. OFF-AXIS ASTIGMATISM

The principal aberration for nonzero field angles in the Ritchey–Chretien design is astigmatism, with the angular astigmatism according to geometric theory given in Table 6.7. Substituting the values of m and β in Table 11.2 into AAS in Table 6.7 gives AAS $= \Gamma\theta^2/2F = 2B_1 a$, where $\Gamma = 8.609$, B_1 is the astigmatism coefficient, θ is the field angle, and a is the radius of the aperture stop.

From Eq. (10.2.1) we find that $\lambda A_2 = B_1 a^2$, where A_2 is the coefficient of astigmatism in the diffraction theory of aberrations. We now find the rms wavefront error at the diffraction focus of the astigmatic image by putting $\varepsilon = 0.33$ and $A_2 = \Gamma\theta^2 D/8\lambda F$ into ω in Table 10.5. The result, with θ expressed in arc-minutes, is

$$\omega(\mu m) = 0.00197\theta^2 \text{ (arc-min)},$$
$$\omega(\text{waves at 633 nm}) = 0.0031\theta^2 \text{ (arc-min)}. \quad (11.2.2)$$

From Section 10.II.B we take the convention for diffraction-limited as $\omega \leq \lambda/14$ and find from Eq. (11.2.2) that $\theta \leq 4.8$ arc-min at 633 nm. For example, the FOS apertures at 3.6 arc-min off-axis are illuminated by images that are essentially diffraction-limited for visible and near-ultraviolet wavelengths. The center of the field of the FOC apertures, on the other hand, is approximately 6.6 arc-min off-axis and the residual astigmatism of the HST must be corrected by the transfer optics of the FOC.

III. CONCLUDING REMARKS

These characteristics of the HST images are based on extensive modeling and represent the best estimate of what can be expected once HST is in orbit. Predictions have been made of the effect of dust on the mirrors on the PSF and EE, and the results predict some additional fraction of light scattered into the image wings. This fraction is uncertain because it is sensitive to the size distribution of the dust particles. Definitive image characteristics will be known only after extensive observations in space.

The imaging characteristics of diffraction-limited telescopes at visible wavelengths were of little more than academic interest prior to the start of the space age, with images from large ground-based telescopes dominated by seeing. Fortunately, the diffraction theory of aberrations discussed in Chapter 10 was well established at the time when thoughts turned to designing a space observatory as large as HST.

The theory, as presented in Chapters 10 and 11, is only an introduction to the main features of the diffraction theory of aberrations, but for many systems this is sufficient for design purposes. Relations for orthogonal aberrations of higher order have been derived, and these are used in designing optical systems of the highest precision. The reader should consult the literature for information on these refinements.

REFERENCES

Mahajan, V. (1978). *Appl. Opt.* **17**, 3329.
O'Neill, E. (1963). "Introduction to Statistical Optics." Addison-Wesley, Reading, Massachusetts.
Wetherell, W. (1980). See reference listed in Chapter 10.

BIBLIOGRAPHY

Goodman, J. (1968). "Introduction to Fourier Optics." McGraw-Hill, New York.
Hecht, E., and Zajac, A. (1974). "Optics," Chap. 11. Addison-Wesley, Reading, Massachusetts.
Steward, E. (1983). "Fourier Optics: An Introduction." Wiley, New York.
Hubble Space Telescope
"Instrument Handbooks" (1985). Space Telescope Science Institute, Baltimore, Maryland.
"The Space Telescope Observatory" (1982). NASA CP-2244. NASA, Washington, D.C.
Leckrone, D. (1980). *Publ. Astron. Soc. Pac.* **92**, 5.
Schroeder, D. (1985). "Advances in Space Research: Astronomy from Space," Vol. 5, No. 3, p. 157. Pergamon, Oxford.

Chapter 12 | Spectrometry: Definitions and Basic Principles

Spectral analysis of celestial objects is probably the most important method for learning about the physics of these sources, and a large fraction of telescope time is used to get spectral data. In this chapter we begin to consider the characteristics of spectrometers used with telescopes to obtain these data.

We use the term *spectrometer* to refer to any of several types of spectroscopic instruments. A spectrograph is an instrument in which many spectral elements are recorded simultaneously with an area detector having many resolution elements. A monochromator is an instrument in which single spectral elements are recorded sequentially in time by a detector with a single resolution element. Many of the results that follow apply to either type of instrument, but if there is a difference the distinction is noted.

A simple method of getting spectral information is filter photography, with broad- or narrowband filters placed in the beam ahead of the telescope focal surface. For point sources the result is one piece of spectral information per photograph for each source in the field, while for extended objects there is one piece of spectral information for each resolution element on the two-dimensional detector.

More detailed spectral information is obtained if the light is sent through a dispersing element, such as a prism or diffraction grating. In this case a spectrum is obtained for each source whose light passes through the disperser, with the number of pieces of spectral information per source determined by the mode in which the disperser is used. In the so-called slitless mode,

12. Spectrometry: Definitions and Basic Principles

a prism or grating acts as a dispersing filter and gives a spectrum for each source in the field. When the disperser is part of a slit spectrometer, a spectrum is obtained for each source whose light passes through the slit. The characteristics of these modes are discussed in more detail in the following sections.

In succeeding sections we present the basic principles that govern the operation of all spectrometric devices and define such terms as spectral purity, spectral resolution, and luminosity. The approach parallels that given by the author in a reference cited at the end of the chapter. Another source of information about the principles of spectrometry is the excellent book by Meaburn.

I. SPECTROMETER TYPES AND MODES

Each type of spectrometer is denoted by the kind of dispersing element that is used, hence prism, grating, or Fabry-Perot spectrometer. Each of these elements is characterized by its *angular dispersion*, defined as $d\beta/d\lambda$, where $d\beta$ is the angular difference between two rays of wavelength difference $d\lambda$ emerging from the disperser. This is shown schematically in Fig. 12.1 for a single ray incident on the dispersing element. Relations for angular dispersion of different elements are given in Chapter 13.

The angular dispersion is clearly a parameter associated with the dispersing element, independent of the configuration in which it is used. When the element is part of an optical system, the characteristics of both are combined to define the *linear dispersion* $dl/d\lambda$, where dl is the linear separation on a focal surface between two rays of wavelength difference $d\lambda$.

If collimated light is incident on the disperser, then the linear dispersion is given by

$$dl/d\lambda = f\, d\beta/d\lambda = fA, \quad (12.1.1)$$

where f is the focal length of the optics following the dispersing element and A is the angular dispersion. This case is shown in Fig. 12.2.

Spectrometer configurations to which Eq. (12.1.1) applies include the

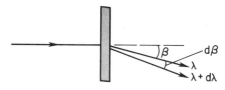

Fig. 12.1. Schematic of dispersive element. Angular dispersion $A = d\beta/d\lambda$.

I. Spectrometer Types and Modes

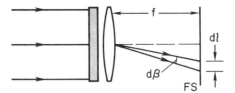

Fig. 12.2. Spectrum in focus on focal surface FS. Linear dispersion $= f\, d\beta/d\lambda$.

slitless mode where a prism or grating is placed in front of a telescope, in which case f is the focal length of the telescope. This type of slitless mode is often called the objective mode. Equation (12.1.1) also applies to most prism and grating slit spectrometers in which a separate collimator gives a collimated beam and f is the focal length of the camera optics. Fabry–Perot spectrometers also have separate collimator and camera optics.

If a convergent beam of light is incident on the disperser, the linear dispersion is

$$dl/d\lambda = s\, d\beta/d\lambda = sA, \qquad (12.1.2)$$

where s is the distance from the disperser to the focal surface. This case is shown in Fig. 12.3.

Configurations to which Eq. (12.1.2) applies include the slitless mode where a disperser, usually a grating or grism, is placed in the telescope beam ahead of the focal surface. The grism is a combination of a grating and a prism, with the grating as the main dispersing element. This type of slitless mode has been dubbed the "nonobjective" mode by Hoag. Equation (12.1.2) also applies to the so-called Monk–Gillieson spectrometer, in which a mirror preceding the grating is both collimator and camera.

For any of the spectrometer modes noted above it is convenient to define P, the reciprocal linear dispersion or *plate factor*, where

$$P = (fA)^{-1}, \qquad (12.1.3a)$$

$$P = (sA)^{-1}, \qquad (12.1.3b)$$

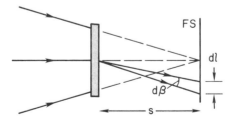

Fig. 12.3. Spectrum in focus on focal surface FS with convergent light incident on disperser. inear dispersion $= s\, d\beta/d\lambda$.

for the modes in Figs. 12.2 and 12.3, respectively. The units of P are usually given as angstroms per millimeter.

The terms *angular* and *linear* dispersion above do not apply to another type of spectrometer used to get spectral information, the so-called Fourier spectrometer. This instrument is basically a Michelson interferometer whose output is an interferogram from which the spectral information is derived by Fourier analysis. Because the Fourier spectrometer is not a dispersive device, the definitions in the following section that include dispersion do not apply to this type of spectrometer. A brief introduction to Fourier spectrometers is given in Section 13.V.

II. PARAMETERS FOR SLIT SPECTROMETERS

A general layout of a slit spectrometer in the most commonly used arrangement is shown in Fig. 12.4. Elements of the spectrometer include an entrance slit of width ω and height h at the telescope focus, collimator and camera optics to reimage the entrance slit, and a disperser whose angular dispersion is A. Collimator and camera optics have focal lengths f_1 and f_2, respectively, with a reimaged slit of width ω' and height h' at the camera focus.

The entrance slit subtends angles ϕ and ϕ' on the sky and $\delta\alpha$ and $\delta\alpha'$ at the collimator, where $\phi = \omega/f$, $\phi' = h/f$, $\delta\alpha = \omega/f_1$, and $\delta\alpha' = h/f_1$. The collimated beam incident on the disperser has diameter d_1, with the direction of dispersion parallel to the slit width, or in the plane of the diagram in Fig. 12.4.

The size of the projected slit image depends on f_1, and f_2 and on the characteristics of the disperser. Figure 12.5 shows the collimator and camera represented by equivalent thin lenses, with an object of length l subtending an angle γ at the collimator and its image of length l' subtending an angle γ' at the camera. For a system with no dispersing element between the lenses, $\gamma' = \gamma$ and $l' = l(f_2/f_1)$. Because a system without a disperser is

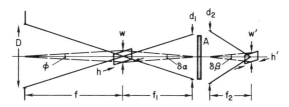

Fig. 12.4. Schematic layout of slit spectrometer with dispersing element A. See text, Section 12.II, for definitions of parameters.

II. Parameters for Slit Spectrometers

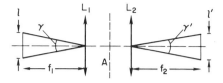

Fig. 12.5. Transverse magnification $|m| = l'/l = f_2/f_1$ in direction perpendicular to dispersion. In direction parallel to dispersion $|m| = r(f_2/f_1)$, where r is the anamorphic factor.

rotationally symmetric about the z axis, this relation is true for any orientation of the object.

If a dispersing element is placed between the lenses, rotational symmetry about the z axis is lost and the equality of the subtended angles is not necessarily preserved for different orientations of the object. In the direction perpendicular to the dispersion the beam passing through the disperser is unchanged, and $\gamma' = \gamma$ holds as before. This is not the case in the direction along the dispersion, where it is necessary to take $\gamma' = r\gamma$ to account for possible magnification effects due to the disperser. In terms of the subtended angles in Fig. 12.4 we have $r = \delta\beta/\delta\alpha$.

The parameter r, called the *anamorphic magnification*, depends on the type and orientation of the dispersing element. At this point we note, without derivation, that $r = d_1/d_2$, the ratio of the beamwidths at the collimator and camera. This relation is derived below, and the form of r in terms of the parameters of a specific disperser is given in Chapter 13.

Applying these results to the slit dimensions in Fig. 12.4 gives

$$\omega' = r\omega(f_2/f_1) = r\phi DF_2, \qquad (12.2.1\text{a})$$

$$h' = h(f_2/f_1) = \phi' DF_2, \qquad (12.2.1\text{b})$$

where $F_2 = f_2/d_1$. Note that these relations also follow directly by substitution into Eq. (2.2.6) with $n' = n$.

The relation in Eq. (12.2.1a) is important in establishing the proper value of F_2 for a detector whose pixel size Δ is properly matched to ω'. If we take a proper match to be one in which two pixels cover the width ω', then $2\Delta = r\phi DF_2$. If we choose $\Delta = 20$ μm, $\phi = 1$ arc-sec, and $D = 4$ m, then from Eq. (12.2.1a) we find $rF_2 = 2$.

From Eq. (12.2.1a) we also see that a constant ω' for a given angular slit width ϕ implies $rDF_2 = $ constant. Thus a spectrometer on a larger telescope requires a camera with a smaller focal ratio, if the ratio ω'/ϕ is to remain constant.

A. SPECTRAL PURITY

Consider a spectrometer entrance slit of width ω illuminated by light of two wavelengths λ and $\lambda + \Delta\lambda$. The slit image in each wavelength has width

ω' and, from Eq. (12.1.1), the separation between the centers of the images is $\Delta l = f_2 A \Delta\lambda$. We define the *spectral purity* $\delta\lambda$ as the wavelength difference for which $\Delta l = \omega'$, hence the spectral images are on the verge of being resolved. Putting this condition on Δl into Eq. (12.2.1) gives

$$\delta\lambda = P\omega' = \frac{r\phi}{A}\frac{D}{d_1}, \qquad (12.2.2)$$

where from Fig. 12.4 we find $f_1/d_1 = f/D$.

For a given telescope diameter and angle on the sky, it is evident from Eq. (12.2.2) that the key factors which determine the spectral purity are the angular dispersion and collimator beam diameter. We also see that a given spectrometer on a larger telescope gives a larger $\delta\lambda$ for a given angle on the sky.

For a Monk-Gillieson spectrometer, shown schematically in Fig. 12.6, the relations in Eq. (12.2.1) apply if f_1 and f_2 are replaced by s_1 and s_2. Combining Eqs. (12.1.3b) and (12.2.1a), we find the spectral purity is given by Eq. (12.2.2), provided d_1/r is replaced by d_2, where d_2 is the beam size at the disperser. Hence for this spectrometer the spectral purity is set by the angular dispersion and beam size at the disperser, for a given ϕ and D.

It is important to note here that the definition of spectral purity is a geometric one that does not take into account the limit on image size set by diffraction. The spectral purity cannot be smaller than $\delta\lambda_0$, the limit set by diffraction, where this limit depends on the angular dispersion, collimator beam diameter, and wavelength. The expression for $\delta\lambda_0$ is derived in a later section.

B. SPECTRAL RESOLUTION

The *spectral resolution* R, a dimensionless measure of the spectral purity, is defined as

$$R = \frac{\lambda}{\delta\lambda} = \frac{\lambda A}{r\phi}\frac{d_1}{D}. \qquad (12.2.3)$$

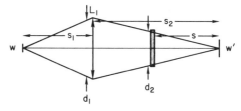

Fig. 12.6. Schematic of Monk-Gillieson spectrometer in direction parallel to dispersion, with dispersing element in convergent light.

II. Parameters for Slit Spectrometers   227

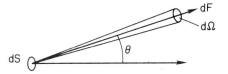

Fig. 12.7. Surface element of area dS and brightness B radiating into solid angle $d\Omega$. See Eq. (12.2.4).

It is clear from Eq. (12.2.3) that a larger telescope requires a larger beam diameter for a given type of disperser, if the resolution is to be kept constant. Because $\delta\lambda = \delta\lambda_0$ at the limit set by diffraction, there is also a largest possible spectral resolution R_0, although in most applications in astronomy the resolution given by Eq. (12.2.3) is considerably smaller than R_0.

C. FLUX, LUMINOSITY, AND ÉTENDUE

We now seek a relation for the energy flux transmitted by the telescope-spectrometer combination, with this relation expressed in terms of the system parameters. This analysis leads to a quantity called the *étendue*, whose importance in the analysis of spectrometric instruments was first emphasized by Jacquinot. A second and related quantity introduced is the *luminosity*, defined as the product of the étendue and net transmittance of the optics.

The derivation of étendue given here makes use of basic photometric definitions given by Born and Wolf. Consider a small, uniformly radiating surface element of area dS and photometric brightness B, as shown in Fig. 12.7. The flux dF radiated into a small cone of solid angle $d\Omega$ in a direction θ from the normal to dS is given by

$$dF = I\,d\Omega = B\cos\theta\,dS\,d\Omega, \qquad (12.2.4)$$

where I is the intensity or flux per unit solid angle in the direction θ. For a spectrometer we take dS as the area of the entrance slit.

The flux passing through the entrance pupil of the spectrometer, taken at the collimator, is the integral of Eq. (12.2.4) over the aperture of the pupil. For the annular cone shown in Fig. 12.8, the solid angle between θ

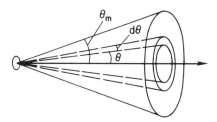

Fig. 12.8. Flux entering spectrometer entrance pupil. See Eq. (12.2.5).

and $\theta + d\theta$ is given by $d\Omega = 2\pi \sin\theta\, d\theta$, and therefore

$$F = 2\pi B\, dS \int_0^{\theta_m} \cos\theta \sin\theta\, d\theta = BU, \qquad (12.2.5)$$

$$U = \pi \sin^2 \theta_m\, dS, \qquad (12.2.6)$$

where $\tan \theta_m = d_1/2f_1$, U is the étendue, and $dS = \omega h$.

Assuming θ_m is small, we can replace $\sin \theta_m$ by θ_m. Substituting for ω, h, and θ_m in terms of the telescope and spectrometer parameters gives

$$U = \frac{\pi d_1^2}{4}\, \delta\alpha\, \delta\alpha' = \frac{\pi D^2}{4}\, \phi\phi' = S\Omega. \qquad (12.2.7)$$

In this form we see that U is the product of the collimator (telescope) area S and the solid angle Ω subtended by the slit at the collimator (or by the slit on the sky).

It is also instructive to write $U = \omega\theta_m \cdot h\theta_m$. When it is written in this form, we see from the discussion following Eq. (2.2.6) that U is the product of two Lagrange invariants with $n = 1$. Because the Lagrange invariant is unchanged through an optical system, the étendue is also a constant of the system. Assuming $n = 1$ for the space surrounding the image, we can write U for the image as

$$U = \frac{\pi d_1 d_2}{4}\, \delta\beta\, \delta\beta', \qquad (12.2.8)$$

where $\delta\beta$ and $\delta\beta'$ are the angles subtended by the projected slit at the camera. Given $\delta\beta = r\,\delta\alpha$ and $\delta\beta' = \delta\alpha'$, equating Eqs. (12.2.8) and (12.2.7) gives the anamorphic magnification $r = d_1/d_2$, as stated above.

Taking τ as the transmittance of the telescope–spectrometer combination, the luminosity L of the system is given by

$$L = \tau U = \tau S\Omega. \qquad (12.2.9)$$

If F is the monochromatic flux incident on the telescope, the flux F' in the projected image of the slit is

$$F' = \tau F = \tau BU = B'U. \qquad (12.2.10)$$

From Eq. (12.2.10) we see that the brightness B' of the image is less than the source brightness B by the factor τ. It is worth emphasizing that Eq. (12.2.10) is restricted to the case where $n = 1$ for the image space. If the image space index is n, as it is in solid or semisolid cameras, then $B' = \tau n^2 B$. As shown in Section 7.V, the focal length of solid cameras is n times smaller than that of the equivalent air camera, and therefore $\delta\beta$ and $\delta\beta'$ in Eq. (12.2.8) are each n times smaller. Thus the image area is n^2 smaller, and the relation between F and F' in Eq. (12.2.10) is unchanged, as must be true for conservation of energy.

D. LUMINOSITY-RESOLUTION PRODUCT

The importance of the concepts of étendue and luminosity in evaluating spectrometer performance is particularly evident in the product of either with the spectral resolution. Taking the product of Eqs. (12.2.3) and (12.2.7) gives

$$L \cdot R = (\tau\pi/4)(D\phi')(\lambda A d_2), \qquad (12.2.11)$$

where d_2 has replaced d_1/r and the factors in the right-hand parentheses are specific to the spectrometer. For stellar sources ϕ' is the diameter of the seeing disk, whereas for extended sources ϕ' is the angular height of the entrance slit.

For constant ϕ' the product $L \cdot R$ is a constant for a given telescope-spectrometer combination. Increasing this product for a given telescope requires either a larger spectrometer beam diameter or a disperser with higher angular dispersion. Note that the width of the entrance slit does not appear in Eq. (12.2.11), hence higher resolution implies lower luminosity, and conversely, as the slit width is changed. Meaburn has evaluated the $L \cdot R$ product for a wide variety of spectrometers, including prism, grating, and Fabry-Perot instruments. We evaluate this product for a selected set of instruments in Chapter 13.

If Eq. (12.2.11) is multiplied by B, the source brightness, the result is a flux-resolution product $F \cdot R$. This product is more useful than Eq. (12.2.11) when ϕ' is not a constant.

For extended sources we take B constant over the slit area. Although the $L \cdot R$ and $F \cdot R$ products increase as ϕ' increases, the brightness B' of the image is unchanged. If the image covers many detector elements, the exposure time is also the same.

For stellar sources the flux in the image of a given star at the entrance slit is constant and $B = C'/\phi'^2$, where C' is a constant and ϕ' is the diameter of the seeing disk. Therefore

$$F \cdot R = C\tau (D/\phi')(\lambda A d_2), \qquad (12.2.12)$$

where $C = \pi C'/4$. When the star image overfills the entrance slit, $\phi < \phi'$ and better seeing means a larger $F \cdot R$ product. When the star image is entirely within the slit, F' is constant and R is inversely proportional to ϕ'.

E. SPECTROMETER SPEED

The exposure time required to record a spectrum depends on the rate at which energy in a given spectral band is collected in a given area on the detector. For a spectrometer the illumination E of an image is defined as the spectral flux received at the detector per unit area. If the area on the detector is that of the reimaged slit, the spectral band is $\delta\lambda$ given by Eq. (12.2.2) and

$$E = F'\delta\lambda/\omega'h' = F'P/h'. \qquad (12.2.13)$$

This relation for E is identical to one for a quantity Bowen called the "speed." It is obvious that greater speed or illumination means shorter exposure times.

Taking $B = C'$ for an extended source and $B = C'/\phi'^2$ for a star, we find F' from Eqs. (12.2.7) and (12.2.10). The results are

$$F'(\text{extended}) = C\tau D^2\phi\phi', \qquad (12.2.14)$$

$$F'(\text{stellar}) = C\tau D^2(\phi/\phi'), \qquad (12.2.15)$$

where $C = \pi C'/4$ and $\phi = \phi'$ if the stellar source is entirely within the slit. Substituting F' into Eq. (12.2.13) and using Eqs. (12.2.1) and (12.2.2) gives

$$E_e = \frac{C\tau\delta\lambda}{rF_2^2} = \frac{C\tau\omega'P}{rF_2^2}, \qquad (12.2.16)$$

$$E_s = \frac{C\tau\delta\lambda}{r(F_2\phi')^2} = \frac{C\tau\phi DP}{F_2\phi'^2} \quad \text{(slit-limited)}, \qquad (12.2.17)$$

$$E_s = \frac{C\tau\delta\lambda}{r(F_2\phi')^2} = \frac{C\tau D^2 P}{r\omega'} \quad \text{(seeing-limited)}, \qquad (12.2.18)$$

where E_e and E_s denote extended source and star, respectively. If the spectrum of a stellar source is widened by trailing the image on the entrance slit, the relations for E_s must be multiplied by h'/H, where H is the height of the widened spectrum at the detector.

For unwidened stellar spectra, it is evident from Eq. (12.2.17) that better seeing means greater speed in the slit-limited case, in part because more light passes through the slit and in part because the image height h' is shorter. In the seeing-limited case all of the light passes through the slit and, given $\omega' \propto \phi'$, speed increases only in inverse proportion to improved seeing. For extended sources it is evident from Eq. (12.2.16) that seeing has no effect on speed.

It is important to note the dependence of speed on telescope diameter and camera focal ratio. For extended sources we see that speed is independent of diameter, and greater speed requires a faster camera. In the seeing-limited case for stellar sources, the speed is proportional to the telescope area. For stellar sources in the slit-limited case, the most usual situation with spectrometers on large telescopes, the speed is proportional to diameter and inversely proportional to the camera focal ratio.

It was pointed out by Bowen that scaling the size of a spectrometer in direct proportion to the telescope diameter does not change the speed, at the same spectral purity. This is easily shown by noting that $F'/\omega'h'$ is independent of D. Hence an increase in speed can be achieved only by using a spectrometer camera with a smaller focal ratio. This is one of the major reasons why much effort has gone into the design and construction of fast cameras.

In our discussion above we have assumed a stellar image with uniform brightness rather than a more realistic one with a bright center and fainter surrounding halo. The results found with a realistic profile are essentially the same as those above, and little is gained by introducing this refinement. For further discussion of spectrometer speed, the reader should consult the reference by Bowen.

F. CONCLUDING COMMENTS

We noted above that many spectroscopic observations of stellar sources, especially with large telescopes, are made in the slit-limited mode with the star image wider than the slit. To recover most of the light intercepted by the slit jaws, different types of so-called image slicers have been devised. Such a device, in effect, slices the image into several strips and places these end to end along the length of the slit. For an excellent discussion of the principles of image slicers and their practical realization, the reader should consult the reference by Hunten.

III. PARAMETERS FOR SLITLESS SPECTROMETERS

Based on the discussion in the preceding section, the relations for slitless spectrometers are easily found. In this section we give the important relations without extensive discussion. The major differences for the slitless mode are (1) the image size of a stellar source is set by atmospheric seeing or diffraction rather than a slit, (2) the anamorphic magnification r is one in all practical configurations, and (3) the diameter d_1 is the beam size at the dispersing element.

Thus the relations in Section II apply to slitless configurations if ϕ is replaced by ϕ' and r is set equal to one. With these changes the spectral purity is given by

$$\delta\lambda = \frac{\phi'}{A}\frac{D}{d_1} = \frac{\phi'}{A}\frac{f}{s}, \qquad (12.3.1)$$

where f is the telescope focal length and d_1 is the diameter of the dispersing element. In the objective mode, as shown in Fig. 12.2, $s = f$ and $d_1 = D$. The illumination of an image with spectral band $\delta\lambda$ is given by Eq. (12.2.18).

IV. PARAMETERS IN DIFFRACTION LIMIT

We noted in Section II that the relation for the spectral purity did not take into account the limit on image size set by diffraction. In this section we determine the form of the spectrometric parameters in the case where a monochromatic image in the focal plane of a spectrometer is diffraction-limited. The basic diffraction relations used in this discussion can be found in any introductory optics text and are given here without derivation.

We begin by noting that the normalized intensity of the Fraunhofer diffraction pattern for a slit of width b and length l, where $b \ll l$, is given by

$$I = (\sin \gamma / \gamma)^2,$$

where γ is the phase difference between two waves, one from the center of the slit and one from the edge. The path difference Δ between these two waves is shown in a cross section of the slit in Fig. 12.9, where $\Delta = (b/2) \sin \theta$ and θ is the angle between the z axis and the direction of the waves. In all cases of interest $\theta \ll 1$ and $\sin \theta$ is replaced by θ. The corresponding phase difference $\gamma = (2\pi/\lambda)\Delta = \pi\theta b/\lambda$.

For a square aperture of side d with collimated light incident normally on the aperture, the normalized intensity of the Fraunhofer diffraction pattern is the product of single-slit intensities,

$$I = \left(\frac{\sin \gamma}{\gamma}\right)^2 \left(\frac{\sin \gamma'}{\gamma'}\right)^2, \qquad (12.4.1)$$

where $\gamma = \pi\theta d/\lambda$, $\gamma' = \pi\theta' d/\lambda$, and θ' and θ are measured in perpendicular

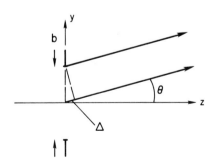

Fig. 12.9. Cross section of slit of width b with path difference $\Delta = (b/2) \sin \theta$ between center and marginal ray.

IV. Parameters in Diffraction Limit

directions from the z axis. A profile of the intensity pattern, in either the xz or yz plane, is shown by the solid curve in Fig. 12.10. The first minimum in the pattern is at $\gamma = \gamma' = \pi$, hence $\theta = \theta' = \lambda/d$.

Using Eq. (12.4.1), it is easy to show that $I = 0.405$ when $\gamma' = 0$ and $\theta = 0.5\lambda/d$, hence two images of equal intensity whose angular separation is λ/d are resolved according to the Rayleigh criterion. The net intensity at the point midway between the peaks is about 20% less than that at either peak. Note that the condition for resolution with a square aperture is similar to that given in Section 11.I for a circular aperture.

We now assume a perfect telescope-spectrometer system and assume the aperture of the spectrometer collimator is square. The FWHM of the diffraction image at the telescope focus is approximately λF. In the notation of Fig. 12.4 we take $\omega = \lambda F$ and $h = 2\lambda F$ as the effective slit dimensions. Substituting ω and h into Eq. (12.2.1) gives

$$\omega' = r\lambda F_2, \qquad h' = 2\lambda F_2. \tag{12.4.2}$$

The first relation in Eq. (12.4.2) can be used to find rF_2 for a detector whose pixel size Δ is matched to ω'. Assuming $2\Delta = \omega'$, as done following Eq. (12.2.1), we find $rF_2 = 65$ for a pixel size of 20 μm with $\lambda = 500$ nm. It is evident from this value of rF_2 that a spectrometer on a diffraction-limited telescope has no need for a Schmidt-type camera.

Following the procedure in Section II.A we take $\Delta l = \omega' = r\lambda F_2$ as the separation between two monochromatic images that are just resolved by

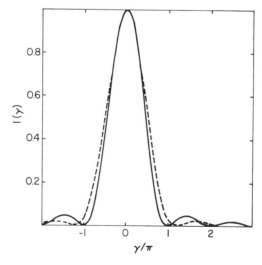

Fig. 12.10. Intensity profiles of diffraction image from square aperture (solid curve) and circular aperture (dashed curve).

the Rayleigh criterion. The difference in wavelength $\delta\lambda_0$ between these two images, according to Eq. (12.1.1), is given by $\delta\lambda_0 = \Delta l/f_2 A$. Therefore

$$\delta\lambda_0 = \frac{r\lambda F_2}{f_2 A} = \frac{r\lambda}{Ad_1} = \frac{\lambda}{Ad_2}, \tag{12.4.3}$$

$$R_0 = \frac{\lambda}{\delta\lambda_0} = \frac{Ad_1}{r} = Ad_2. \tag{12.4.4}$$

Note that the change in Eqs. (12.2.2) and (12.2.3) to get the relations for the diffraction-limited case is simply a substitution of λ for ϕD.

Following this procedure, and noting that $\phi' = 2\lambda/D$, we can transform the remaining parameters in Section II into their diffraction-limited counterparts. The results are

$$U = \pi\lambda^2/2, \tag{12.4.5}$$

$$L \cdot R = \tau(\pi\lambda^2/2)Ad_2, \tag{12.4.6}$$

$$F \cdot R = (C\tau D^2/2)Ad_2, \tag{12.4.7}$$

$$E_s = \tau FP/h' = C\tau D^2 \delta\lambda_0/r(2\lambda F_2)^2. \tag{12.4.8}$$

In writing Eq. (12.4.7) we take $B = C'(D/2\lambda)^2$ and $C = C'\pi/4$. This form of B assumes, in effect, that the image brightness is constant over the core of the diffraction-limited image.

The relations in Eqs. (12.4.5)–(12.4.8) apply only to stellar sources. For extended sources ϕ' is the angular height of the entrance slit and Eqs. (12.2.11) and (12.2.16) are correct as given, provided the brightness over the slit is uniform in the diffraction limit.

It is possible to make various comparisons between the relations for diffraction-limited and seeing-limited cases, and we give two such comparisons. Consider the 2.4-m Hubble Space Telescope and a ground-based 4-m telescope. Taking the ratio of Eqs. (12.4.7) and (12.2.12) gives

$$\frac{F \cdot R(\text{HST})}{F \cdot R(4\text{-m})} = \frac{\tau_0 D_0^2 \phi'}{2\tau\lambda D},$$

where the parameters with subscript 0 refer to HST and ϕ' is the diameter of the seeing-limited image. Taking $\tau_0 = \tau$, $\phi' = 1$ arc-sec, and $\lambda = 500$ nm gives an $F \cdot R$ ratio of about 7. Note that a ground-based telescope in the 25-m class has an $F \cdot R$ product comparable to that of the space telescope. A complete analysis of the detectability of a faint source takes into account the sky background and detector characteristics; we defer this discussion to Chapter 16.

Comparing Eqs. (12.2.3) and (12.4.4), we find

$$R/R_0 = \lambda/\phi D = K. \tag{12.4.9}$$

Assuming $\phi = 1$ arc-sec we find that $K = \lambda(\mu m)/5D(m)$. For $\lambda = 500$ nm = 0.5 μm and $D = 1$ m, $K = 0.1$. Thus the resolution at visible wavelengths at which spectrometers on ground-based telescopes are often used is well below the resolution that is theoretically possible. At infrared wavelengths, where λ is larger and seeing is typically better, it is possible that R can approach R_0.

REFERENCES

Bowen, I. (1952). *Astrophys. J.* **116**, 1.
Hoag, A., and Schroeder, D. (1970). *Publ. Astron. Soc. Pac.* **82**, 1141.
Jacquinot, P. (1954). *J. Opt. Soc. Am.* **44**, 761.
Schroeder, D. (1974). "Methods of Experimental Physics: Astrophysics," Vol. 12, Part A: "Optical and Infrared," Chap. 10. Academic Press, New York.

BIBLIOGRAPHY

For discussions of Fraunhofer diffraction see any of the texts listed in the bibliography in Chapter 2.

Bowen, I. (1962). "Stars and Stellar Systems II, Astronomical Techniques," Chap. 2. Univ. of Chicago Press, Chicago, Illinois.
Hunten, D. (1974). "Methods of Experimental Physics: Astrophysics," Vol. 12, Part A: "Optical and Infrared," Chap. 4. Academic Press, New York.
James, J., and Sternberg, R. (1969). "The Design of Optical Spectrometers." Chapman & Hall, London.
Meaburn, J. (1976). "Detection and Spectrometry of Faint Light." Reidel, Dordrecht, Netherlands.

Chapter 13 | Dispersing Elements and Systems

We now turn our attention to specific dispersing elements, discussing in turn in this and following sections the prism, diffraction grating, and Fabry-Perot interferometer. We give expressions for the angular dispersion and resolution for each of these and point out other characteristics that are important in their application. We also discuss briefly the characteristics of the Fourier spectrometer.

I. DISPERSING PRISM

The angular dispersion A of a prism used at minimum deviation is derived in Section 3.II. A dispersing prism is generally used at or near minimum deviation, defined as that orientation at which θ in Fig. 3.6 is a minimum and rays inside the prism are parallel to the base. From Eq. (3.2.8) we have

$$A = \frac{d\beta}{d\lambda} = \frac{t}{a}\frac{dn}{d\lambda}, \quad (13.1.1)$$

where t is the base length, a the beamwidth into and out of the prism, and $dn/d\lambda$ the rate of change of index with wavelength. Because a is the same on either side of the prism, there is no anamorphic magnification and $r = 1$.

I. Dispersing Prism

Substituting A into Eq. (12.4.4) and noting that $d_2 = a$ gives the limiting spectral resolution of a single prism in a square beam as

$$R_0 = t\, dn/d\lambda. \tag{13.1.2}$$

If there are k identical prisms in series, then Eqs. (13.1.1) and (13.1.2) are each multiplied by k.

Dispersion curves for three glasses selected from the Schott glass catalog are shown in Fig. 13.1. The form of these curves is typical of those for all transparent glasses, with $dn/d\lambda$ going approximately as the inverse cube of the wavelength. Taking UBK7 at $\lambda = 500$ nm as an example, we find $dn/d\lambda = 0.066\ \mu\text{m}^{-1}$, and thus $R_0 = 10\,000$ for $t = 150$ mm. This base length corresponds to a beam diameter of about 100 mm for a 60° prism.

Prisms are often used in the objective mode on Schmidt telescopes in the 1-m class to obtain spectra suitable for classification. In this configuration K from Eq. (12.4.9) is approximately 0.1 and therefore $R = 1000$ for a UBK7 prism whose base is 150 mm. This resolution is sufficient for this mode, where the plate factor P is typically about 150 Å/mm.

As we show in Section II, the resolution of a prism is low compared to what is possible with a grating large enough to accept the same beam

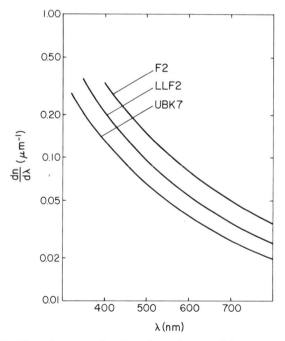

Fig. 13.1. Dispersion curves for three glasses from the Schott glass catalog.

diameter. For this reason, prisms are little used as primary dispersing elements in slit spectrometers, having been largely replaced by gratings. Prisms are, however, often used as cross-dispersers in spectrometers that use gratings for the primary dispersion. For the objective mode, a prism is much more efficient than a grating, for reasons made clear when we discuss grating efficiency.

II. DIFFRACTION GRATING; BASIC RELATIONS

The diffraction grating is the primary dispersing element in most astronomical spectrometers, where it has the advantage of significantly larger limiting resolution than a prism of comparable size. A grating is also versatile in the spectral formats it can provide and can be quite efficient over a reasonable spectral range, though usually not as efficient as a prism. We discuss these, and other, characteristics of gratings in this section and the following one. A number of additional references which discuss gratings are given at the end of the chapter.

A. GRATING EQUATION

The starting point for our discussion of gratings is the well-known grating equation, the derivation of which is found in any introductory optics text. We also derive this equation from the point of view of Fermat's principle in Chapter 14. The grating equation is

$$m\lambda = \sigma(\sin \beta \pm \sin \alpha), \qquad (13.2.1)$$

where m is the order number, σ is the distance between successive, equally spaced grooves or slits, and α and β are angles of incidence and diffraction, respectively, measured from the normal to the grating surface. The parameter σ is also called the grating constant. The plus sign in Eq. (13.2.1) applies to a reflection grating, the minus sign to a transmission grating.

Schematics of grating cross sections are shown in Fig. 13.2 for both a reflection and a transmission grating. The plane defined by the incident ray and normal to the grating surface is in the plane of the paper in Fig. 13.2, with the grating grooves perpendicular to this plane. The angles α and β are governed by the same sign convention as that given for i and i' in Chapter 2. For a reflection grating α and β have the same signs if they are on the same side of the grating normal, while for a transmission grating they have the same signs if the diffracted ray crosses the normal at the point of diffraction. In each of the diagrams in Fig. 13.2, α and β have the same

II. Diffraction Grating; Basic Relations

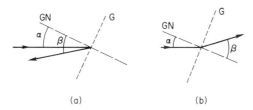

Fig. 13.2. Schematic showing angles of incidence α and diffraction β for (a) reflection grating and (b) transmission grating. See the discussion following Eq. (13.2.1) for the sign convention.

signs. Note that $m = 0$ for a reflection grating when $\alpha = -\beta$, while for a transmission grating this condition holds when $\alpha = \beta$.

B. ANGULAR DISPERSION

The angular dispersion follows directly from Eq. (13.2.1) by holding α constant and differentiating with respect to β, with the result

$$A = \frac{d\beta}{d\lambda} = \frac{m}{\sigma \cos \beta}, \qquad (13.2.2a)$$

or

$$A = \frac{\sin \beta + \sin \alpha}{\lambda \cos \beta}. \qquad (13.2.2b)$$

Unless otherwise noted, the following discussion refers to reflection gratings, hence the plus sign in Eq. (13.2.2b).

From Eq. (13.2.2a) we see that angular dispersion in a given order m is a function of σ and β. Changing A in this case means choosing a grating with a different grating constant and/or using the grating at a different angle of diffraction.

We see from Eq. (13.2.2b) that A, at a given wavelength, is set entirely by the angles α and β, independent of m and σ. Thus a given angular dispersion can be obtained with many combinations of m and σ, provided the angles at the grating are unchanged and m/σ is constant. Recognition of this fact led to the development of coarsely ruled reflection gratings specifically designed to achieve high angular dispersion by making α and β large, typically about 60°. Such gratings, called echelles, have groove densities in the range of 300 to 30 per millimeter with values of m in the range of 10 to 100 for visible light. On the other hand, typical first-order gratings have groove densities in the range 300 to 1200 per millimeter. A typical echelle and a grating used in first order are compared in a later section.

C. ANAMORPHIC MAGNIFICATION

The relation for anamorphic magnification is derived from Eq. (13.2.1) by holding λ constant and finding the change in β for a change in α. The result is $\cos \beta \, d\beta + \cos \alpha \, d\alpha = 0$, and it follows that

$$r = \frac{|d\beta|}{|d\alpha|} = \frac{\cos \alpha}{\cos \beta} = \frac{d_1}{d_2}. \tag{13.2.3}$$

The relation between the beamwidths and angles is derived from the geometry in Fig. 13.3, with $d_1/\cos \alpha = d_2/\cos \beta$.

Because r is in the denominator of Eq. (12.2.3), it is clear that the choice $r < 1$, hence $\beta < \alpha$, gives higher resolution. This condition, in turn, means that the grating normal is more nearly in the direction of the camera than of the collimator. If the grating is to accept all of the light from the collimator, it follows that $W = d_1/\cos \alpha$ is required, where W is the width of the grating.

D. SPECTRAL RESOLUTION

Substituting Eq. (13.2.2a) into Eq. (12.4.4) gives the limiting spectral resolution as

$$R_0 = \frac{md_2}{\sigma \cos \beta} = \frac{mW}{\sigma} = mN, \tag{13.2.4a}$$

where N is the total number of grooves in the grating width W. The relation between W, d_2, and β is shown in Fig. 13.3. Writing Eq. (13.2.4a) in terms of angles, we get

$$R_0 = W(\sin \beta + \sin \alpha)/\lambda. \tag{13.2.4b}$$

From Eq. (13.2.4b) we see that R_0 is directly proportional to the grating width for a given pair of angles. Using the geometry in Fig. 13.3, we also see that the numerator in Eq. (13.2.4b) has a simple geometric interpretation; it is the total path difference between the marginal rays spanning the grating width, and R_0 is the number of wavelengths in this path difference.

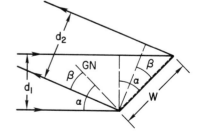

Fig. 13.3. Change in beamwidth due to anamorphic magnification of grating. See Eq. (13.2.3).

II. Diffraction Grating; Basic Relations

Table 13.1
Equations for Reflection Gratings

$$m\lambda = \sigma(\sin\beta + \sin\alpha)$$

$$A = \frac{d\beta}{d\lambda} = \frac{m}{\sigma\cos\beta} = \frac{\sin\beta + \sin\alpha}{\lambda\cos\beta}$$

$$R_0 = \frac{W(\sin\beta + \sin\alpha)}{\lambda}$$

$$R = \frac{W(\sin\beta + \sin\alpha)}{\phi D}$$

The resolution for the seeing-limited case is

$$R = [W(\sin\beta + \sin\alpha)]/\phi D. \qquad (13.2.5)$$

It is evident from this relation why there has been a concerted effort to produce larger reflection gratings and echelles as the size of telescopes has increased.

For ease of reference, the important grating relations are brought together in Table 13.1.

E. FREE SPECTRAL RANGE

For a given pair of α and β, the grating equation is satisfied for all wavelengths for which m is an integer. Thus there are two wavelengths in successive orders, λ and λ', for which we have $m\lambda' = (m+1)\lambda$. The wavelength difference $\Delta\lambda = \lambda' - \lambda$ is called the *free spectral range*, where

$$\Delta\lambda = \lambda/m. \qquad (13.2.6)$$

The two wavelengths are diffracted in the same direction, and confusion is the result unless one is rejected by a filter or they are separated with a cross-disperser. Both techniques are used to eliminate the wavelength overlap, with a cross-disperser most often used when m is large and a filter when m is small.

F. COMPARISON OF GRATING AND ECHELLE

As illustration of the relations given above, we now consider two specific gratings and give their characteristics at a common wavelength near 500 nm. The reflection gratings chosen are assigned the following parameters:

	m	$1/\sigma$ (mm)	δ (°)
First-order grating	1	1200	17.5
Echelle	45	79	63.5
$\alpha = \delta + \theta,$	$\beta = \delta - \theta,$	$\theta > 0$ if $\alpha > \beta$	

The reader can verify for each that the grating equation is satisfied for a wavelength near 500 nm when $\theta = 0$. The parameters chosen are typical of those for a first-order grating and echelle.

The angle δ, the so-called blaze angle, is introduced here because it is one of the key grating parameters; its significance is discussed fully in the next section. At this point we simply note that grating efficiency is a maximum when α and β are chosen as given. A sketch of the geometry showing the relation between these angles is shown in Fig. 13.4.

It is useful at this point to write the relations in Table 13.1 for the special case where $\theta = 0$, hence $\alpha = \beta = \delta$. This defines the so-called Littrow configuration, for which we get

$$m\lambda = 2\sigma \sin \delta, \tag{13.2.7}$$

$$A = 2 \tan \delta / \lambda, \tag{13.2.8}$$

$$R_0 = 2W \sin \delta / \lambda = 2d_1 \tan \delta / \lambda, \tag{13.2.9}$$

$$R = 2d_1 \tan \delta / \phi D. \tag{13.2.10}$$

Although the Littrow configuration is not a practical one for a reflection grating, the angle θ is small in most grating spectrometers. Thus these relations are useful for calculating good first approximations to the true values, and comparisons of different gratings are easily made. For our chosen grating and echelle, R and R_0 in the Littrow mode are 6.36 times larger for the echelle, at the same d_1 and λ. It is also worth noting that W is 2.14 times larger for the echelle in the same configuration.

Values derived from relations in Table 13.1 are given in Table 13.2 for our chosen grating and echelle, with $d_1 = 100$ mm, $\phi = 1$ arc-sec, and $D = 4$ m. From the entries there we see that R of the grating changes very little with changing θ, while for the echelle the change is proportionately much larger. The trend in the values of W/d_1 follows a similar pattern.

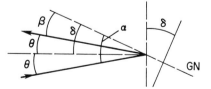

Fig. 13.4. Relation between blaze angle δ, grating normal GN, and angles of incidence and diffraction.

III. Grating Efficiency

Table 13.2

Characteristics of Typical Grating and Echelle[a]

	Resolution in Littrow configuration	
	Grating	Echelle
R_0	1.3E5	8.0E5
R	3.1E3	2.0E4

	Parameters in non-Littrow configuration			
	Grating		Echelle	
θ (°)	$\phi DR/d_1$	W/d_1	$\phi DR/d_1$	W/d_1
0	0.630	1.05	4.01	2.24
2	0.637	1.06	4.31	2.41
4	0.644	1.07	4.67	2.61
6	0.652	1.09	5.08	2.86

[a] Parameters of grating and echelle are given in the text; $d_1 = 100$ mm, $\phi = 1$ arc-sec, $D = 4$ m.

Compared to R_0 of the UBK7 prism given in the previous section, we also see from Table 13.2 that the grating and echelle have values that are 12.5 and 80 times larger. The increases in resolution potential and achievable resolution for a given slit width are obviously significant.

The increase in resolution R with increasing positive θ, and thus $\alpha > \beta$, is a result that merits further comment. This result is a bit surprising at first glance because the angular dispersion, according to Eq. (13.2.2a), decreases with increasing θ. At the same time, however, the anamorphic magnification r decreases at a higher rate. Hence R, which is proportional to A/r, has the behavior noted above. We discuss the consequences of this further in terms of the $L \cdot R$ product in a later section.

III. GRATING EFFICIENCY

The absolute efficiency of a grating is defined as the fraction of the energy at wavelength λ incident on the grating directed into a given diffracted order, where the fraction of energy diffracted into a given order is determined by the so-called blaze function. In this section we examine the characteristics of the grating blaze function.

An exact treatment of grating efficiency is carried out by applying Maxwell's equations of electromagnetic theory to the interaction of light with a grating surface. For our purposes the exact theory is not necessary and a scalar approximation is used instead. The polarization of the incident light is ignored in the scalar theory, but the results obtained from it are a good first approximation to the exact results. Important differences between the two approaches are noted. For a discussion of the exact theory, the reader should consult the book by Hutley and other references given at the end of the chapter. Good introductions to the results of the scalar theory can be found in handbooks by Bausch and Lomb and Jobin Yvon.

A. BLAZE FUNCTION

Consider a reflection grating consisting of a number N of equally spaced facets, or grooves, of width b with center-to-center spacing σ, as shown in Fig. 13.5. For a beam of collimated light incident at angle α, the normalized intensity of the diffracted wave is

$$I = \mathrm{IF} \cdot \mathrm{BF} = \frac{\sin^2 N\gamma'}{\sin^2 \gamma'} \cdot \frac{\sin^2 \gamma}{\gamma^2}, \quad (13.3.1)$$

where $2\gamma'$ is the phase difference between the centers of adjacent grooves and γ is the phase difference between the center and edge of one groove. The derivation of Eq. (13.3.1) is found in any introductory optics text.

The relations for the phase differences are

$$\gamma' = \frac{\pi\sigma}{\lambda}(\sin \beta + \sin \alpha), \quad \gamma = \frac{\pi b}{\lambda}(\sin \beta + \sin \alpha).$$

The interference factor (IF) is a maximum when $\gamma' = m\pi$, where m is an integer and IF(max) = 1. Substituting $m\pi$ for γ' gives the grating equation (13.2.1).

The factor BF is the intensity of a single-slit diffraction pattern and is called the *blaze function*. The blaze function is a maximum when $\gamma = 0$, hence $\alpha = -\beta$. This corresponds to $m = 0$ in the grating equation. As noted in Section 12.IV, the first minimum in this pattern occurs at $\gamma = \pi$. We also noted there that BF = 0.405 when γ is $\pi/2$, and from Fig. 12.10 we see that the angular width of BF at this intensity level is λ/b.

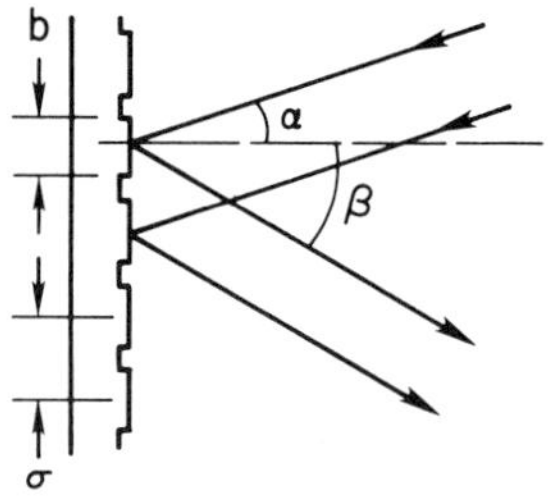

Fig. 13.5. Schematic of unblazed reflection grating with groove width b and separation σ.

III. Grating Efficiency

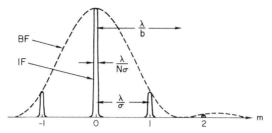

Fig. 13.6. Intensity pattern of single diffracted wavelength for grating in Fig. 13.5. BF, Blaze function; IF, interference factor.

A sketch of the intensity pattern for a single wavelength is shown in Fig. 13.6, where it is evident that the pattern is simply the interference (IF) term modulated by the blaze function BF. From Fig. 13.6 we see that this grating directs most of the light to zero order and thus the efficiency in any other order is low.

In order to increase the efficiency in a dispersed order it is necessary, in effect, to move the blaze function along the axis in Fig. 13.6 until its peak coincides with an interference maximum in the dispersed order. This is done by tilting each facet of the grating by an angle δ relative to the surface, as shown in Fig. 13.7. For a tilted groove the phase difference from center to edge is

$$\gamma = \frac{\pi \sigma \cos \delta}{\lambda} (\sin \theta - \sin \theta')$$

$$= \frac{\pi \sigma \cos \delta}{\lambda} [\sin(\beta - \delta) + \sin(\alpha - \delta)]. \qquad (13.3.2)$$

where the width $b = \sigma \cos \delta$ for a groove profile with right-angle corners.

The shifted blaze function BF is a maximum when $\gamma = 0$, and from Eq. (13.3.2) we get $\alpha + \beta = 2\delta$ at the blaze peak. From Fig. 13.7 we get $\alpha = \delta + \theta$ and $\beta = \delta - \theta'$, where $\theta' = \theta$ at the blaze peak. The wavelength is called the *blaze wavelength* λ_0, and Eq. (13.2.1) at this wavelength is $m\lambda_0 = 2\sigma \sin \delta \cos \theta$.

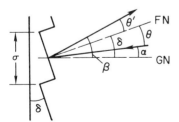

Fig. 13.7. Reflection grating of Fig. 13.5 with tilted facets to shift blaze function BF by angle 2δ.

The net effect of "blazing" a grating is to get maximum efficiency in the direction in which light would be reflected by specular reflection in the absence of diffraction. Note that in a Littrow configuration $\alpha = \beta = \delta$, and the light returns on itself. The groove profile for a transmission grating is similar to that shown in Fig. 13.7, where each facet is a long narrow prism. In this case the blaze peak is in the direction in which light would be refracted in the absence of diffraction.

The diffracted efficiency for a wavelength not at the blaze peak is determined by the value of BF for that wavelength in the diffracted direction. Figure 13.8 shows two wavelengths, λ_+ and λ_-, on opposite sides of λ_0. The grating equation for these two wavelengths is

$$m\lambda_\pm = \sigma \sin \alpha + \sigma \sin(\beta_0 \pm \varepsilon_\pm)$$
$$= m\lambda_0 - \sigma \sin \beta_0 (1 - \cos \varepsilon_\pm) \pm \sigma \cos \beta_0 \sin \varepsilon_\pm, \quad (13.3.3)$$

where β_0 is the diffraction angle at the blaze peak.

We now choose $\varepsilon_\pm$ as that pair of angles for which BF = 0.405, noting again that the width of a monochromatic single-slit diffraction peak between these points is λ/b. For a grating groove $b = \sigma \cos \delta$ and $\varepsilon_\pm = \lambda_\pm / 2\sigma \cos \delta$. This relation for $\varepsilon_\pm$ is accurate only in the small-angle approximation, but for most gratings of interest this accuracy is sufficient.

Assuming $\varepsilon_\pm$ is small and $\cos \beta_0 \simeq \cos \delta$, we drop the middle term in Eq. (13.3.3) and find the wavelength limits

$$\lambda_\pm = \frac{m\lambda_0}{m \mp \frac{1}{2}}, \quad (13.3.4)$$

$$\lambda_+ - \lambda_- = \frac{m\lambda_0}{m^2 - \frac{1}{4}} = \frac{\lambda_0}{m}, \quad \text{for large } m. \quad (13.3.5)$$

In first order, $\lambda_+ = 2\lambda_0$ and $\lambda_- = 2\lambda_0/3$. The asymmetry in these wavelengths about the peak is expected because the blaze function is broader for the longer wavelength.

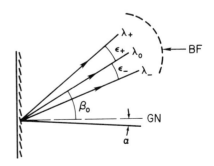

Fig. 13.8. Schematic of blaze function envelope with blaze peak at $\lambda = \lambda_0$. See Eq. (13.3.3).

III. Grating Efficiency   247

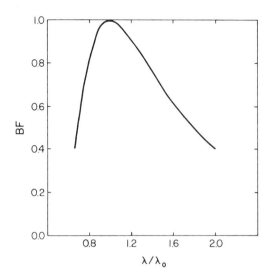

Fig. 13.9. Blaze function for grating with $m = 1$. See Eq. (13.3.4) and following discussion.

The overall blaze function for wavelengths λ_- to λ_+ in first order is shown in Fig. 13.9. The values of γ used to calculate BF at other wavelengths are given by $\gamma = m\pi(\lambda_0 - \lambda)/\lambda$. At the wavelengths $\lambda_\pm$ given by Eq. (13.3.4), the reader can verify that $\gamma = \pm \pi/2$, as expected. Published efficiency curves of first-order gratings with σ greater than a few wavelengths are similar to the curve in Fig. 13.9, though there is a difference between different polarizations of light. As examples, see efficiency curves in the Bausch and Lomb Grating Handbook.

For an echelle with large m, the asymmetry in the wavelengths given by Eq. (13.3.4) is much smaller and usually can be ignored. The blaze function spanning many echelle orders has the form shown in Fig. 13.10 for the Littrow configuration. The width of each peak in the scalloped curve is given by Eq. (13.3.5).

One important feature of the efficiency curve in Fig. 13.10 is that each peak covers one free spectral range of the echelle spectrum, as is seen by

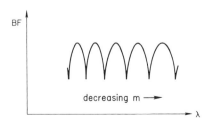

Fig. 13.10. Blaze function for echelle with $m \gg 1$ in Littrow configuration. See Eq. (13.3.5).

comparing Eqs. (13.2.6) and (13.3.5). Hence each wavelength over an extended range is diffracted with an efficiency no less than 40% that of the blaze wavelength.

When an echelle with rectangular grooves is illuminated at an angle $\alpha > \delta$, part of the width of each facet is not used and the effective groove width is less than $\sigma \cos \delta$. From the geometry in Fig. 13.11 we find that the effective groove width b' is

$$b' = \sigma \cos \alpha / \cos \theta. \qquad (13.3.6)$$

Because the groove width is smaller, the angular width of the blaze function is larger. At the same time, according to Eq. (13.2.2a), the angular dispersion is smaller and one free spectral range has a smaller angular spread. Combining these two effects, we find that the fraction of BF spanned by one free spectral range is a factor $\cos \alpha / \cos \beta$ smaller than in the Littrow mode. Thus variation in the blaze function across a free spectral range is less pronounced when $\alpha > \beta$ than in the Littrow mode.

At the same time, however, the efficiency at the peak is reduced because some light at the blaze peak is now diffracted into neighboring orders. As shown by Bottema, the peak efficiency is reduced by a factor of $\cos \alpha / \cos \beta$. The net result of all these effects is a broader, but lower, efficiency profile. Profiles of a single blaze peak are shown in Fig. 13.12 for $\theta = 0$ and $\theta = 5°$, at a blaze angle of 63.5°. Examination of these curves shows that the efficiency at the ends of a free spectral range is somewhat higher at $\theta = 5°$ than at $\theta = 0°$. The average efficiency across the range, however, decreases as θ increases from zero.

The effects of groove shadowing for $\alpha > \beta$ are present for all gratings with rectangular grooves, but are much less significant for gratings with small blaze angle. For our examples in Section II.F, $\cos \alpha / \cos \beta$ for $\theta = 5°$ is 0.95 for the grating and 0.70 for the echelle.

When light is incident on an echelle at an angle $\alpha < \beta$, each facet is fully illuminated but a fraction of the light is sent back to the collimator. The effective groove width in this case is again given by Eq. (13.3.6), as is evident from Fig. 13.11 when the arrows on the rays are reversed. Thus the blaze peak is again broadened. In this case, however, the angular dispersion is

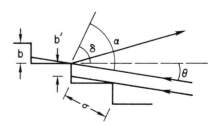

Fig. 13.11. Cross section of echelle showing effective facet width b' for $\alpha > \delta$. See Eq. (13.3.6).

III. Grating Efficiency

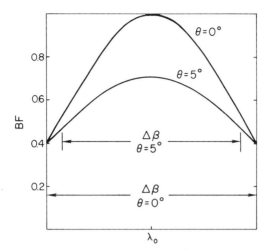

Fig. 13.12. Blaze function for single echelle order at two values of θ. Angular width of the free spectral range is $\Delta\beta$. See the discussion following Eq. (13.3.6).

larger for increasing β and one free spectral range spans the same portion of the blaze peak. The peak efficiency in this case is reduced by the factor $\cos\beta/\cos\alpha$.

A final point to be made on grating efficiency is the extent to which polarization effects are important. Let P and S represent the diffracted efficiency for light polarized parallel and perpendicular, respectively, to the length of the grating grooves. We define fractional polarization as $|P-S|/(P+S)$. The size of this fraction is determined by the size of λ/b, hence the effect is smaller for a grating with larger grooves.

Unpublished measurements at $\lambda = 480$ nm on a tan $\delta = 2$ echelle with 79 grooves/mm give a fractional polarization of 0.02, where $\lambda/b = 0.085$. The conclusion is that the curves in Figs. 13.10 and 13.12 are valid for either plane of polarization, to a good approximation. For a grating with 600 grooves/mm, the difference in the efficiency curves is significant, as seen from examples in the Bausch and Lomb Grating Handbook.

B. LUMINOSITY-RESOLUTION PRODUCT

The luminosity-resolution product for a grating follows from a substitution of A into Eq. (12.2.11). The result in terms of d_1 and the blaze angle δ is

$$L \cdot R = \tau \frac{\pi D \phi' d_1}{2} \frac{\sin\delta \cos\theta}{\cos(\delta+\theta)}. \qquad (13.3.7)$$

For a grating with a small blaze angle, this product is essentially constant over a range of θ of many degrees. With our grating example above, this product changes by less than 6% when θ is changed from 0 to 10°. For a typical grating, therefore, it is sufficient to evaluate Eq. (13.3.4) at δ and ignore the dependence on θ.

For an echelle with a large blaze angle, on the other hand, the dependence of $L \cdot R$ on θ is an important one. Over the range of θ in Table 13.2, this product changes by about 27% for our echelle example. The size of this change indicates that a closer look at $L \cdot R$ for echelles is in order.

It appears from Eq. (13.3.7) that $L \cdot R$ can be made as large as desired for a given d_1 by choosing $\alpha = \delta + \theta$ near $\pi/2$. This is not feasible in practice, however, because the width W needed to collect all the light is larger than the width of any practical grating. An additional complication at these large angles is that the diffracted beamwidth d_2 is significantly greater than d_1. This extended width makes the design of Schmidt cameras more difficult at the focal ratios usually required.

As an illustration of these comments, an R4 ($\tan \delta = 4$) echelle with $\theta = 5°$ has $W/d_1 = 6.39$ and $d_2/d_1 = 2.08$. By comparison, an R2 ($\tan \delta = 2$) echelle has $W/d_1 = 2.73$ and $d_2/d_1 = 1.43$ at the same θ. At the present time, given the state of large-grating technology and camera design, the available R2 echelles are the best choices for getting large $L \cdot R$ with a grating.

Large, commercially available, R2 echelles have a width-to-length ratio of two. From the data in Table 13.2, we see that $W/d_1 > 2$, hence a beam unvignetted by the echelle can cover its width but not its length. This is illustrated in Fig. 13.13, where W_c is the projected echelle width seen from the collimator and H is the height of the grooves. In terms of W, height $H = W/2$ and $W_c = W \cos \alpha$.

Given this fixed width-to-length ratio, the designer of an echelle spectrometer has three options:

(1) Choose $d_1 = W_c$, hence no vignetting at the echelle.
(2) Choose $d_1 > W_c$, but not larger than H, and accept some vignetting.
(3) Choose $\alpha < \beta$ in a way to make $d_1 = H = W_c$.

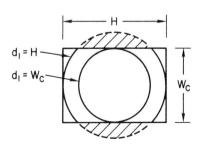

Fig. 13.13. Overfilling of echelle to increase $L \cdot R$ product. See Section 13.III.B.

In options (1) and (2), we assume α is larger than β.

Option (1) is acceptable, but does not make full use of the echelle surface. Option (3) has the same $L \cdot R$ as option (1), but is constrained to cos $\alpha = 0.5$, hence $\alpha < \beta$. As noted in Section VI.G, the variation in efficiency across a free spectral range is larger when $\alpha < \beta$, hence (1) is preferable even though its $L \cdot R$ is the same.

Option (2) is a trade-off between larger d_1 than (1) and smaller τ due to vignetting. A larger d_1 translates into a wider entrance slit for the same projected slit width ω', a result that follows from Eq. (12.2.1a) written as $\omega' = rDf_2\phi/d_1$. A full analysis shows that the gain in light at the entrance slit more than offsets the light loss at the echelle, up to the point where $d_1 = H$. For angle $\theta = 5°$ the net gain amounts to about 17%. It is worth noting that this is true for a circular beam but not for a square one. If the beam is square, the two effects cancel one another and $L \cdot R$ for option (2) is the same as for (1).

C. CONCLUDING COMMENTS

In the preceding sections we discussed efficiency as it applies to classically ruled gratings with rectangular groove shape. Another type of grating now available is the holographic grating, produced by recording interference fringes from two laser beams in a photosensitive material. This type of grating has a different groove shape than the ruled type, with efficiency a sensitive function of groove shape and number of grooves per millimeter.

Although holographic gratings are competitive with ruled gratings when used in first order, they have not been made at the groove densities characteristic of echelles. First-order gratings are typically used at spectral resolutions of 1E3 to 1E4, and in this range either type of grating can be used. For moderately high resolution in the range 3E4 to 1E5, the echelle is clearly superior for broad spectral coverage. References for holographic gratings are given at the end of the chapter.

IV. FABRY–PEROT INTERFEROMETER

The Fabry–Perot spectrometer is an important instrument for astronomical observations that require very high spectral resolution of limited spectral ranges and/or large angular fields. As we show, the Fabry–Perot is superior to a grating or echelle instrument in these cases.

The basic theory of the Fabry-Perot can be found in any optics text, and extensive discussions of its application in astronomy are given by Roesler and Meaburn. Our discussion of the Fabry-Perot is brief, with results given without derivation, and the reader should consult the references for details. Although relations for the Fabry-Perot are often given in terms of wave number, the reciprocal of the wavelength, we choose to give all results in terms of wavelength.

A schematic diagram of a Fabry-Perot spectrometer is shown in Fig. 13.14. For a Fabry-Perot with material of index n, usually air, between the interferometer plates, the wavelengths transmitted with maximum intensity are given by the relation

$$m\lambda = 2nd \cos \theta, \tag{13.4.1}$$

where m is an integer order number, d the spacing between the plates, and θ the angle of incidence. Because the Fabry-Perot is axially symmetric, the pattern in the focal plane of a camera for a broad monochromatic source is a set of concentric rings. Each ring is a separate order, and the usual mode of operation is one in which a circular aperture isolates the central order. All the results to follow assume $\theta = 0$.

The free spectral range $\Delta\lambda$ is found by the procedure used in Section II.E, with a similar result,

$$\Delta\lambda = \lambda/m = \lambda^2/2nd. \tag{13.4.2}$$

Because m is usually large, a filter or predisperser is required to eliminate all unwanted orders.

The spectral purity is determined by the characteristics of the interferometer transmission function, the so-called Airy function, and the result is

$$\delta\lambda = \Delta\lambda/N = \lambda^2/2Nnd, \tag{13.4.3}$$

where N is called the finesse. The parameter N depends on the plate quality and the reflectance of the coatings. For plates of high quality with good coatings, a typical value of N is in the range 30-50. From Eq. (13.4.3) the spectral resolution is

$$R = \lambda/\delta\lambda = N\lambda/\Delta\lambda = 2Nnd/\lambda. \tag{13.4.4}$$

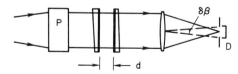

Fig. 13.14. Schematic of Fabry-Perot spectrometer. P, Prefilter-collimator combination; D, detector.

IV. Fabry–Perot Interferometer 253

The angular diameter $\delta\beta$ of an axial hole that accepts all of the light in the full width at half-maximum (FWHM) of the central order is given by

$$\delta\beta = (8/R)^{1/2}. \quad (13.4.5)$$

Substituting Eq. (13.4.5) into Eq. (12.2.8) gives the étendue for a circular hole as

$$U = \frac{\pi d_1^2}{4} \frac{\pi(\delta\beta)^2}{4} = \frac{\pi d_1^2}{4} \frac{2\pi}{R}, \quad (13.4.6)$$

where the anamorphic magnification is one, hence $d_2 = d_1$. Therefore the luminosity-resolution product for the Fabry–Perot is given by

$$L \cdot R = 2\pi\tau \cdot \pi d_1^2/4. \quad (13.4.7)$$

The transmittance τ is the product of the transmittance τ_0 of the optical elements and the average of the Airy function over the FWHM of the central order, where the latter factor is about 0.8.

We now want to compare $L \cdot R$ for the Fabry–Perot with that for a typical echelle with the same collimator beam diameter. With the substitution of $d_1\delta\alpha'$ for $D\phi'$ in Eq. (13.3.7), we get

$$\frac{L \cdot R(\mathrm{FP})}{L \cdot R(\mathrm{E})} = \frac{\tau_{\mathrm{FP}}}{\tau_{\mathrm{E}}} \frac{\pi}{\delta\alpha' \tan\delta}, \quad (13.4.8)$$

with the echelle in the Littrow configuration. Note that $\delta\alpha'$, the angle subtended by the slit length at the collimator, is D/d_1 times larger than the slit length projected on the sky.

For a well-defined projected slit, assume the largest practical $\delta\alpha'$ is about 1°. Taking $\tan\delta = 2$, the ratio in Eq. (13.4.8) is then about 100 for the same transmittances and beam diameters in each instrument. If we assume the largest echelle beam is twice that of the largest Fabry–Perot beam, the $L \cdot R$ ratio is reduced to 25, still substantially larger for the Fabry–Perot. The value of $\delta\alpha'$ assumed here is obviously appropriate for a large extended source. If $D = 4$ m and $d_1 = 200$ mm, the corresponding angle on the sky is 180 arc-sec.

We now compare the Fabry–Perot and echelle for a stellar source, with each spectrometer on a 4-m telescope. The beam diameters are 100 and 200 mm, respectively, for the Fabry–Perot and echelle, and the stellar seeing disk is taken as 1 arc-sec. For an entrance slit width equal to the seeing disk diameter, the resolution of the echelle is 4.0E4. This resolution is far below the limiting resolution of 1.6E6 and the echelle could be used to get higher resolution, say 10 times higher, but only at the expense of luminosity.

For the Fabry–Perot we assume $R = 4.0\mathrm{E}5$, a resolution appropriate for the observation of interstellar absorption lines, for example. With this R

we find $\delta\beta = 15$ arc-min, which projects to about 22 arc-sec on the sky. Light loss at the entrance hole is clearly not a problem with this telescope-spectrometer and the system is far from any reasonable seeing limit at this resolution.

These two examples illustrate the kinds of observations for which the Fabry-Perot is especially well suited, very high spectral resolution for sources of small angular size or moderate to high resolution for sources of large angular size.

The principal limitation of the Fabry-Perot is set by its mode of operation, that of scanning successive spectral elements in sequence. This scanning is achieved by varying either n or d in Eq. (13.4.1), with the scan rate set by the desired signal-to-noise ratio. The echelle spectrometer, on the other hand, is most often used in a mode in which the flux in all spectral elements over a wide wavelength range is recorded simultaneously. With solid-state area detectors of high efficiency, the echelle spectrograph, with its multiplex advantage, can overcome much of its etendue disadvantage.

V. FOURIER TRANSFORM SPECTROMETER

Although a Fourier transform spectrometer, hereafter denoted by FTS, is not a dispersing system in the sense defined in Section 12.I, it gives an output from which the spectrum can be derived. Because the FTS is used to get spectral data, especially in the infrared, a brief discussion of its characteristics is in order. Results are given without derivation; for discussions of the theory of the FTS the reader should consult the references at the end of the chapter.

An FTS is basically a scanning Michelson interferometer with collimated light as the input, as shown schematically in Fig. 13.15. The input beam is divided by a beamsplitter with approximately one-half going to each of the

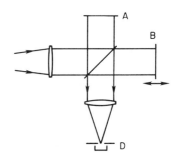

Fig. 13.15. Schematic of Fourier transform spectrometer. A, Fixed mirror; B, movable mirror; D, detector.

V. Fourier Transform Spectrometer

mirrors A and B. The light reflected from the mirrors is again divided by the beamsplitter, with approximately one-half of the original beam recombined and sent to the detector D.

For a collimated monochromatic beam the intensity at D is determined by the path difference between the two arms of the interferometer. Let x_a and x_b denote the respective distances from the center of the beamsplitter to the center of each mirror. The OPD between the recombined beams, assuming $n = 1$, is then $2(x_b - x_a) = 2\Delta x$, and the phase difference is $2k\Delta x$, where $k = 2\pi/\lambda$. Note that Eq. (13.4.1) for the Fabry–Perot applies to the Michelson if d is replaced by Δx.

The fraction $T(k, \Delta x)$ of the incident beam in the output beam is given by

$$T(k, \Delta x) = \tfrac{1}{2}[1 + \cos(2k\Delta x)]. \qquad (13.5.1)$$

If the beams from the two arms are in phase $[\cos(2k\Delta x) = 1]$ we have $T = 1$; if the beams are π out of phase $[\cos(2k\Delta x) = -1]$ then $T = 0$. The relation in Eq. (13.5.1) is a direct consequence of two-beam interference.

Given an incident beam whose spectrum is $I(k)$, the flux F in the output beam is

$$F(\Delta x) = C \int I(k) T(k, \Delta x)\, dk$$

$$= \text{constant} + \frac{C}{2} \int I(k) \cos(2k\Delta x)\, dk, \qquad (13.5.2)$$

where C is a constant. The output $F(\Delta x)$ for all Δx from a minimum value, usually zero, to the maximum value is called the interferogram. The integral in the second line of Eq. (13.5.2) is the Fourier cosine transform of the spectrum. From the theory of Fourier transforms, the transform of the recorded flux $F(\Delta x)$ is the spectrum.

The spectral resolution R_0 achievable with the FTS is directly proportional to the maximum Δx in the scan producing the interferogram and is given by

$$R_0 = \lambda/\delta\lambda_0 = 4\Delta x(\max)/\lambda. \qquad (13.5.3)$$

If, for example, $\Delta x(\max) = 10$ cm, then $R_0 = 4\text{E}5$ at $\lambda = 1000$ nm.

Because Eq. (13.4.1) applies to both the Michelson and Fabry–Perot interferometers, the relations for $\delta\beta$, U, and $L \cdot R$ in Eqs. (13.4.5)–(13.4.7) also apply to an FTS. As a consequence, the resolution achieved in practice is smaller than that given in Eq. (13.5.3) by about a factor of two, when the angular diameter of the exit aperture is set according to Eq. (13.4.5).

Given the similarities between an FTS and a Fabry-Perot spectrometer and the discussion in the preceding section, it is evident that the FTS is well suited to observations requiring very high spectral resolution. It is also evident that an FTS has a significant $L \cdot R$ advantage over an echelle. The principal advantage of an FTS over a scanning Fabry-Perot is that all of the light in the passband of interest is being recorded all of the time. Thus the étendue advantage of an FTS over an echelle is maintained at high resolution.

REFERENCES

Bottema, M. (1980). *Proc. Conf. Gratings Periodic Structures*, SPIE **240**, 171.
"Diffraction Grating Handbook" (1970). Bausch & Lomb, Rochester, New York.
"Handbook of Diffraction Gratings, Ruled and Holographic." Jobin-Yvon, Metuchen, New Jersey.
"Optical Glass." Schott Optical Glass, Duryea, Pennsylvania.

BIBLIOGRAPHY

Diffraction gratings

Hutley, M. (1982). "Diffraction Gratings." Academic Press, New York.
Loewen, E. (1983). "Applied Optics and Optical Engineering," Vol. 9, Chap. 2. Academic Press, New York.
Maystre, D. (1984). "Progress in Optics," Vol. 21, Chap. 1. North-Holland, Amsterdam.
Richardson, D. (1969). "Applied Optics and Optical Engineering," Vol. 5, Chap. 2. Academic Press, New York.
Stroke, G. (1963). "Progress in Optics," Vol. 2, Chap. 1. North-Holland, Amsterdam.

Fabry-Perot spectroscopy

Meaburn, J. (1976). "Detection and Spectrometry of Faint Light." Reidel, Dordrecht, Netherlands.
Roesler, F. (1974). "Methods of Experimental Physics: Astrophysics," Vol. 12, Part A: "Optical and Infrared," Chap. 12. Academic Press, New York.
Vaughan, A. (1967). "Annual Review of Astronomy and Astrophysics," Vol. 5, p. 139. Annual Reviews, Palo Alto, California.

Fourier transform spectroscopy

Connes, P. (1970). "Annual Review of Astronomy and Astrophysics," Vol. 8, p. 209. Annual Reviews, Palo Alto, California.
Ridgeway, S., and Brault, J. (1984). "Annual Review of Astronomy and Astrophysics," Vol. 22, p. 291. Annual Reviews, Palo Alto, California.
Schnopper, H., and Thompson, R. (1974). "Methods of Experimental Physics: Astrophysics," Vol. 12, Part A: "Optical and Infrared," Chap. 11. Academic Press, New York.

Chapter 14 | Grating Aberrations; Concave Grating Spectrometers

Most astronomical spectrometers use a grating or echelle as the primary dispersing element. As noted in Chapter 13, this choice is made for flexibility in choice of spectrum format and resolution to match modern detectors and for the ability to get broad spectral coverage at good efficiency. In this chapter we discuss the limitations on grating performance from the point of view of geometric optics, both for the grating itself and for selected concave grating instruments.

The limitations of a grating spectrometer are determined by two principal factors: the aberrations due to the collimator and camera optics, and the aberrations introduced by the grating. Grating aberrations are determined by the type of grating surface, whether it is plane or spherical, and the focal ratio of the incident beam. In most spectrometers the incident light is collimated, but there are systems in which the incident light is a convergent beam.

In this chapter we use Fermat's principle to derive the aberrations of a grating surface, either reflecting or refracting. By combining these results with the aberrations of the other optics in a spectrometer, we can determine overall spectrometer performance. This system analysis is given for concave grating spectrometers in this chapter and plane grating instruments in the following chapter.

The approach followed in deriving the grating aberrations is similar to those used by Beutler, Namioka, and Welford. Each uses a different notation and the reader should note these differences when comparing results given by different authors. Our approach parallels that given in Chapter 5 for a general refracting surface, with those results modified to include the grating characteristics.

I. APPLICATION OF FERMAT'S PRINCIPLE TO GRATING SURFACE

Although a plane reflection grating is the element of choice in most grating spectrometers, we choose to set up a more general case in order to derive the general grating equation. After showing that this equation for a transmission grating is similar to that for a reflection grating, we carry out subsequent derivations of aberrations for the reflection case only. During this discussion it will become evident why plane gratings are preferred over spherical gratings for most applications.

A sketch of a grating surface is shown in Fig. 14.1, with the origin of the coordinate system at the vertex of the surface. The grating grooves are taken parallel to the y axis, with the separation between adjacent grooves, measured perpendicular to the yz plane, given by σ. The medium in front of the surface has index n; the medium on the other side has index n'. The object and image points are at Q and Q', respectively, and an arbitrary ray from Q intersects the grating surface at $B(x, y, z)$. The equation of the surface is given by Eq. (5.1.1) with K and b each set equal to zero. With this simplification we restrict our analysis to plane or spherical grating surfaces.

Unlike the situation in Chapter 5, the grating surface is not rotationally symmetric about the z-axis. Thus we locate Q, as shown in Fig. 14.1, distances h and w from the xz and yz planes, respectively. The chief ray from Q to 0 makes angle γ with the xz plane, and its projection on the xz plane makes

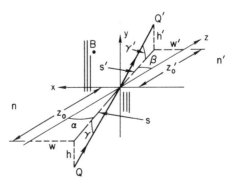

Fig. 14.1. Coordinate system for grating with center at origin and rulings parallel to y axis. Q and Q' are object and image points, respectively; point B is on the grating surface.

I. Application of Fermat's Principle to Grating Surface

angle α with the z axis. The image point Q' and the refracted chief ray are defined in a similar way with primed quantities, except that β is used as the counterpart of α.

It is important to note here that most spectrometers used on telescopes have $\gamma = 0$ at the center of the entrance slit. This type of spectrometer is the so-called in-plane design, in which the incident and diffracted chief rays are in the xz plane. For an in-plane spectrometer with a long slit there is a range of γ and, as we show below, the result is spectral line curvature. If γ is not zero at the slit center, then the spectrometer is an off-plane design. In this case the spectral lines from a long slit source are both tilted and curved. The origin of these effects is discussed below.

Proceeding now with Fermat's principle applied to the arbitrary ray shown in Fig. 14.1, we write the optical path length OPL between Q and Q' as

$$\text{OPL} = n[QB] + n'[BQ'] + (m\lambda/\sigma)x, \tag{14.1.1}$$

where Fermat's principle is satisfied when $\delta(\text{OPL}) = 0$ for any change in B on the surface. The wavelength λ in Eq. (14.1.1) is the vacuum wavelength.

The first two terms in Eq. (14.1.1) are the same as those in Eq. (5.1.2); the additional term is what makes the surface a grating. This is most easily seen by assuming x changes by σ, a step from one groove to an adjacent one. Because of this change in x there is an accompanying $\delta(\text{OPL}) = m\lambda$, and a corresponding phase difference of $2\pi m$ between the two points. If m is an integer, diffracted rays from these points are in phase and the effective change in OPL is zero. Thus these two rays interfere constructively at the image. Note that this conclusion is based on the assumption that the sum of the first two terms is the same for both rays, as required by Fermat's principle for a nongrating surface.

The procedure from this point on is similar to that carried out in Chapter 5. The line segments in Eq. (14.1.1) are written in terms of the parameters in Fig. 14.1 and each, in turn, is expanded in powers of x and y.

From the geometry in Fig. 14.1 we find

$$[QB] = [(x-w)^2 + (y-h)^2 + (z_0 - z)^2]^{1/2},$$
$$[BQ'] = [(x-w')^2 + (y-h')^2 + (z_0' - z)^2]^{1/2}, \tag{14.1.2}$$

where

$$w = s \sin \alpha, \quad h = s \tan \gamma, \quad z_0 = s \cos \alpha,$$
$$w' = s' \cos \beta, \quad h' = s' \tan \gamma', \quad z_0' = s' \cos \beta. \tag{14.1.3}$$

The distances s and s' in Eqs. (14.1.3) are measured along the projections of $[QB]$ and $[BQ']$ on the xz plane, respectively. The usual sign conventions apply to all parameters, with all angles in Fig. 14.1 positive, s, w, h, and z_0 negative, and the primed distances positive.

A. GENERAL GRATING EQUATION

To derive the grating equation we need only the linear terms in the expansion of Eq. (14.1.1). Substituting Eqs. (14.1.3) into Eqs. (14.1.2), we find that the OPL can be written as

$$\text{OPL} = n's' - ns + y(n \sin \gamma - n' \sin \gamma')$$
$$+ x[(m\lambda/\sigma) + n \cos \gamma \sin \alpha - n' \cos \gamma' \sin \beta]$$
$$+ \text{terms in higher powers of } x \text{ and } y. \qquad (14.1.4)$$

Terms in higher powers lead to aberrations and their forms are given in a subsequent section.

Taking the partial derivatives of OPL with respect to x and y and setting each equal to zero gives

$$n \sin \gamma = n' \sin \gamma', \qquad (14.1.5)$$

$$m\lambda = \sigma(n' \cos \gamma' \sin \beta - n \cos \gamma \sin \alpha). \qquad (14.1.6)$$

These equations are the grating counterpart of Snell's law for a general surface of revolution. Equation (14.1.5) applies in the yz plane and is simply Snell's law, while Eq. (14.1.6) is the grating equation.

For a reflection grating $n' = -n$, hence $\gamma' = -\gamma$ and

$$m\lambda = -n\sigma \cos \gamma (\sin \beta + \sin \alpha), \qquad (14.1.7)$$

where $n = 1$ for light incident in the $+z$ direction. Note that β is a function of γ for constant α and λ, hence the image of a long straight slit parallel to the y axis is curved. The details of this curvature are discussed in the next section.

For a transmission grating surface we choose $\gamma = 0$ and get $m\lambda = \sigma(n' \sin \beta - n \sin \alpha)$. Applying this relation at the grating surface and Snell's law at the other gives, assuming plane-parallel surfaces,

$$m\lambda = n\sigma(\sin \beta - \sin \alpha), \qquad (14.1.8)$$

where α is the angle of incidence at the first surface, β the angle of diffraction following the second surface, and n the index of the medium, usually air, in which the grating is located. The index n is positive for light directed in the $+z$ direction. Note that the index of the blank is absent from Eq. (14.1.8). Thus the transmission grating is, in effect, a diffracting element of negligible thickness.

There is also an element called a grism in which a transmission grating is put on one surface of a prism. The relation in Eq. (14.1.8) applies to the grating if the combination is treated as a prism nearly in contact with a grating of negligible thickness.

B. SPECTRUM LINE CURVATURE AND TILT

The curvature of the image of a straight slit is a consequence of the dependence of β on γ noted following Eq. (14.1.7). At constant α and λ we get

$$\frac{d\beta}{d\gamma} = \frac{\sin \gamma(\sin \alpha + \sin \beta)}{\cos \gamma \cos \beta} = \tan \gamma \cdot \lambda A, \quad (14.1.9)$$

where A is the angular dispersion of the grating for nonzero γ. For small γ the total change in β between $\gamma = 0$ and some largest γ_0 is found by integrating Eq. (14.1.9) between these limits. The result, expressed as $\Delta\beta$, is

$$\Delta\beta = (\gamma_0^2/2)\lambda A.$$

Note that $\Delta\beta > 0$, hence the change in β is toward longer wavelengths. In the camera focal plane, as shown in Fig. 14.2, the linear displacement from a straight image is $f_2\Delta\beta$, where

$$f_2\Delta\beta = \frac{\lambda A}{2f_2}(f_2\gamma_0)^2 = \frac{(f_2\gamma_0)^2}{2\rho}, \quad (14.1.10)$$

where $\rho = f_2/\lambda A$ is the radius of curvature of the image.

The slope at a point on the curved image, relative to a straight image, is $d\beta/d\gamma$ as given in Eq. (14.1.9). Hence a short entrance slit for which γ is not zero at the center is imaged as a tilted, though nearly straight, line. This assumes, of course, that the entrance slit is not tilted out of the yz plane. Line tilt of this type is present in all off-plane spectrometer designs.

Note that line curvature and tilt are present even if the grating and spectrometer are otherwise free of aberrations. If the instrument has large astigmatism, this curvature is superposed on any curvature of the astigmatic image. The reader can consult the reference by Welford for details of this combination.

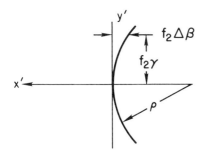

Fig. 14.2. Spectrum line curvature in spectrometer focal plane with radius of curvature ρ. See Eq. (14.1.10).

It is also important to note that curvature and tilt are larger for an echelle than for a typical grating of small blaze angle, in direct proportion to the angular dispersion. This factor is an important one to take into account in the design of any spectrometer, but especially of an echelle instrument.

II. GRATING ABERRATIONS

The aberrations of a grating are found from the higher-order terms in Eq. (14.1.4). In the general case, these terms contain factors depending on $\sin \gamma$ and $\sin \gamma'$ to various powers. Most spectrometers, however, are of the in-plane design and so we choose to set γ and γ' to zero. The factors left out of the aberration coefficients with this choice are of order γ^2 and γ'^2 smaller than those that remain.

Another simplification results if we consider only the case of a reflection grating. This is sufficient because the aberration coefficients for a plane reflection grating also apply to a transmission grating, given the discussion leading to Eq. (14.1.8).

We also give only the squared and cubed terms in the OPL expansion; thus the only aberration coefficients we derive are those of astigmatism and coma. Spherical aberration, which comes from the fourth-power terms, is negligible in most cases. If it is significant for a specific grating type, its value is given in our discussion for that type.

Following our discussion in Section 5.I, we define G as the optical path difference between the general and chief rays and find

$$G = -\frac{nx^2}{2}\left[\frac{\cos^2 \beta}{s'}+\frac{\cos^2 \alpha}{s}-\frac{\cos \beta + \cos \alpha}{R}\right]$$
$$-\frac{ny^2}{2}\left[\frac{1}{s'}+\frac{1}{s}-\frac{\cos \beta + \cos \alpha}{R}\right]$$
$$-\frac{nx^3}{2}\left[\frac{\sin \beta}{s'}\left(\frac{\cos^2 \beta}{s'}-\frac{\cos \beta}{R}\right)+\frac{\sin \alpha}{s}\left(\frac{\cos^2 \alpha}{s}-\frac{\cos \alpha}{R}\right)\right]$$
$$-\frac{nxy^2}{2}\left[\frac{\sin \beta}{s'}\left(\frac{1}{s'}-\frac{\cos \beta}{R}\right)+\frac{\sin \alpha}{s}\left(\frac{1}{s}-\frac{\cos \alpha}{R}\right)\right]$$
$$= A_1 x^2 + A'_1 y^2 + A_2 x^3 + A'_2 xy^2. \tag{14.2.1}$$

Note that Eq. (14.2.1) does not contain the linear terms from Eq. (14.1.4); these are zero from the grating equation.

A comparison of terms in Eq. (14.2.1) with corresponding ones in Eq. (5.1.5) shows that they are the same provided x and y are interchanged, θ and θ' are replaced by α and β, respectively, and n' is replaced by $-n$.

Because of this correspondence we can use many of the results derived in Chapter 5. One important difference in the results to follow is that the small-angle approximation is not applied, except in selected cases.

The locations of the astigmatic images are found by setting either A_1 or A_1' in Eq. (14.2.1) to zero. Setting $A_1 = 0$ gives

$$\frac{\cos^2 \beta}{s_t'} + \frac{\cos^2 \alpha}{s} = \frac{\cos \beta + \cos \alpha}{R}, \quad (14.2.2)$$

where s_t' is the location of the tangential astigmatic image. This is a line image perpendicular to the xz plane, hence parallel to the grating grooves, as shown in Fig. 14.3. The detector must be located at this image if grating astigmatism is not to degrade the spectral resolution. For a plane transmission grating the right side of Eq. (14.2.2) is zero and the plus on the left side is changed to minus.

Setting A_1' to zero gives s_s', the location of the sagittal astigmatic image. Taking this expression and Eq. (14.2.2) we find that the separation $\Delta s'$ between the astigmatic image is

$$\frac{\Delta s'}{s_s' s_t'} = \left[\frac{(\sin^2 \beta)(\cos \beta + \cos \alpha)}{R \cos^2 \beta} + \frac{\sin^2 \alpha - \sin^2 \beta}{s \cos^2 \beta} \right]$$

$$= -\frac{2A_1'}{n} = \frac{2A_1'}{n'}, \quad (14.2.3)$$

where the relation between $\Delta s'$ and A_1' is the grating counterpart of Eq. (5.2.5) and follows by substituting s_t' for s' in A_1'. The analog of Eq. (5.2.6) for the transverse astigmatism is

$$\text{TAS} = \frac{\Delta s'}{s_s'} y = \frac{2A_1' y s_t'}{n'}, \quad (14.2.4)$$

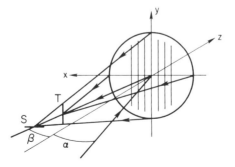

Fig. 14.3. Tangential (T) and sagittal (S) astigmatic images. Maximum spectral resolution is achieved with the detector at the T image.

where the total length of the tangential image is $2|\text{TAS}|$. The length of the line image in units of the grating groove length is $|\text{TAS}|/y = 2A_1's_t'/n'$. We now return to Eq. (14.2.2) and select several combinations of s_t' and s which satisfy that equation. Commonly used names are given for each combination, or mounting as it is usually called, with the results given in Table 14.1.

The characteristics of each mounting in Table 14.1 are easily described in terms of the object and image locations. For the Rowland mounting the entrance slit and image lie on the Rowland circle, a circle of diameter R tangent to the concave grating at its vertex. The grating in the Wadsworth mounting is illuminated with collimated light and the curved focal surface is roughly a distance $R/2$ from the grating vertex. A convergent or divergent light bundle is incident on the grating in the Monk-Gillieson mounting. The object and image lie on opposite sides of a reflection grating, and the minus sign in Table 14.1. applies. For a transmission grating the plus sign in Table 14.1 applies and object and image are on the same side.

We get the transverse astigmatism for each mounting in Table 14.1 by substituting R, s, and s_t' into Eqs. (14.2.3) and (14.2.4). The results are given in Table 14.2, where TAS is expressed in units of grating groove length. There is no entry for a plane grating in collimated light because its A_1' is zero.

From the entries in Table 14.2 we see that astigmatism is zero for the Rowland mounting only when $\alpha = \beta = 0$, corresponding to the zero order. The Wadsworth mounting has zero astigmatism on the grating normal and small astigmatism over a limited range on either side. For the Monk-Gillieson mounting, the astigmatism is zero when $\beta = \pm\alpha$, where the minus sign gives the zero order for a plane reflection grating and the plus sign is the zero order for a transmission grating. For either plane grating, therefore, there is a direction in which astigmatism is zero at a diffracted wavelength given by $m\lambda = 2\sigma \sin \beta$.

Table 14.1

Grating Mountings

Grating	s	s_t'	Name
Concave	$R \cos \alpha$	$R \cos \beta$	Rowland
Concave	∞	$\dfrac{R \cos^2 \beta}{\cos \beta + \cos \alpha}$	Wadsworth
Plane	∞	∞	
Plane	s	$\pm s \left(\dfrac{\cos^2 \beta}{\cos^2 \alpha} \right)$	Monk-Gillieson

II. Grating Aberrations

Table 14.2

Astigmatism of Grating Mountings

Rowland: $\dfrac{TAS}{y} = \sin^2 \beta + \sin^2 \alpha \left(\dfrac{\cos \beta}{\cos \alpha} \right)$

Wadsworth: $\dfrac{TAS}{y} = \sin^2 \beta$

Monk-Gillieson: $\dfrac{TAS}{y} = \dfrac{\sin^2 \beta - \sin^2 \alpha}{\cos^2 \alpha}$

$A_1' = \dfrac{n'}{2s_t'} \dfrac{TAS}{y}$

The coefficients A_2 and A_2' in Eq. (14.2.1), with $-n'$ substituted for n, are

$$A_2 = \dfrac{n'}{2} \left(\dfrac{\cos^2 \alpha}{s} - \dfrac{\cos \alpha}{R} \right) \left(\dfrac{\sin \alpha}{s} - \dfrac{\sin \beta}{s'} \right), \qquad (14.2.5)$$

$$A_2' = \dfrac{n'}{2} \left[\dfrac{\sin \beta}{s'} \left(\dfrac{1}{s'} - \dfrac{\cos \beta}{R} \right) + \dfrac{\sin \alpha}{s} \left(\dfrac{1}{s} - \dfrac{\cos \alpha}{R} \right) \right], \qquad (14.2.6)$$

where Eq. (14.2.2) is used to simplify Eq. (14.2.5). Relations analogous to Eqs. (5.3.4) and (5.3.5) for these coefficients are

$$TA_x = \dfrac{s_t'}{n'} (3A_2 x^2 + A_2' y^2), \qquad TA_y = \dfrac{s_t'}{n'} 2A_2' xy,$$

where TA denotes the transverse aberration. The expressions for the transverse tangential and sagittal coma are

$$TTC = 3A_2 x^2 s_t'/n', \qquad TSC = A_2' y^2 s_t'/n'. \qquad (14.2.7)$$

With reference to Fig. 5.5 with x and y interchanged, the relations in Eqs. (14.2.7) give the coma in the direction parallel to the x axis. If coma is present, its effect is to degrade the spectral resolution.

We now evaluate Eq. (14.2.5) for each of the grating mountings in Table 14.1. It turns out that A_2' is comparable in size to A_2 and we give its value only for the one mounting in which $A_2 = 0$. The results are shown in Table 14.3.

From the entries in Table 14.3 we see that TTC for the Rowland mounting is zero and TSC is small because A_2' goes as the cube of factors that are usually small. Coma is zero for the Rowland mounting only in zero order. The Wadsworth mounting has zero coma on the grating normal and small coma over a limited range on either side. For the Monk-Gillieson mounting

14. Grating Aberrations; Concave Grating Spectrometers

Table 14.3

Coma of Grating Mountings

Rowland:	$A_2 = 0$
	$A_2' = \dfrac{n'}{2R^2}(\sin\beta\tan^2\beta + \sin\alpha\tan^2\alpha)$
Wadsworth:	$A_2 = \dfrac{n'\sin\beta}{2R^2}\left[\dfrac{\cos\alpha(\cos\beta + \cos\alpha)}{\cos^2\beta}\right]$
Monk-Gillieson:	$A_2 = \pm\dfrac{n'\cos^2\alpha}{2s^2}\left[\sin\alpha \pm \sin\beta\left(\dfrac{\cos^2\alpha}{\cos^2\beta}\right)\right]$

the plus and minus signs apply to reflection and transmission gratings, respectively, and coma is zero only in zero order.

Spherical aberration is negligible in all practical configurations of the Rowland and Monk-Gillieson mountings. The spherical aberration coefficient for the Wadsworth mounting on the grating normal is given by

$$A_3 = \frac{n}{8R^3}\cos^2\alpha(1 + \cos\alpha), \qquad (14.2.8)$$

with the transverse spherical aberration given by Eq. (5.4.2). Its size, to a good approximation, is the same as that of a sphere in collimated light.

The aberration coefficients for the concave grating mountings are derived assuming the pupil is at the grating. Given the limited sizes of concave gratings, this is the only practical way of using the grating. A plane grating has no preferred axis and the aberration coefficients are independent of the pupil location.

For each of the grating mountings, we now determine the curvature of the image surface on which the tangential astigmatic images are located. This is easily done by applying Eqs. (5.7.1) and (5.7.2) to the relations for s_t' in Table 14.1. The curvatures obtained are given in Table 14.4, where the

Table 14.4

Tangential Image Surface Curvatures

Rowland:	$\kappa_t = -2/R$
Wadsworth:	$\kappa_t = -(2/R)(1 + \tfrac{3}{2}\cos\alpha)$
Monk-Gillieson:	$\kappa_t = \pm(3\cos^2\alpha)/s$

plus and minus signs for the Monk–Gillieson mounting apply to the reflection and transmission cases, respectively.

Note that the curvature for the Rowland mounting is exact because the images lie on the Rowland circle. For the other mountings in Table 14.4 the curvatures are only approximations to the exact curvatures, though the relations given are adequate for most configurations of these mountings. The image surface for each mounting is concave as seen from the grating.

All of the aberration relations needed for analysis of the different mountings are now in hand. The plane grating in collimated light has no aberrations, and aberration analysis of spectrometers with this mounting is reduced to considering the collimator and camera optics and any anamorphic magnification of the grating. The rest of this chapter is a discussion of the characteristics of selected concave grating spectrometers. Discussion of the characteristics of plane grating instruments follows in Chapter 15.

III. CONCAVE GRATING MOUNTINGS

In this section we look further at the characteristics of the two concave grating mountings introduced in the previous sesction. Although these mountings are used little, if at all, for stellar spectroscopy with ground-based telescopes, they are used often for ultraviolet spectroscopy from space. In this spectral region they are practical alternatives to plane grating spectrometers. For information on other concave grating mountings, such as the Seya–Namioka monochromator, the references should be consulted.

A. ROWLAND MOUNTINGS

The Rowland mounting was probably the first type of grating spectrometer used for astronomical spectroscopy. Although it was adequate for getting solar spectra, it was a failure in stellar spectroscopy, primarily because of its astigmatism. As an example of the size of the astigmatism, we take $\lambda = 500$ nm for a first-order grating with 600 grooves/mm and a diameter of 100 mm. From the relation in Table 14.2 we get $TAS/y = 0.090$ for $\alpha = 0$ and $TAS/y = 0.045$ for $\alpha = \beta$; hence image lengths are 9 and 4.5 mm, respectively, for a point source at the entrance slit.

Note that these lengths depend only on the grating size and are independent of the output beam focal ratio. Given these image lengths, the speed of a Rowland spectrometer is much less than that of a stigmatic spectrometer at the same camera focal ratio. This is evident from Eq. (12.2.17), with the widening factor h'/H included for the Rowland instrument.

Rowland spectrometers are used in the extreme ultraviolet, where reflection efficiencies are low and the number of optical surfaces must be kept to an absolute minimum. In this spectral range both near-normal and grazing incidence mountings are used. Because of the shorter wavelengths, astigmatism is tolerable for near-normal mountings. For the grating example above, the astigmatism is smaller by a factor of 25 at $\lambda = 100$ nm for the same grating diameter. A comparison of different Rowland mountings suitable as scanning monochromators in the ultraviolet has been given by Namioka.

Grazing incidence spectrometers based on the Rowland circle are used at still shorter wavelengths, and their large astigmatism is reduced by using ellipsoidal and toroidal concave gratings. For further discussion of grazing incidence instruments, the reader should consult the references at the end of the chapter.

B. WADSWORTH MOUNTING

The Wadsworth mounting was a successful replacement for the Rowland mounting in the early days of stellar spectroscopy, primarily because it has significantly smaller astigmatism. With the same visible grating used above, the Wadsworth can be made stigmatic at $\lambda = 500$ nm by choosing $\sin \alpha = 0.3$. At wavelengths of 400 and 600 nm we find that TAS/y is 0.0036, and the image length is 0.36 mm for the same grating diameter. This image length increases for wavelengths farther from the stigmatic wavelength, roughly as the square of the wavelength difference.

Comparing the astigmatic image lengths of the Wadsworth and Rowland mountings, it is evident that any decrease in transmittance of the Wadsworth is more than compensated by its greater speed. This advantage of the Wadsworth over the Rowland mounting holds at shorter wavelengths, provided high-efficiency reflecting films are available.

With $\sin \alpha$ given above, it is straightforward to calculate the transverse coma and spherical aberration. Using A_2 from Table 14.3 and Eqs. (14.2.7) we find TTC $= 0.175(d/F) \sin \beta$, where d is the grating diameter and F the grating focal ratio. With $d = 100$ mm, as above, and $F = 10$, we find the magnitude of TTC $= 0.1$ mm at 400 and 600 nm. Because coma is proportional to $\sin \beta$, its size is linearly dependent on the difference between the actual and corrected wavelengths.

Using Eqs. (14.2.8) and (5.4.2) we find TSA $= 0.014d/F^2$, hence for our chosen d and F we get TSA $= 14$ μm. Compared to the size of the coma, it is evident that spherical aberration is negligible over most of the spectral region centered on 500 nm.

Because the grating in the Wadsworth mounting has collimated light incident, a separate collimator is required. A schematic diagram of a Wadsworth spectrometer is shown in Fig. 14.4, with a flat fold mirror in series with an on-axis collimator mirror. In a properly designed system for a Cassegrain telescope, the fold mirror is entirely inside the shadow of the secondary and does not vignette the collimated beam.

The collimator has no off-axis aberrations in this arrangement and spherical aberration is also zero if the mirror is a paraboloid. If the mirror is spherical, its spherical aberration adds to that of the grating. Other possible collimators include an off-axis parabolic mirror and a tilted concave mirror. The former is free of on-axis aberrations, while the latter has off-axis aberrations. The optical properties of a tilted spherical collimator and a concave grating have been calculated by Namioka and Seya, and these references can be consulted for details. They consider both collimated and noncollimated light incident on the grating.

C. INVERSE WADSWORTH

In the standard Wadsworth mounting the grating is both the disperser and the camera. In the inverse Wadsworth mounting the slit is at the focus of the grating, which is both the disperser and the collimator, and a separate camera is required. The general aberration relations, not given here, are derived using the procedure given above. From these relations it turns out that coma and astigmatism are smallest at a given wavelength when α is zero. This is not surprising in view of Fermat's principle; compared to the standard Wadsworth, whose aberrations are zero on the grating normal, the light rays are simply reversed in direction.

Assume a camera of focal length f that is aberration-free and a concave grating for which the collimator focal ratio is F. The transverse aberrations of the inverse Wadsworth at $\alpha = 0$ are given by

$$\text{TAS} = \frac{f}{2F}\left[\frac{\sin^2\beta \sin^2(\beta/2)}{\cos^2\beta}\right], \quad \text{TTC} = \frac{3f}{16F^2}\left[\frac{\sin\beta \sin^2(\beta/2)}{\cos^2\beta}\right],$$

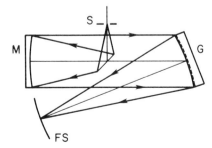

Fig. 14.4. Schematic of Wadsworth spectrometer. S, Entrance slit; M, collimator; G, concave grating; FS, focal surface.

where the total length of the tangential image is 2TAS. Given that $\sin \beta = m\lambda/\sigma$ when $\alpha = 0$, we see that coma and astigmatism increase roughly in proportion to the cube and fourth power of the wavelength, respectively. With $F = 10$, $f = 1000$ mm, and the same 600 grooves/mm grating used above, the reader can verify that the image length and TTC are approximately 0.22 mm and 14 μm, respectively, at a wavelength of 500 nm. Compared to a standard Wadsworth of the same focal length, the average coma and astigmatism are significantly less in the visible and ultraviolet spectral regions. The spherical aberration coefficient of the inverse Wadsworth is the same as that of a standard Wadsworth.

From the aberration relations above, we see that there is freedom to choose a shorter camera focal length and thereby reduce the transverse aberrations. A Schmidt camera, with its corrector plate close to the grating, is a practical choice for a fast camera in an inverse Wadsworth. Both the incident and diffracted light from the grating pass through the corrector in this arrangement and, with the stop at the grating, the corrector size is essentially the same as that of the grating.

D. CONCLUDING COMMENTS

From our discussion of concave grating mountings, it is evident that there are useful configurations, especially for short wavelengths, in which aberrations are tolerable. None of these mountings, however, has found favor on ground-based telescopes. One reason is that, except for the inverse Wadsworth, there is little freedom in choice of camera focal ratios and match of projected slit to pixel size.

Another shortcoming of the concave grating is its efficiency relative to that of the plane grating. A concave grating ruled by conventional methods has lower overall efficiency than a plane grating because grooves near one side of the concave grating have a different effective blaze angle than those near the other side. To overcome this defect, at least in part, concave gratings with bipartite and tripartite rulings have been made. The Bausch and Lomb Grating Handbook is a good reference for more information about these types of rulings and grating efficiency in general.

REFERENCES

Beutler, H. (1945). *J. Opt. Soc. Am.* **35**, 311.
Namioka, T. (1959). *J. Opt. Soc. Am.* **49**, 446.
Welford, W. (1965). "Progress in Optics," Vol. 4, Chap. 6. North-Holland, Amsterdam.

BIBLIOGRAPHY

Seya-Namioka monochromator
Namioka, T. (1959). *J. Opt. Soc. Am.* **49**, 951.

Concave grating mountings
James, J., and Sternberg, R. (1969). "The Design of Optical Spectrometers." Chapman & Hall, London.
Meltzer, R. (1969). "Applied Optics and Optical Engineering," Vol. 5, Chap. 3. Academic Press, New York.
Namioka, T. (1961). "Space Astrophysics," p. 228. McGraw-Hill, New York.

Wadsworth mounting
Seya, M., and Namioka, T. (1967). *Sci. Light* **16**, 158.

Chapter 15 | Plane Grating Spectrometers

Plane grating spectrometers have distinct advantages over concave grating instruments and have been the almost universal choice for large telescopes. Spectrometers in which the grating is illuminated by collimated light have only the aberrations of the collimator and camera degrading the spectrum quality. With careful attention to the design of these optical subsystems, aberrations can be reduced to an insignificant level. In addition, the freedom to choose collimator and camera focal ratios to get the best possible match between projected slit and pixel size is a significant advantage.

Another important advantage of a plane grating is its adaptability to configurations in which many orders are arranged to cover a convenient two-dimensional format suitable for modern detectors. Such modes are discussed later in this chapter.

I. PLANE GRATING MOUNTINGS

In this section we consider two types of plane grating mountings with the grating in collimated light. One type in common use has a camera from the family of Schmidt cameras with either an off-axis paraboloid or on-axis paraboloid with fold mirror as collimator. This type, which we choose to call "fast," is used exclusively in the spectrograph mode. The other type discussed is the Czerny–Turner mounting, used in either the spectrograph

I. Plane Grating Mountings

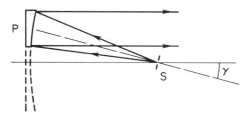

Fig. 15.1. Off-axis paraboloidal collimator P with angle γ between axes of telescope and mirror. S, Entrance slit of spectrometer at telescope focus.

or monochromator mode. We also give the characteristics of a version of the Monk–Gillieson mounting, a type used only in the monochromator mode.

A. FAST PLANE GRATING SPECTROMETERS

The most common collimator arrangements for a spectrometer of this type are a folded on-axis paraboloid and an off-axis paraboloid. The former is shown schematically in Fig. 14.4, the latter in Fig. 15.1. Either type has zero spherical aberration, but both have coma and astigmatism at off-axis points on a long slit. The off-axis aberrations are of no consequence for a stellar source but do set a limit when extended sources are observed. In the latter case it is important to know the size of these aberrations as a function of position on the slit.

Consider first an on-axis paraboloid without the fold mirror, as shown in Fig. 15.2. The chief ray at height y on the entrance slit comes from the center of the telescope exit pupil at angle ψ with the z axis. The exit pupil is a distance $f_p \delta$ from the slit, where f_p is the telescope primary focal length and δ is given in Eq. (2.5.4). The pupil is distance $W = -(f_p \delta + f_1)$ from

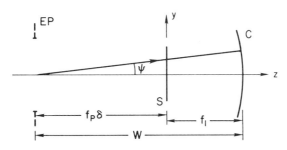

Fig. 15.2. Chief ray from center of telescope exit pupil at angle ψ with telescope axis. S, Spectrometer entrance slit; C, on-axis collimator mirror.

the collimator, where $W < 0$ by the sign convention and f_1 is the collimator focal length. From Eq. (2.5.5) we get $\psi = \theta(m/\delta)$, where θ is the angle on the sky.

Substituting these results in the coefficients in Table 5.6 for a paraboloid, with $n = 1$, we find

$$B_1 = -\frac{\theta^2}{R}\frac{mD}{\delta d}, \quad B_2 = \frac{\theta}{R^2}\frac{D}{d}, \quad (15.1.1)$$

where R is the collimator radius of curvature, and d and D are the diameters of the collimator and telescope, respectively. In arriving at Eqs. (15.1.1) we used the fact that $f_p\delta/f_1$ is the ratio of the exit pupil diameter to the collimator diameter. As expected, given our discussion in Section 5.V, the coma coefficient does not depend on pupil position.

The transverse aberrations are found by substituting Eqs. (15.1.1) into (5.5.9), with s' replaced by the camera focal length f_2. The angular aberrations, projected on the sky, are found by dividing each transverse aberration by the system focal length f_s, where $f_s = f(f_2/f_1) = F_2D$. With these substitutions we get

$$\text{AAS} = \frac{\theta^2}{2F}\frac{m}{\delta}, \quad \text{ATC} = \frac{3\theta}{16F^2}, \quad (15.1.2)$$

where F is the focal ratio of the collimator or telescope.

As an example, we take parameters for a Cassegrain telescope in Table 6.8 and evaluate Eqs. (15.1.2) at a field angle of 10 arc-min. The results are 0.28 and 1.13 arc-sec for the collimator astigmatism and coma, respectively. From these results we see that coma sets the limit on allowable slit length; for typical seeing a slit 20 arc-min long gives negligible loss in spatial resolution along the slit.

Analysis including the telescope aberrations shows that the astigmatism of a Ritchey–Chretien is opposite that of the collimator, and the net astigmatism is smaller than that from Eqs. (15.1.2) by nearly a factor of 10. With a classical Cassegrain, collimator and telescope aberrations cancel one another and spatial resolution along the slit is determined entirely by seeing.

For the off-axis paraboloid shown in Fig. 15.1, γ is the angle between the axes of the telescope and paraboloid. The usual arrangement has the slit length perpendicular to the plane defined by telescope and paraboloid axes. For this collimator we give results derived from ray traces. With $\gamma = 10°$ we find blur diameters of approximately 0.5 and 1.5 arc-sec at $\theta = 1$ and 3 arc-min, respectively. For this example the total slit length is limited to about 4 arc-min.

Ray-trace results show that the blur diameter for different γ and θ is approximately proportional to the product of the angles. An empirical relation for blur diameter is blur (arc-sec) $= 0.05\gamma$ (deg)θ (arc-min). This relation is an approximate one, and ray-trace results are required for more exact measures of blur. The results given here also hold for a slit in the plane defined by the axes of the telescope and paraboloid.

In summary, for slit spectroscopy of extended objects the on-axis paraboloid is the better choice for the collimator if a slit more than a few arc-minutes in length is required. This is achieved at the expense of an additional optical element, the folding flat. The echelle spectrograph on the 4-m Mayall telescope at Kitt Peak has an on-axis paraboloid collimator; the RC spectrograph on the same telescope has an off-axis paraboloid.

The design considerations for the camera end of a fast plane grating spectrometer, assuming a Schmidt-type camera, are covered in Chapter 7. As noted there, requirements for cameras for direct photography are different in some respects from those for spectrometer cameras. For example, chromatic focal shifts in solid and semisolid cameras, intolerable in direct photography, are accommodated in spectrometer cameras by tilting the detector to match the focal surface.

Spectrometers also require a wider camera in the direction of primary dispersion to accommodate beam expansion due to anamorphic magnification and to avoid vignetting of the dispersed beams. This is true for both gratings used in a single order and cross-dispersed echelles. Unlike the case of a direct camera, the pupil in a spectrometer is usually displaced from the corrector, and this means a somewhat larger corrector and camera mirror to cover the same field. It is possible to place the corrector close to the grating, as suggested by Bowen, but this makes camera interchange more difficult.

A final significant difference between the modes of operation is the location of the focal surface. It is usually internal in a Schmidt telescope but almost always external in a spectrometer camera. A schematic of a folded Schmidt camera is shown in Fig. 15.3, with the fold mirror roughly midway between the corrector and spherical mirror. A study of the layout in Fig. 15.3 shows that careful placement of the fold mirror is required to ensure an efficient camera. It is essential that the detector does not see collimated light, and this requires that the detector be in the shadow of the fold mirror, as seen from the corrector. If the distance from the fold mirror vertex to detector is large, the size of the hole in the fold mirror must also be large to avoid vignetting the beam from the spherical mirror. But the hole in the fold mirror should be as small as possible to minimize vignetting of the beam from the corrector. The trade-off between these competing requirements sets the design of a folded Schmidt.

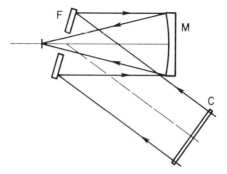

Fig. 15.3. Folded Schmidt camera. C, Corrector plate; F, folding flat; M, spherical mirror.

B. CZERNY–TURNER MOUNTINGS

The Czerny–Turner design is a widely used laboratory spectrometer suited for either the monochromator or spectrograph mode. This design is also used for astronomical spectrometers; for example, the main spectrograph at the McMath solar telescope at Kitt Peak is of this type. We discuss characteristics of both modes of operation, starting with the monochromator mode.

The optical layout for a Czerny–Turner mounting, hereafter denoted CZ, is shown in Fig. 15.4. The spherical collimator and camera mirrors are M_1 and M_2, respectively, G is the grating, and the entrance and exit slits are at Q and Q', respectively. The axis of each mirror is tilted with respect to its incident chief ray, and thus each has both on- and off-axis aberrations.

To ensure that the tangential fan of rays incident on the grating is strictly collimated, the slit at Q is at the tangential focus of the collimator. The consequence of this choice is that the distance from the camera vertex to Q' is independent of the grating angles and separation between the grating and each mirror. The spectrum is scanned by rotating the grating about an axis along its central groove.

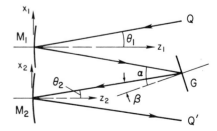

Fig. 15.4. Optical arrangement of Czerny–Turner mounting. M_1, Collimator; G, grating; M_2, camera; Q (Q'), entrance (exit) slit. Dispersion direction is in the plane of the diagram.

I. Plane Grating Mountings

The aberrations of the CZ mounting are easily found using the results in Chapter 5. In the monochromator mode, the position of the stop is of no consequence because the beam location at the camera is always the same. Thus we can take the coefficients for each mirror from Table 5.2, where the magnification m is infinite for the collimator and zero for the camera. One change is needed in these coefficients before writing the system coefficients. The coefficient A_1 denotes the astigmatism at the sagittal image, but for a grating instrument we require the astigmatism at the tangential image. Hence we take A'_1 instead of A_1, where, as seen in Eqs. (5.2.3) and (5.2.4), these coefficients have opposite signs. This sign change has no effect on the final relations that give the length of the astigmatic image.

With this change we take the coefficients from Table 5.2 and substitute them into Eq. (5.6.7) to get the system coefficients. The results are

$$A_{1s} = \frac{\theta_1^2}{R_1} + \frac{\theta_2^2}{R_2}\left(\frac{y_2}{y_1}\right)^2, \quad (15.1.3)$$

$$A_{2s} = \frac{\theta_1}{R_1^2} - \frac{\theta_2}{R_2^2}\left(\frac{x_2}{x_1}\right)^3, \quad (15.1.4)$$

where $y_1 = y_2$, $n_1 = n_2 = -1$ from Fig. 15.4, and subscripts 1 and 2 refer to collimator and camera mirrors, respectively.

Because R_1 and R_2 have the same sign, the astigmatisms of the two mirrors add to give the system astigmatism. A similar result is found for the spherical aberration of the system. With a proper choice of angles, however, the coma coefficient can be made zero.

The relation between x_1 and x_2 is derived from the geometry in Fig. 15.4. If x is the coordinate of a marginal ray at the grating, then $x_1 \cos \theta_1 = x \cos \alpha$ and $x_2 \cos \theta_2 = x \cos \beta$. Putting these results into Eq. (15.1.4) and setting $A_{2s} = 0$ gives

$$\frac{\theta_2}{\theta_1} = \left(\frac{R_2}{R_1}\right)^2 \left(\frac{\cos \alpha_0}{\cos \beta_0}\right)^3 \left(\frac{\cos \theta_2}{\cos \theta_1}\right)^3. \quad (15.1.5)$$

The relation in Eq. (15.1.5) is an approximation to the exact third-order equation, in which the left side is the ratio of the sines of the angles. As shown shortly, it is necessary to have θ_1 and θ_2 small in order to keep the astigmatism small. For most purposes, therefore, the paraxial approximation is adequate and $\cos \theta$ factors can be replaced by one.

Note that a choice of θ_1 and θ_2 to satisfy Eq. (15.1.5) is possible for any set of grating angles, but once they are chosen there is a small residual coma at other wavelengths. This residual, a result of wavelength-dependent anamorphic magnification, has been called subsidiary coma by Welford. The size of the subsidiary coma is given in an example to follow.

278   15. Plane Grating Spectrometers

Putting Eq. (15.1.5) into Eq. (15.1.3), with factors in $\cos \alpha_0$ and $\cos \beta_0$ retained, we find

$$A_{1s} = \frac{\theta_1^2}{R_1}\left[1 + \left(\frac{R_2}{R_1}\right)^3\left(\frac{\cos \alpha_0}{\cos \beta_0}\right)^6\right]. \qquad (15.1.6)$$

The transverse aberrations are found by substituting the system coefficients into Eq. (5.5.9). The results are given in Table 15.1, with d and F_1 denoting the collimator diameter and focal ratio, respectively.

The relations in Table 15.1 describe a spectrometer in which the coma is zero at one wavelength and negligible over an extended range of wavelengths. It is useful at this point to determine the characteristics of a representative example, that of a 1-m, $f/10$ spectrometer with $R_1 = R_2 = 2$ m. We choose $\alpha - \beta = 8°$ and $\theta_1 = 3°$ to provide clearance between the optical elements and beams. With a 600 groove/mm grating used in first order, and choosing zero coma at $\lambda_0 = 500$ nm, we find $\theta_2 = 2.814°$.

With these parameters we find that the image length is 0.52 mm at the zero-coma wavelength and TSA = 31 μm. At ±200 nm from the corrected wavelength, the transverse coma is 2.6 μm and the length of the astigmatic image is unchanged. For all practical purposes over this range, coma is negligible, astigmatism is constant, and spherical aberration determines the spectral resolution. With a plate factor of 16.7 Å/mm, the resolution at the minimum width of the spherical aberration blur is 0.26 Å.

From this example it is evident that a CZ monochromator gives good spectral resolution, provided the focal ratios of the camera and collimator are sufficiently large. An instrument of this type is suitable for stellar sources, but its astigmatism limits its usefulness for extended sources if spatial

Table 15.1

Transverse Aberrations of Czerny–Turner Monochromator[a]

$$\text{TAS} = \frac{\theta_1^2 d}{2}\left(\frac{R_2}{R_1}\right)\left[1 + \left(\frac{R_2}{R_1}\right)^3\left(\frac{\cos \alpha_0}{\cos \beta_0}\right)^6\right]$$

$$\text{TTC} = \frac{3\theta_1 d}{16 F_1}\left(\frac{R_2}{R_1}\right)\left[1 - \left(\frac{\cos \alpha_0}{\cos \beta_0}\frac{\cos \beta}{\cos \alpha}\right)^3\right]$$

$$\text{TSA} = \frac{d}{64 F_1^2}\left(\frac{R_2}{R_1}\right)\left[1 + \left(\frac{R_1}{R_2}\right)^3\right]$$

[a] Angle θ_2 is chosen to give zero coma at $\lambda = \lambda_0$. All $\cos \theta_1$ and $\cos \theta_2$ factors are omitted. Length of astigmatic image is 2TAS.

resolution along the slit is required. The angle corresponding to the image length projected on the sky can be calculated from Eq. (12.2.1b).

We now consider the CZ in the spectrograph mode, with the grating and camera mirror shown schematically in Fig. 15.5. The z axis of the mirror is the normal to the mirror at its center, with the angle of the chief diffracted ray measured with respect to this axis. An arbitrary chief ray makes angle θ_2 with the axis, with θ_{20} used for the chief ray at the mirror center.

The effective position of a pupil centered on the z axis is shown in Fig. 15.5 as a distance W' from the mirror vertex. Denoting the grating-mirror separation measured parallel to the z axis by W, we find, in the paraxial approximation, $W'\theta_2 = W(\theta_2 - \theta_{20})$.

The aberration coefficients for the camera mirror are now found by substitution of W' for W and θ_2 for ψ in the relations in Table 5.6. Combining these with the coefficients of the collimator gives

$$B_{1s} = \frac{\theta_1^2}{R_1} + \frac{1}{R_2}\left[\theta_2 - \frac{W}{R_2}(\theta_2 - \theta_{20})\right]^2, \qquad (15.1.7)$$

$$B_{2s} = \frac{\theta_1}{R_1^2} - \frac{1}{R_2^2}\left[\theta_2 - \frac{W}{R_2}(\theta_2 - \theta_{20})\right], \qquad (15.1.8)$$

where the paraxial approximation is used for all angles. This approximation is quite adequate for calculating the image aberrations in the spectrograph mode and is used in the discussion and example that follow.

Examination of Eqs. (15.1.7) and (15.1.8) shows that the coefficients are independent of θ_2 when $W = R_2$. The transverse aberrations of the spectrograph are then given by the relations in Table 15.1. An instrument with this choice of W has been described by Willstrop. The principal disadvantages of a spectrograph with this W are its longer length and curved focal surface.

Returning to the geometry of the chief rays in Fig. 15.5, we see that $\beta - \beta_0 = -(\theta_2 - \theta_{20})$. Applying the paraxial approximation to the grating

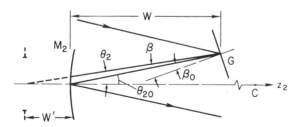

Fig. 15.5. Diffraction and camera angles for Czerny-Turner spectrograph. C, Center of curvature of mirror M_2; W', distance of apparent stop from M_2 with grating as stop.

Table 15.2
Transverse Aberrations of Czerny-Turner Spectrograph[a]

$$\text{TAS} = \frac{d}{2}\left(\frac{R_2}{R_1}\right)\left\{\theta_1^2 + \left(\frac{R_1}{R_2}\right)\left[\theta_{20} + \left(1 - \frac{W}{R_2}\right)\frac{m}{\sigma}(\lambda - \lambda_0)\right]^2\right\}$$

$$\text{TTC} = \frac{3d}{16F_1}\left(\frac{R_1}{R_2}\right)\left(1 - \frac{W}{R_2}\right)\frac{m}{\sigma}(\lambda - \lambda_0)$$

[a] Angle θ_{20} is chosen to give zero coma at $\lambda = \lambda_0$.

equation gives $\beta - \beta_0 = m(\lambda - \lambda_0)/\sigma$. With this substitution in Eqs. (15.1.7) and (15.1.8), the system coefficients are expressed in terms of the wavelength difference with respect to the wavelength at the center. The transverse aberrations, given in this form, are shown in Table 15.2, where θ_{20} is chosen to give zero coma.

The relations in Table 15.2 show that the coma is linearly proportional to the wavelength difference, while the astigmatism depends on this difference in a more complicated way. As an example, we take the same grating and mirror parameters used for the monochromator, with $\lambda_0 = 500$ nm and $W/R_2 = 0.5$. The coma for $\Delta\lambda = \pm 50$ nm is 28 μm, and the image lengths are 0.41, 0.53, and 0.73 mm at $\lambda = 550$, 500, and 450 nm, respectively. The results obtained from ray traces are within a few percent of these values and justify the approximations used.

In contrast to the monochromator mode, neither the coma nor the astigmatism is constant, and both the coma and spherical aberration determine the spectral resolution. Nevertheless, the CZ mounting is practical in the spectrograph mode, provided a modest spectral range is covered.

The final item of interest for the spectrograph mode is the curvature of the tangential focal surface, found by using the geometry in Fig. 15.6. The

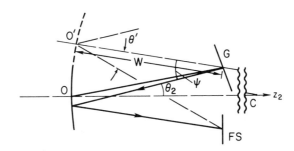

Fig. 15.6. Geometry for finding orientation and curvature of focal surface for Czerny-Turner spectrograph.

line $O'C$ passes through the center of the grating, which is the stop, and the center of curvature C of the mirror. A chief ray that makes angle θ_2 with the line OC makes angle ψ with the line $O'C$. Because all rays from the grating are parallel, an imaginary ray reflected at O' from an extension of the mirror makes angle $\theta' = -\psi$ with the line $O'C$.

Relative to the line $O'C$, which is effectively a z axis of the mirror, the astigmatism coefficient is B_1 in Table 5.6. This coefficient cannot be used to find the transverse astigmatism because the actual mirror is not centered at O', but it can be used to find the image surface curvature. Substituting B_1 in κ_t in Table 5.7 and noting that $\theta' = \psi$, we find

$$\kappa_t = \frac{2}{R_2} - \frac{6}{R_2}\left(1 - \frac{W}{R_2}\right)^2, \qquad (15.1.9)$$

where W is the distance $O'G$ in Fig. 15.6. The center of curvature of the image surface is on the line $O'C$, and a flat image surface is perpendicular to this line.

Setting $\kappa_t = 0$ we find that the condition for a flat image surface is $W/R_2 = 1 \pm 1/\sqrt{3}$, with the minus sign a more convenient choice. With this value for W/R_2, the coma measures for the example above are increased by about 15%, but this is a small price to pay for the convenience of a flat field.

The monochromator and spectrograph modes of the Czerny–Turner design have been thoroughly analyzed in far more detail than given here. Higher-order aberrations and detailed ray-trace analyses lead to refinements of the relations given here, and the interested reader should consult the references at the end of the chapter for more information. Included in these references are treatments of off-plane designs and instruments with curved slits for achieving the highest possible spectral resolution of broad sources. These treatments are not included here because, for astronomical applications, the in-plane design with a short slit is best suited to observations of point sources.

Before concluding this section, it is worth noting a particular version of the Czerny–Turner design called the Ebert–Fastie mounting. In the Ebert–Fastie mounting a single spherical mirror serves as both collimator and camera, thus eliminating the need for alignment of separate mirrors. This mounting is almost always used in the monochromator mode with the grating slightly closer to the mirror than the slits. If the incident and emergent chief rays are parallel, the angle $\alpha - \beta$ is nearly equal to $\theta_1 + \theta_2$ in Fig. 15.4 redrawn with a single mirror. The aberrations of the Ebert–Fastie monochromator are given in Table 15.1 with $R_1 = R_2$. References on this mounting are listed at the end of the chapter.

C. MONK-GILLIESON MOUNTINGS

We now consider the Monk-Gillieson mounting in which a plane grating is illuminated with a convergent light bundle. In this mounting the grating contributes to the system aberrations, but with proper arrangement of the optical elements these aberrations can be made zero at one wavelength. At other wavelengths, as we show below, coma and astigmatism are present in amounts directly proportional to the wavelength difference with the corrected wavelength. Because of these wavelength-dependent aberrations, this mounting is best suited to the monochromator mode.

A schematic layout of a Monk-Gillieson monochromator, hereafter called MG, is shown in Fig. 15.7. The entrance and exit slits are at Q and Q', respectively, and M is a spherical mirror. If the grating G were absent, point Q would be imaged at Q''. The distances from the grating vertex to Q' and Q'', respectively, are s' and s, where s' refers to the tangential astigmatic image. The distance from the mirror vertex to Q'' is $r = s + r'$.

In the discussion that follows, all angles are assumed small enough that the paraxial approximation applies, hence the grating equation is $m\lambda = -n\sigma(\beta + \alpha)$. We also define $\gamma = \alpha - \beta$ and note that $s' = -s$ in this approximation. In terms of these parameters, we find the system aberration coefficients by the usual method, in this case referenced to the grating. The results are

$$B_{1s} = \frac{1}{s}\left[\gamma\frac{m\lambda}{2\sigma} - \frac{n_1\theta^2 r^2}{sR}\right],$$

$$B_{2s} = \frac{1}{s^2}\left[\frac{m\lambda}{2\sigma} - \frac{n_1\theta r^3}{sR^2}\left(\frac{M+1}{M-1}\right)\right], \quad (15.1.10)$$

where M is used for the magnification of the mirror to avoid confusion with the order number m.

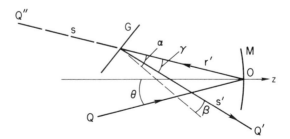

Fig. 15.7. Optical arrangement of Monk-Gillieson mounting. Q (Q'): entrance (exit) slit; M: spherical mirror; G: plane grating.

With two free parameters, θ and γ, both coefficients in Eqs. (15.1.10) can be made zero at any desired wavelength λ_0. Applying these conditions gives

$$\theta_0 = -\frac{m\lambda_0}{n_1\sigma}\frac{sR}{r^2(M+1)}, \qquad \gamma_0 = -\frac{2\theta_0}{M+1}. \qquad (15.1.11)$$

For the layout shown in Fig. 15.7, $M < -1$ and $n_1 = 1$; therefore θ_0 and γ_0 are positive. Substitution of Eqs. (15.1.11) into Eqs. (15.1.10) gives the aberrations at an arbitrary wavelength λ, where $\gamma = \gamma_0$ in the monochromator mode. Using Eq. (5.5.9) we get the transverse coma and astigmatism as follows:

$$\text{TTC} = \frac{3s'}{8F_2^2}\frac{m}{\sigma}|\lambda - \lambda_0|, \qquad (15.1.12)$$

$$\text{TAS} = \frac{\theta_0 d}{|M+1|}\frac{m}{\sigma}|\lambda - \lambda_0|, \qquad (15.1.13)$$

where d is the beam diameter at the grating and F_2 is its focal ratio. Both aberrations increase linearly with the wavelength difference. Because astigmatism depends on θ_0, its value is also proportional to λ_0. The spectral resolution is set by the size of TTC at any wavelength. It is convenient to define the *spectral coma* $\delta\lambda_c$ as the spectral width of an image, where $\delta\lambda_c = \text{TTC} \cdot P$ and the plate factor $P = \sigma/ms'$ for small β. Multiplying Eq. (15.1.12) by P, we see that the spectral coma depends only on the focal ratio of the beam at the grating for a given wavelength difference.

As an example we take a 600 groove/mm grating in first order, with $QO = 500$ mm and $M = -2$ for the mirror. We assume a beam diameter of 50 mm at the mirror, hence $F_2 = 20$, and place the grating to give $s' = 545$ mm. Choosing $\lambda_0 = 320$ nm gives angles $\theta_0 = 3.6°$ and $\gamma_0 = 7.2°$, and at ± 100 nm from the corrected wavelength we find TTC = 31 μm, $\delta\lambda_c = 0.95$ Å, and an image length of 0.2 mm. Ray traces of this system show that the actual λ_0 is about 10% smaller and, with this correction, the spectral coma and astigmatism from the ray traces agree with the calculated values. The difference in λ_0 is a consequence of the paraxial approximation.

Unlike the Czerny–Turner monochromator, the grating in the Monk–Gillieson cannot be rotated about an axis on its face and maintain focus at a fixed exit slit. From Table 14.1 we see that s'_t is a function of α and β for a given s. Noting that both α and β are variables in the scanning mode, with $d\alpha = d\beta$, we find in the paraxial approximation

$$ds'_t = -2s\gamma_0\, d\alpha = -s\gamma_0(m\, d\lambda/\sigma),$$

where $d\lambda$ is the wavelength interval scanned. For the example above, $ds'_i = -2.05$ mm for $d\lambda = 50$ nm. To keep the spectrum in focus on a fixed exit slit, it is necessary to simultaneously translate and rotate the grating. It is not difficult to show that a rotation about an appropriate axis displaced from the grating face gives the required motion. Details on the location of this axis are given in the reference by Schroeder.

It is evident from Eq. (15.1.12) that the Monk–Gillieson scanner is limited to relatively slow beams to keep the coma small, particularly if a large spectral range about the zero-coma wavelength is scanned. A scanning Czerny–Turner or Ebert–Fastie is clearly superior in this regard. One attractive feature of the Monk–Gillieson is its two reflections, hence the entrance and exit slits are on opposite ends of the instrument.

D. CONCLUDING COMMENTS

Plane grating instruments have traditionally been used to observe stellar or near-stellar sources one at a time with the object centered on the entrance slit. For an observing program requiring spectra of many objects of comparable brightness in close proximity on the sky, the observing time can be reduced significantly by recording the spectra of many sources in the same exposure. This is accomplished by using flexible glass fibers to transfer the light from separate sources in the telescope focal plane to the spectrometer slit. Each fiber is positioned at a source on one end and aligned along the slit at the other end. This technique of multiobject spectroscopy is applicable to objects within a cluster of stars or galaxies, as demonstrated by a number of observers.

The discussion in this section is intended to illustrate the general characteristics of different types of plane grating spectrometers in both the spectrograph and monochromator modes. Each of the types is well suited to certain kinds of observations, and the choice of one over another clearly depends on its intended application. From a comparison of plane and concave grating instruments, however, it should be evident why plane grating spectrometers are the choice for ground-based telescopes.

II. ECHELLE SPECTROMETERS

For high-resolution astronomical spectroscopy, $R \approx 1E5$, an echelle is the preferred choice over a grating used in low order. The principal reason for this, larger luminosity, is discussed in Section 13.III.B. A second important advantage of the echelle is the two-dimensional format of the spectrum, which permits broad spectral coverage. Because the echelle has a large groove spacing it is used at high-order numbers, as the example in Section 13.II.F illustrates. Thus it is necessary to provide cross-dispersion to separate the orders or use a filter to isolate a single order.

In this section we discuss the form of the two-dimensional format with different cross-dispersers, the location of the cross-disperser within an echelle spectrometer, and different types of collimator and camera possibilities. We also give an example of parameters appropriate for an echelle instrument on a 4-m telescope.

A. SPECTRUM FORMATS

If wavelengths in different orders are to be recorded without confusion, a cross-disperser must be put in series with the echelle. A cross-disperser is simply another element, usually a prism or another grating, whose dispersion is at right angles to that of the echelle and whose function is to separate the orders. The angular dispersion of the order separator is usually many times smaller, and the combination of elements gives a two-dimensional spectrum format. An example of a typical echelle spectrum is shown in Fig. 15.8. The format outline is set by the relative dispersions of the two elements; in this section we discuss the factors that determine a spectrum format.

We consider first the factors that determine the length of the spectrum in a given echelle order. With a camera of focal length f_2, the spectrum in the focal plane has length $f_2 \Delta\beta$, where $\Delta\beta$ is the angular width of one free spectral range $\Delta\lambda$. Combining Eqs. (13.2.2a) and (13.2.6) we get

$$\Delta\beta = \frac{d\beta}{d\lambda} \Delta\lambda = \frac{\lambda_0}{\sigma \cos \beta_0}, \qquad (15.2.1)$$

where λ_0 is the blaze wavelength in the mth order and β_0 is its angle of diffraction. This relation is not exact but for $m > 10$ is a good approximation to the exact angular width. The free spectral range within this $\Delta\beta$ is

$$\Delta\lambda = \frac{\lambda_0}{m} = \frac{\lambda_0^2}{2\sigma \sin \delta \cos \theta}. \qquad (15.2.2)$$

As noted in Section 13.III.A, $\Delta\lambda$ is the spectral range between the approximate half-intensity points of the blaze function; it is also the separation between blaze wavelengths in adjacent orders. For the echelle example in Section 13.II.F, $\Delta\beta = 4.33°$ and $\Delta\lambda = 111$ Å with $\theta = 5°$ in the forty-fifth order.

286 15. Plane Grating Spectrometers

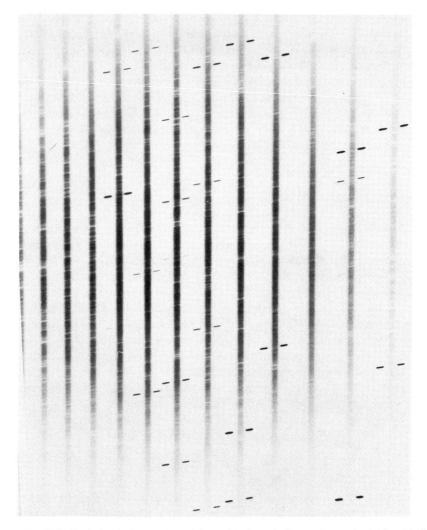

Fig. 15.8. Typical echelle spectrum taken with the echelle spectrograph at Pine Bluff Observatory. (Photograph courtesy of C. M. Anderson.)

From Eqs. (15.2.1) and (15.2.2) it is evident that σ is the controlling parameter for a given blaze angle and wavelength. Recall also that the parameters that set the resolution are the blaze angle and diameters of the collimator and telescope. Because the camera design depends on both beam diameter and focal length, it is clear that the order length $f_2\Delta\beta$ also depends on parameters other than the groove spacing.

II. Echelle Spectrometers

Table 15.3

Equations for Echelle at Blaze Peak

$m\lambda_0 = \sigma(\sin \beta_0 + \sin \alpha) = 2\sigma \sin \delta \cos \theta,$	$\beta_0 = \delta - \theta, \quad \alpha = \delta + \theta$
$A = \dfrac{2 \sin \delta \cos \theta}{\lambda_0 \cos \beta_0},$	$\Delta\lambda = \dfrac{\lambda_0^2}{2\sigma \sin \delta \cos \theta}$
$R = \dfrac{2d_1 \sin \delta \cos \theta}{\phi D \cos \alpha},$	$\Delta\beta = A\Delta\lambda = \dfrac{\lambda_0}{\sigma \cos \beta_0}$

For convenient reference, we give the important relations for an echelle in Table 15.3, including those for order width and free spectral range. Note that the angular dispersion and resolution are not constant over a single order, but their values at the blaze peak are an average over the order.

Assuming a cross-disperser with angular dispersion A_c, the separation Δy between adjacent orders is given by $\Delta y = f_2 A_c \Delta \lambda$. If A and A_c are assumed constant, each echelle order is tilted by an angle ψ with respect to the direction of echelle dispersion, where $\tan \psi = A_c/A$. This tilt of orders is shown in Fig. 15.9.

For a first-order grating cross-disperser $A_c = 1/\sigma_c \cos \beta_c$. Although A_c is essentially constant over any echelle order, A for the echelle is not strictly constant along an order. Therefore $\tan \psi$ changes by a few percent from one end to the other and the order is slightly curved. Ignoring this curvature, the tilt of echelle orders with a grating is given by $\tan \psi = \text{constant} \cdot \lambda_0$. For a prism cross-disperser $A_c \propto \lambda^{-3}$, as shown in Section 3.II.D, hence each order has significant curvature. In this case the tilt of an order at its center is given by $\tan \psi = \text{constant} \cdot \lambda_0^{-2}$.

Given the angular dispersions of the grating and prism, we can write the order separation Δy in terms of the blaze wavelength. The results are

$$\Delta y(\text{grating}) \propto \lambda_0^2, \quad \Delta y(\text{prism}) \propto \lambda_0^{-1}. \quad (15.2.3)$$

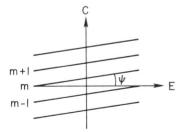

Fig. 15.9. Tilt of echelle orders relative to directions of echelle dispersion (E) and cross-dispersion (C).

Thus gratings give larger order separation at longer wavelengths, and the reverse is true for prisms. When detector area is limited, as for a charge-coupled device (CCD), a prism cross-disperser makes better use of the available area.

To illustrate the formats given by different cross-dispersers, we take a 31.6 groove/mm echelle with $\tan \delta = 2$ and $\theta = 5°$. For the grating cross-disperser, we assume one with 158 grooves/mm used in first order; for the other we assume two 45° UBK7 prisms in series. The format outlines for these two cases, to the same scale, are shown superposed in Fig. 15.10. The outlines cover a wavelength range from 400 to 750 nm, with $m = 141$ and 76 at these wavelengths, respectively.

The linear size of the formats shown in Fig. 15.10 is proportional to the camera focal length, as is the linear separation between orders. The angular separation between orders, projected on the sky, is proportional to the beam diameter and is independent of the camera focal length. We give numerical values for these formats in our discussion below of a particular echelle spectrometer design.

B. CROSS-DISPERSION MODES

Many different methods of order separation are possible. The cross-dispersion can be done with a separate prism or grating spectrometer following an echelle instrument. Because each has its own set of collimator and camera optics, such a system is less efficient and more prone to misalignment than an echelle spectrometer with internal cross-dispersion.

In this section we consider only internal cross-dispersion modes. Possibilities include (1) a prism(s) or plane grating between the echelle and

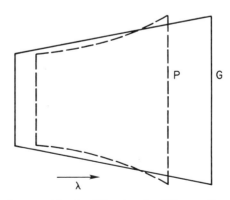

Fig. 15.10. Echelle formats with prism (P) and grating (G) cross-dispersers. See text, Section 15.II.A, for parameter values of the echelle and cross-disperser.

camera optics and used single-pass, (2) a concave grating that is both camera and cross-disperser, (3) a concave grating that is also the collimator, and (4) a prism or transmission grating close to the echelle and used double-pass.

We first consider mode(1), in which the disperser follows the echelle, oriented with $\alpha > \beta$, as shown schematically in Fig. 15.11. The direction of the echelle dispersion is in the plane of Fig. 15.11, and the prism or grating must be clear of the collimator beam and large enough to accept the dispersed light in each echelle order. In the direction perpendicular to the echelle dispersion, the width of the cross-disperser is the diameter of the collimator beam.

It is evident from Fig. 15.11 that the cross-disperser can be placed closer to the echelle if the angle θ is larger, hence the height of the cross-disperser is less. Larger θ, however, also means larger dispersed beam height because of anamorphic magnification. The latter effect largely cancels the reduction in size obtained by putting the cross-disperser closer to the echelle; the net effect is that the size of the cross-disperser depends only weakly on the choice of θ, except for θ near zero.

We pointed out in Section 13.III.A that the efficiency at the blaze peak and the average efficiency over each order decrease as θ increases. In order to keep the efficiency as high as possible, θ is chosen as small as possible, within the constraint of having the dispersed beam clear the collimator beam. Given these competing effects, it turns out that θ in the range 4–6° is a good compromise between beam clearance and efficiency for this arrangement.

The choice of a grating versus a prism depends on the required order separation and spectral range to be covered. Gratings usually have higher dispersion and, if a camera with a short focal length is used, may be the only practical option. One disadvantage of a grating is its changing efficiency over a wide spectral range, as shown in Fig. 13.9, so that several may be needed. Another disadvantage of a grating is the less than optimum use of detector area, as shown by Eq. (15.2.3). The main advantage of plane gratings is a wide selection of available groove spacings to give a range of cross-dispersions.

A prism has high and constant efficiency over its range of transmittance but lower dispersion than a grating; hence more than one prism may be

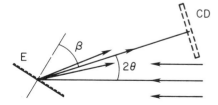

Fig. 15.11. Echelle (E) with postdisperser (CD), denoted as mode (1) in text.

required to get the necessary order separation in a spectrometer with a fast camera. Large prisms are also more expensive than gratings of comparable size, and the choice of transparent glasses for near-ultraviolet wavelengths is limited. A significant advantage of a prism over a grating is that it gives a format that makes better use of detector area.

Mode (2) has a concave grating postdisperser/camera subject to the same size constraints noted for mode (1). The selection of concave gratings, especially in larger sizes, is much more limited than that of plane gratings. The aberrations introduced by the concave grating are the same as those described for the Wadsworth mounting in Section 14.III.B; hence this cross-disperser mode is limited to focal ratios of 10 or larger.

When a concave grating doubles as cross-disperser and collimator, the echelle is illuminated with beams at different angles γ. The setup is shown in Fig. 15.12, where the direction of the grating dispersion is in the plane of the diagram. The consequence of nonzero γ is tilted spectral lines on the image surface, with the tilt proportional to γ and the slope given by Eq. (14.1.9). Substituting A from Table 15.3 into Eq. (14.1.9), the slope of the spectral lines is, to a good approximation, $d\beta/d\gamma = 2 \tan \gamma \tan \delta$.

As an example, assume a concave grating with 150 grooves/mm and an R2 echelle. Over the spectral range 400-700 nm the lines are tilted over the angular range $\pm 5°$, with maximum tilt at the wavelength extremes. This tilt is not a problem for a stigmatic instrument with a point source at the entrance slit, but the concave grating in this mounting has the same aberrations as the inverse Wadsworth mounting discussed in Section 14.III.C. Therefore the images are astigmatic, line tilt is present, and reduction of the spectral data is made more difficult.

For a stigmatic instrument used only for point sources, mode (4), using a double-pass cross-disperser, is a practical arrangement. Of the two options given above for this mode, the prism is the only viable choice because of its constant efficiency. The efficiency curve for a double-passed transmission grating is sharply peaked, and the grating option is not practical.

From our discussion it is clear that mode (1) with either a plane grating or prism following the echelle is the preferred mode. Line tilt is absent and, if the instrument is stigmatic, spectra of either stellar or extended sources can be taken. Unlike mode (2), with its concave grating camera, mode (1) is adaptable to different camera types. We discuss two possible configurations with an internal postdisperser in the following section.

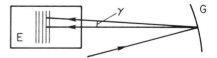

Fig. 15.12. Echelle (E) with concave grating predisperser and collimator (G), denoted as mode (3) in text.

C. ECHELLE SPECTROMETER CONFIGURATIONS

Echelle spectrometers with mode (1) cross-dispersion have been built in two different configurations, one based on the Czerny–Turner mounting and the other with a folded Schmidt camera. The former is limited to camera focal ratios of 10 or larger; the focal ratio of the latter is set by the camera and may be as fast as $f/2$ for a monochromatic beam. We discuss briefly the characteristics of each, followed by a design example for a fast echelle spectrometer.

The optical configuration for a modified Czerny–Turner echelle instrument is shown in Fig. 15.13. The dispersing elements are located between spherical collimator and camera mirrors, with each mirror tilted to provide beam clearance. With this mirror arrangement, the relations for coma and astigmatism given in Section 15.I.B for the Czerny–Turner spectrograph mode apply.

The principal difference between the design in Fig. 15.13 and the conventional Czerny–Turner design in Fig. 15.5 is the orientation of the primary disperser with respect to the camera mirror. In the standard Czerny–Turner the grating grooves are perpendicular to the plane defined by the mirror normal and chief ray; in the echelle design the echelle grooves are parallel to this plane. Therefore the sagittal focal surface of the camera mirror is

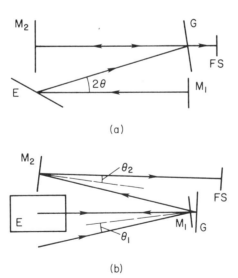

Fig. 15.13. Echelle spectrograph layout based on modified Czerny–Turner mounting. E, Echelle; G, grating; M_1, M_2, collimator, camera mirrors, respectively; FS, focal surface. Echelle dispersion is in the plane of diagram (a) and cross-dispersion in plane of diagram (b).

the one on which the astigmatism does not degrade the spectral resolution. Although this echelle system can be configured along the lines of the standard Czerny-Turner, the angle θ_2 at the camera mirror must be larger to provide beam clearance and thus the aberrations are also larger.

Versions of the modified Czerny-Turner design are used with a number of Cassegrain telescopes, including the 0.9-m telescope at the Pine Bluff station of Washburn Observatory and the 1.5-m telescope at the Mount Hopkins station of Whipple Observatory. More details about this design are found in the reference by Schroeder and Anderson.

The optical layout of an echelle spectrometer with a Schmidt camera is shown in Fig. 15.14. This particular design is the basis for the echelle instruments used at the Cassegrain foci of the 4-m telescopes at Kitt Peak and Cerro Tololo Observatories. The collimator is a folded paraboloid, and the system aberrations are those of the Schmidt camera. The beam diameter at the collimator is approximately 130 mm, and the echelle is overfilled along its longer dimension. With interchangeable cameras and reflection

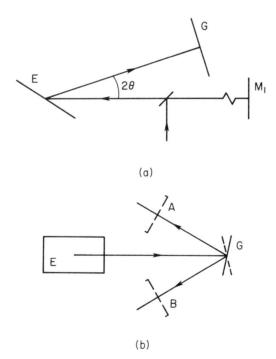

Fig. 15.14. Optical layout of echelle spectrograph for the 4-m Mayall telescope at Kitt Peak National Observatory. (A and B) Entrance apertures of Schmidt cameras. Other symbols are defined in the caption of Fig. 15.13.

grating cross-dispersers, this type of spectrometer can provide a variety of plate factors and two-dimensional formats.

D. ECHELLE DESIGN EXAMPLE

To illustrate the results in the previous sections, we take some specific parameters and determine the characteristics of the instrument and spectrum format. The assumed parameters include a telescope diameter $D = 4$ m and an echelle with 31.6 grooves/mm and $\delta = 63.5°$ used at $\theta = 5°$. We require that the system be capable of $R = 5E4$ when the slit width $\phi = 1$ arc-sec and the projected slit width $\omega' = 30$ μm.

Defining $\Gamma = 2 \sin \delta \cos \theta / \cos \alpha$, we can rewrite R from Table 15.3 and Eq. (12.2.1a) in suitable units as

$$R\phi \text{ (arc-sec)} = 200\Gamma \frac{d_1 \text{ (mm)}}{D \text{ (m)}}, \qquad (15.2.4)$$

$$\omega' \text{ (μm)} = 5r\phi \text{ (arc-sec)} D \text{ (m)} F_2, \qquad (15.2.5)$$

where $\Gamma = 4.87$ and $r = 0.70$ for the echelle angles chosen. Solving for d_1 and F_2 gives $d_1 = 205$ mm and $F_2 = 2.14$, hence the camera focal length $f_2 = 440$ mm.

From these results we see that the only possible instrument configuration is one with a Schmidt camera. Note that the results derived so far are independent of the echelle groove spacing; this parameter is used to find the length $f_2 \Delta \beta$ of each echelle order. With our chosen echelle, the order lengths are 10.7 and 19.8 mm, respectively, at 400 and 750 nm, with corresponding plate factors of 2.67 and 5.0 Å/mm.

The order separation and format width are determined by the cross-disperser, so far unspecified. If a first-order grating with 158 grooves/mm is used, the orders at 400 and 750 nm are separated by 23.3 mm; if two 45° UBK7 prisms are used, the orders at these wavelengths are separated by 15.7 mm. The echelle and cross-disperser parameters used in this example are the same as those used for the format outlines in Fig. 15.10.

The spacings of the orders within the formats depend on the wavelength according to Eq. (15.2.3). For the grating the angular separations between adjacent orders, projected on the sky, are 4.7 and 16.1 arc-sec at 400 and 750 nm, respectively. The corresponding separations for the prism cross-disperser are 7.6 and 4.5 arc-sec, with smaller separation at the longer wavelength.

Although this example illustrates the approach in the design of a particular echelle instrument, the same procedure can be applied to the design of any echelle or grating spectrometer. Once the basic outline of a design is found, the choice of an optical system that fits this outline can be made.

E. CONCLUDING COMMENTS

Although our discussion of echelle formats and spectrometer configurations is given in terms of echelles with $\tan \delta = 2$, the relations used apply to any blaze angle. A grating with $\tan \delta < 1$ is often called an echellette and can also be used with a cross-disperser to separate orders. Consider, for example, a grating with 300 grooves/mm and $\tan \delta = 0.75$. From the grating equation in Table 15.3 we get $m\lambda_0$ (μm) $= 4 \cos \theta$; hence most of the visible spectrum is covered in four orders, $m = 6$ through $m = 9$. These orders are easily separated with a crossed prism in the configuration shown in Fig. 15.11. Because of the smaller $\sin \delta$ compared to an echelle, such a system is appropriate for medium spectral resolution over a wide wavelength range in a two-dimensional format.

III. NONOBJECTIVE SLITLESS SPECTROMETERS

An important technique for low-dispersion spectroscopy with large telescopes is the nonobjective mode. This mode is one in which a dispersing element, prism or blazed transmission grating, or a combination of the two is placed in the converging beam near the telescope focal surface, as shown in Fig. 12.3. We discuss the characteristics of each of these in the nonobjective mode. The plate factor P and spectral purity $\delta\lambda$ for this mode are given by Eqs. (12.1.3b) and (12.3.1), respectively, and are repeated here for convenient reference.

$$P = (sA)^{-1}, \qquad (15.3.1)$$

$$\delta\lambda = \frac{\phi' f}{A s}, \qquad \delta\lambda_c = \text{TTC} \cdot P, \qquad (15.3.2)$$

where s is the distance from the element to the focus, f is the telescope focal length, A is the angular dispersion, and $\delta\lambda_c$ is the spectral coma as defined in the previous section.

The term "nonobjective" is used to distinguish this type of slitless instrument from the classical objective mode in which a prism or grating covers the aperture of a telescope. The disperser in the objective mode is in collimated light and the spectral resolution is determined by the seeing or telescope aberrations. The discussion in Sections 12.III and 13.I is sufficient for the objective mode.

The advantages of the nonobjective mode include ease of mounting a disperser on any telescope with a minimum of effort and cost, as no auxiliary optics are needed. Hence slitless spectroscopy is not limited to telescopes of modest size. Another advantage is that within broad limits set by aberrations and disperser size any plate factor is possible. A disadvantage of this mode is that aberrations are present when the disperser is placed in a converging beam, but, as we show, their effect can often be reduced to a negligible level compared to the seeing limit. It is also important to note that this mode is not an alternative to the objective mode but is complementary. The objective mode is typically used to give P in the range of 100–300 Å/mm, while the nonobjective mode is suitable for larger plate factors.

We now consider in turn the aberrations of the prism, grating, grism, and prism–grating in a converging beam. In each case we apply the paraxial approximation to all angles, an assumption that is justified for all practical configurations. Any significant deviations from results so derived are noted.

A. NONOBJECTIVE PRISM

Although a prism is rarely used in this mode, its characteristics are important when it is combined with a grating. Thus we determine prism aberrations in anticipation of the discussion of a grism.

Consider a thin prism of index N with apex angle γ, as shown in Fig. 15.15, and angles of incidence θ_1 and θ_2 at the first and second surfaces, respectively. It is convenient to express these angles in terms of the apex angle. If $\theta_1 = \varepsilon\gamma$ then, from Snell's law, we find $\theta_2 = -\gamma(1-\varepsilon/N)$. The parameter ε determines the prism orientation with respect to the chief ray; when $\varepsilon = 0$ the incident chief ray is perpendicular to the first surface, and when $\varepsilon = N/2$ the prism is set for minimum deviation.

The pertinent aberration coefficients for each surface are those of astigmatism and coma, and from Table 5.1 we get

$$A_{11} = -\frac{\theta_1^2(N^2-1)}{2N^2 s_1}, \quad A_{12} = \frac{\theta_2^2 N(N^2-1)}{2s_2}, \qquad (15.3.3)$$

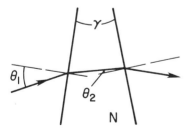

Fig. 15.15. Cross section of prism of index N with apex angle γ. θ_1 and θ_2 are angles of incidence at the first and second faces, respectively.

$$A_{21} = -\frac{\theta_1(N^2-1)}{2N^2 s_1^2}, \quad A_{22} = \frac{\theta_2 N(N^2-1)}{2s_2^2}, \quad (15.3.4)$$

where s_1 and s_2 are the object distances at the first and second surfaces, respectively. The astigmatism coefficients are reversed in sign from those in Table 5.1 to reflect the change from sagittal to tangential image, as discussed in Section 15.I.B.

Assuming the prism thickness is small compared to the distance to the focal surface, we have $s_2 = Ns_1$. This assumption, in turn, implies that the beam size is the same at the two surfaces, hence each prism aberration coefficient is simply the sum of surface coefficients. Substituting for θ_1 and θ_2, with $s_1 = s$, we find

$$A_{1p} = \frac{\gamma^2(N^2-1)}{2s}\left(1 - \frac{2\varepsilon}{N}\right), \quad (15.3.5)$$

$$A_{2p} = -\frac{\gamma(N^2-1)}{2Ns^2}, \quad (15.3.6)$$

where the subscript p denotes prism. Note that the coma coefficient is independent of ε, hence coma does not depend on the prism orientation. For the astigmatism coefficient we see that it is zero when $\varepsilon = N/2$, the prism orientation at minimum deviation.

An analysis including the prism thickness shows that, to a good approximation, the prism coma coefficient is the sum of Eqs. (15.3.6) and (7.2.11), where the latter is the coefficient for a plate of thickness t. The contribution of the thickness term is smaller than that of Eq. (15.3.6) by a factor of $\varepsilon t/N^2 s$ and can be ignored.

The transverse and spectral coma are given by

$$\text{TTC} = 3A_{2p}x^2 s = \frac{3\gamma(N^2-1)}{8NF^2} s, \quad (15.3.7)$$

$$\delta\lambda_c = \frac{3(N^2-1)}{8NF^2} \frac{d\lambda}{dN}, \quad (15.3.8)$$

where F is the focal ratio of the converging beam, and the angular dispersion A of a thin prism is $\gamma \, dN/d\lambda$. Note that the spectral coma is independent of the prism angle and distance from the focal surface.

As an example, consider a thin UBK7 prism in an $f/8$ beam. At a wavelength of 400 nm, $N = 1.53$ and, with $dN/d\lambda$ from Fig. 13.1, we get $\delta\lambda_c = 400$ Å. Thus a prism in the nonobjective mode is useful only for very low resolution, of order 10 in this example.

A final feature to note for a prism is the tilt of the focal surface. Figure 15.16 shows chief rays from sources in different parts of the field passing

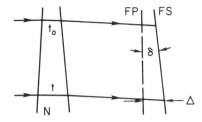

Fig. 15.16. Tilt of focal surface FS, shift from nominal telescope focal plane FP, due to prism in converging beam.

through different prism thicknesses. Neglecting the prism angle, we use Eq. (2.4.5) to find the shift in focus from the nominal telescope focal surface. Because the shift in focus is proportional to t, the average local prism thickness, the surface on which the spectrum is imaged is tilted with respect to the telescope focal surface. Using Eq. (2.4.5) we find that the angle δ between these two surfaces, in terms of the prism angle, is given by $\delta = \gamma(N-1)/N$. This relation is relevant to our discussions of the grism and prism-grating.

B. NONOBJECTIVE TRANSMISSION GRATING

We now consider a transmission grating in a converging beam at distance s from the telescope focal surface. The thickness of the grating blank contributes to the aberrations only if the grating face is not normal to the incident chief ray. Compared to the grating aberrations, the contribution of the thickness is roughly $t/3s$ smaller for practical tilt angles and is ignored.

The relations for this mode are those for the Monk-Gillieson mounting in Tables 14.1-14.4, with the appropriate sign for a transmission grating. In the paraxial approximation these are

$$A_{1g} = (\beta^2 - \alpha^2)/2s, \qquad (15.3.9)$$

$$A_{2g} = (\beta - \alpha)/2s^2 = (1/2s^2)m\lambda/\sigma, \qquad (15.3.10)$$

$$\kappa_t = -3/s, \qquad (15.3.11)$$

where the grating equation is used to rewrite the coma coefficient in Eq. (15.3.10).

Considering first the coma, we find

$$\text{TTC} = 3A_{2g}x^2s = (3s/8F^2)m\lambda/\sigma = 3\lambda/8F^2P, \qquad (15.3.12)$$

$$\delta\lambda_c = 3\lambda/8F^2, \qquad R = 8F^2/3, \qquad (15.3.13)$$

where F is the focal ratio of the converging beam and R is the spectral resolution. Note that the spectral coma and resolution are independent of the grating parameters and distance to the focal surface.

As an example, a grating in an $f/8$ beam has $\delta\lambda_c = 23$ Å at a wavelength of 400 nm, with $R = 170$. Compared to the prism example above, the improvement in spectral resolution is indeed substantial. Given the size of the spectral coma, a grating in this mode can be used at significantly higher dispersion or lower plate factor than a nonobjective prism.

The spectral resolution achievable with a nonobjective grating is set by spectral coma if the dispersion is large and by seeing if the dispersion is small. The boundary between these is found by setting $\delta\lambda_c = \delta\lambda$, where the latter is given in Eq. (15.3.2).

Solving this relation for P gives, in appropriate units,

$$P \text{ (Å/mm)} = 75\lambda \text{ (Å)}/F^3 D \text{ (m) } \phi' \text{ (arc-sec)}, \qquad (15.3.14)$$

where D is the telescope diameter. The seeing blur is larger than the coma blur for any P larger than that given by Eq. (15.3.14). Results from this relation are plotted in Fig. 15.17 for a selected set of focal ratios.

The constraint put on the plate factor by this relation is a conservative one because the entire width of the comatic image is used in the spectral coma. Because about 80% of the light in a comatic image is within a width $TTC/2$, a plate factor limit that is one-half that given in Eq. (15.3.14) and Fig. 15.17 is somewhat more realistic.

We now determine the characteristics of the astigmatic image. It is evident from Eq. (15.3.9) that the astigmatism can be made zero at a wavelength λ_0 by setting $\beta = -\alpha$, hence $2\alpha = -m\lambda_0/\sigma$. If the grating is tilted by angle α, however, the detector is also at angle α with respect to the grating surface. As a result of this tilt the plate factor now varies across the detector, and the fractional change in P across a detector of width W is $|\alpha|W/s$.

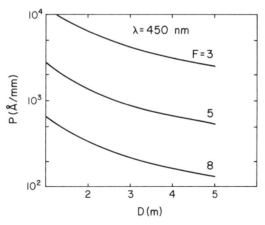

Fig. 15.17. Plate factor P at which spectral coma equals seeing blur of 1 arc-sec for nonobjective transmission grating. See Eq. (15.3.14).

The transverse astigmatism is

$$\text{TAS} = 2A_{1g}ys = \frac{(\beta^2 - \alpha^2)s}{2F} = \frac{\lambda(\lambda - \lambda_0)}{2FP^2 s}, \quad (15.3.15)$$

where the image length is 2TAS. For a given P and F we see that TAS varies inversely as s. To keep P constant for a larger s means choosing a grating with fewer grooves per millimeter. Alternatively, for a given grating at different distances from the focal surface, the size of TAS is linear with s.

As an example, take a first-order grating with 150 grooves/mm in an $f/8$ beam. This grating has a plate factor of 1000 Å/mm when s is 67 mm. At wavelengths of 400 and 600 nm, respectively, the astigmatic image lengths are 30 and 68 μm when $\alpha = 0$. If λ_0 is 500 nm, hence $\alpha = -2.15°$, the image lengths are 7 and 11 μm at 400 and 600 nm, respectively. Choosing a grating with 75 grooves/mm and letting $s = 133$ mm reduces the image lengths by a factor of two. Although astigmatism does not affect spectral resolution, it is evident that it is desirable to keep the image lengths short to maintain spectrographic speed.

From Eq. (15.3.11) we see that the spectrum of each source has its own curved surface with radius $s/3$, with the surface concave as seen from the grating. Because all of the spectra are recorded on a flat detector, this curvature results in a defocus blur and can degrade the spectral resolution.

The image surfaces for the examples with $\alpha = 0$ and $\alpha = -2.15°$ are shown in Fig. 15.18, with a line for each representing the optimum location of a detector for the 400–600 nm range. It is evident from Fig. 15.18 that the detector "fits" the image surface somewhat better for the tilted grating, with the zero-order image also in better focus. Simple geometry can be used in each case to determine the defocus blur at the ends of the spectral range.

A final important feature of the nonobjective grating mode is the presence of a zero-order reference for each spectrum, a reference not present for a prism in any slitless mode. This is a significant advantage in that quantitative measures of line positions are now possible. This is important, for example, for approximate measures of redshifts in emission-line objects. For a typical blazed grating with $P \approx 1000$ Å/mm, it turns out that the brightness in the zero order is comparable to that in the dispersed spectrum. Examples of such spectra are shown in the reference by Hoag and Schroeder.

C. NONOBJECTIVE GRISM

The principal defects in spectra taken with a nonobjective grating are coma across the spectral range and defocus at the ends of the range. These

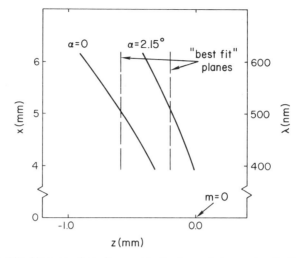

Fig. 15.18. Tilted image surfaces for nonobjective transmission mode with $P = 1000$ Å/mm. Horizontal scale is stretched by 2.5 times relative to vertical scale. Zero order and spectrum are in focus on surface with radius of curvature 22 mm.

defects are significantly reduced with a grism, a prism with a grating replicated on one of its faces. The characteristics of a grism in convergent light were first described by Bowen and Vaughan; in our discussion we reproduce the main results from their treatment.

A cross section of a grism is shown in Fig. 15.19, with the same notation for the angles as in Fig. 15.15. We again assume the paraxial approximation and neglect the prism thickness. In this approximation, the aberration coefficients of the grism are the sum of the corresponding coefficients of the prism and grating. Taking results from the previous sections, we get

$$A_{1s} = \frac{\gamma^2(N^2-1)}{2s}\left(1 - \frac{2\varepsilon}{N}\right) + \frac{\beta^2 - \alpha^2}{2s}, \quad (15.3.16)$$

$$A_{2s} = \frac{1}{2s^2}\left[\frac{m\lambda}{\sigma} - \frac{\gamma(N^2-1)}{N}\right]. \quad (15.3.17)$$

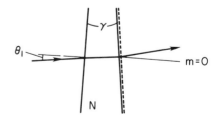

Fig. 15.19. Cross section of grism with apex angle γ. θ_1 is the angle of incidence at the first face; the grating is on the second face.

III. Nonobjective Slitless Spectrometers

The coma coefficient is zero for $\lambda = \lambda_0$ when the prism angle γ is given by

$$\gamma = \frac{m\lambda_0}{\sigma} \frac{N}{N^2 - 1}, \qquad (15.3.18)$$

and therefore

$$\text{TTC} = \frac{3s}{8F^2} \frac{m}{\sigma} |\Delta\lambda| = \frac{3|\Delta\lambda|}{8F^2 P}, \qquad (15.3.19)$$

where $\Delta\lambda = \lambda - \lambda_0$. Comparing the spectral coma derived from Eq. (15.3.19) with that for a grating given in Eq. (15.3.13), we see that coma for the grism is several times smaller. This implies, in turn, that a grism can be used at a plate factor that is smaller by the same amount, or at a faster focal ratio, before the coma and seeing blurs are equal.

The direction of the diffracted chief ray at the zero-coma wavelength is shown in Fig. 15.19. In this direction the dispersions of the prism and grating add; the grism dispersion is typically a few percent larger than that of the grating alone.

The astigmatism coefficient, unlike the coma, depends on ε and the grism orientation. With the grating equation, $m\lambda = \sigma(\beta - \alpha)$, and $\alpha = \gamma(\varepsilon - N)$, we can rewrite Eq. (15.3.16) as

$$A_{1s} = -\frac{\gamma^2(N^2 - 1)}{2N^2 s} [\zeta^2 + 2N\varepsilon(1 - \zeta) - N^2(1 - \zeta)^2],$$

where $\zeta = \lambda/\lambda_0$. Note that A_{1s} is independent of ε when $\zeta = 1$. Setting $dA_{1s}/d\zeta = 0$ and evaluating at $\zeta = 1$ gives $\varepsilon = 1/N$. With this choice of ε we have astigmatism constant near λ_0 and, to a good approximation, constant over a significant range centered on λ_0. Note that $\beta = 0$ at the corrected wavelength for this grism orientation. We show shortly that this choice of ε also significantly reduces defocus due to image surface curvature.

With $\varepsilon = 1/N$, substitution of A_{1s} into Eq. (14.1.14) gives

$$\text{TAS} = \frac{d\gamma^2(N^2 - 1)}{2N^2} [1 - (N^2 - 1)(1 - \zeta)^2], \qquad (15.3.20)$$

where d is the beam diameter at the grating surface. Note that TAS is largest at λ_0 and decreases slowly as λ changes.

We now illustrate these results with an example using a grating with 75 grooves/mm and a fused silica prism with $\gamma = 2.76°$, hence $\lambda_0 = 500$ nm. Assume an $f/8$ beam and $s = 240$ mm, hence $P \approx 555$ Å/mm. With these parameters we get TTC = 11 μm for $\Delta\lambda = \pm 100$ nm, and an image length of 38 μm at the corrected wavelength. Results from ray traces of a 10-mm-thick grism show that the zero-coma wavelength is 510 nm, with a plate factor of 535 Å/mm in this vicinity. Aberrations from ray traces are in excellent agreement with those given by the relations for TTC and TAS.

If this grism is placed in an $f/4$ beam at $s = 120$ mm, the plate factor $P \approx 1100$ Å/mm. With these changes the image length is unchanged and TTC is two times larger. Because the beam is faster, aberrations due to the thickness of the grism are larger and a ray-trace analysis is necessary to determine whether they are significant.

We noted in our discussion of the nonobjective prism that the nominal surface on which the spectra are in focus is tilted by an angle $\delta = \gamma(N-1)/N$ to the telescope focal plane. With $\varepsilon = 1/N$, the second surface of the grism is tilted by the same angle. Because the grating is on this surface, the perpendicular distance between the grating and detector is constant and the plate factor is the same over the field.

The fit between a tilted detector and the curved image surface of a single spectrum is shown in Fig. 15.20 for the grism example above, with tilt $\delta = 0.87°$. Comparing this diagram with Fig. 15.18, it is evident that the fit is better for the grism mode. A better fit is obtained for the grism example if θ_1 is 0.5° smaller, but at the expense of variable P in the dispersion direction. Thus the grism corrects the two major defects of the nonobjective grating, but with the slight added complication of a tilted detector.

In our discussion we have ignored the grism thickness and the variation of grism index with wavelength. Their effects change some of the results

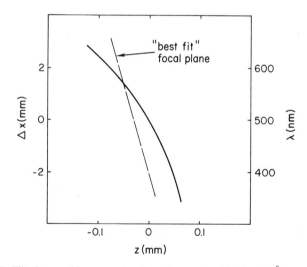

Fig. 15.20. Tilted, curved image surface for grism mode with $P = 550$ Å/mm. Grism parameters are $\alpha = 1.89°$, $\gamma = 2.76°$, $\delta = 0.87°$. Horizontal scale is stretched by 20 times relative to vertical scale.

III. Nonobjective Slitless Spectrometers

above, though not significantly, and were included by Bowen and Vaughan in their analysis.

D. NONOBJECTIVE PRISM-GRATING

The final nonobjective device considered consists of a separate prism and grating, as shown in Fig. 15.21. In this system there are two additional degrees of freedom: the separation between the elements and their relative orientations. With this system it is possible to eliminate both coma and astigmatism at one wavelength in the desired spectral range. This type of device is then suitable for fast beams, and it is used, for example, at the $f/2.7$ prime foci of the 4-m telescopes at the Kitt Peak and Cerro Tololo Observatories.

Because of the several degrees of freedom for the prism-grating, we make no attempt at a thorough analysis but only give selected results without derivation. The aberration coefficients for the prism-grating are easily found by substituting the prism and grating coefficients into Eq. (5.6.7), with the results

$$A_{1s} = \frac{1}{2s_2} \left[\gamma^2 (N^2 - 1) \left(1 - \frac{2\varepsilon}{N} \right) \left(\frac{s_1}{s_2} \right) + \beta^2 - \alpha^2 \right], \quad (15.3.21)$$

$$A_{2s} = \frac{1}{2s_2^2} \left[\frac{m\lambda}{\sigma} - \frac{\gamma(N^2 - 1)}{N} \left(\frac{s_1}{s_2} \right) \right], \quad (15.3.22)$$

where s_1 and s_2 are the distances from the prism and grating, respectively, to the focal surface. Note that these relations reduce to those for the grism when $s_1 = s_2$. Setting A_{2s} to zero gives

$$\gamma = \frac{s_2}{s_1} \frac{m\lambda_0}{\sigma} \frac{N}{N^2 - 1}. \quad (15.3.23)$$

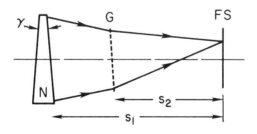

Fig. 15.21. Nonobjective prism-grating. FS, Focal surface.

Because the prism is farther from the focus than the grating in the arrangement in Fig. 15.21, coma correction is achieved with a smaller apex angle compared to the grism.

The way in which astigmatism varies is most easily seen by taking the prism and grating tilt angles of the grism, hence $\varepsilon = 1/N$ and $\delta = \gamma(N-1)/N$, respectively. With these values as starting points, and assuming s_2 and λ_0 are fixed, Eq. (15.3.21) can be evaluated at λ_0 for different s_1. Assigning the grism parameters to the grating and prism, astigmatism is zero when $s_1 = 1.88 s_2$. If ε is made smaller, then astigmatism is zero at a smaller value of s_1. Ray traces show good agreement with the calculated values for an $f/8$ beam. In faster beams, say $f/3$, the relations above are a good first approximation, and ray traces are necessary to optimize the system and find the best grating orientation to fit the curved image surface to the tilted detector.

The slitless modes discussed above assume a dispersing element in the converging beam ahead of the telescope focal surface. It is also possible to place a disperser in the diverging beam behind the focal surface and use a separate camera to focus the spectra. Because of the added optical elements this arrangement is less efficient than those discussed above, but it has the advantage that the focal ratio of the final beam can be chosen different from that of the telescope. A system of this type with a transmission grating and Schmidt camera has been built at the Royal Greenwich Observatory.

IV. CONCLUDING COMMENTS

With the exception of the nonobjective mountings, the discussion in this chapter has made little mention of the appropriate telescope focus for each instrument type. Instruments of smaller size are usually used at the Cassegrain focus, while large-beam spectrometers are usually placed at the fixed coudé focus. Coudé beam diameters as large as 300 mm are used, with the dispersed light most often sent to one of several Schmidt cameras. For large telescopes now in the planning or design phase, large-beam spectrometers will most often be placed at a Nasmyth focus on a platform that rotates with the telescope.

REFERENCES

Bowen, I. (1960). "Stars and Stellar Systems I, Telescopes," Chap. 4. Univ. of Chicago Press, Chicago, Illinois.

Schroeder, D. (1966). *Appl. Opt.* **5**, 545.
Willstrop, R. (1965). *Mon. Not. R. Astr. Soc.* **130**, 16.

BIBLIOGRAPHY

All types of spectrometers
James, J., and Sternberg, R. (1969). "The Design of Optical Spectrometers." Chapman & Hall, London.

Czerny-Turner mounting
Namioka, T., and Seya, M. (1967). *Sci. Light* **16**, 169.
Reader, J. (1969). *J. Opt. Soc. Am.* **59**, 1189.
Rosendahl, G. (1962). *J. Opt. Soc. Am.* **52**, 408 and 412.

Dispersers in convergent light
Bowen, I., and Vaughan, A. (1973). *Publ. Astron. Soc. Pac.* **85**, 174.
Buchroeder, R. (1974). Unpublished study for Kitt Peak National Observatory.
Hoag, A., and Schroeder, D. (1970). *Publ. Astron. Soc. Pac.* **82**, 1141.

Ebert-Fastie mounting
Fastie, W. (1952). *J. Opt. Soc. Am.* **42**, 641 and 647.
Welford, W. (1965). "Progress in Optics," Vol. 4, Chap. 6. North-Holland, Amsterdam.

Echelle spectrometers
Chaffee, F., and Schroeder, D. (1976). "Annual Review of Astronomy and Astrophysics," Vol. 14, p. 23. Annual Reviews, Palo Alto, California.
Schroeder, D. (1970). *Publ. Astron. Soc. Pac.* **82**, 1253.
Schroeder, D., and Anderson, C. (1971). *Publ. Astron. Soc. Pac.* **83**, 438.

Monk-Gillieson mounting
Kaneko, T., Namioka, T., and Seya, M. (1971). *Appl. Opt.* **10**, 367.
Schroeder, D. (1966). See reference listed above.

Chapter 16 | System Noise and Detection Limits

Given a particular telescope and instrument combination, it is essential to know its capabilities for making a specific type of observation. To determine these capabilities requires knowledge not only of the telescope and instrument characteristics but also of the characteristics of other parts of the overall system. The system as used here includes the detector and, in the case of ground-based telescopes, the atmosphere. The ability of a particular system to measure a given signal is then determined by including all of these factors in a system analysis.

In this chapter we discuss the characteristics of detectors that are important for the detection and resolution of point sources. We consider a detector in terms of its modulation transfer function (MTF) and the effect this has on the resolution of images. We do not discuss the physics of detectors, although characteristics of some specific types are used in examples of limiting-magnitude calculations.

Our discussion of the atmosphere considers the characteristics of a distorted wavefront entering a telescope and the effect on the image of a point object. Distorted wavefronts give rise to the phenomena of "scintillation" and "seeing;" we consider the later phenomenon from the point of view of the time-averaged MTF of the atmosphere.

In most cases the output of a detector includes both the desired signal from the source being observed and "noise" from unwanted sources. This noise can arise from many different sources, such as light from the sky background in the vicinity of the object under study, and noise intrinsic to the detector in the form of "dark current" for a photomultiplier or "dark counts" for a photon-counting detector. For detectors such as CCDs there is also "readout" noise introduced during the process of reading out the accumulated information stored on the array of pixels. In the absence of sky background and detector noise, the recorded signal shows fluctuations due to "photon noise." This noise arises from statistical fluctuations in the number of recorded photons about some average value, where the average number of detected photons is found from a large number of identical exposures. The effect of any noise contributor is analyzed in terms of signal-to-noise ratio, the topic of one of the sections in this chapter.

The final section of this chapter is a discussion of the detection limits in the presence of noise for different types of observations. Included are relations appropriate to direct imaging and spectrometric observations for both the slit-limited and slitless modes. Examples of selected modes are given for both ground-based telescopes and HST.

I. DETECTOR CHARACTERISTICS

The important detector parameters for our purposes are pixel size, quantum efficiency, intrinsic noise, and MTF. A pixel or picture element is a single detector element in an array of elements, such as a single AgBr grain on a photographic plate or a single diode on a silicon-based detector. Pixel sizes on panoramic detectors are typically in the range 15–50 μm.

A. QUANTUM EFFICIENCY AND INTRINSIC NOISE

If n_i is the number of photons per second incident on a detector and n_0 is the number of detected photons, the *quantum efficiency Q* is defined as $Q = n_0/n_i$. For a sensitive photographic plate Q is typically of the order of 1%, while for a CCD detector the quantum efficiency approaches unity in the near infrared. The rapid development of CCD technology during recent years has made this detector the one of choice for many spectrometric and direct imaging systems.

The principal noise contribution in the output of most electronic detectors is the dark current or, equivalently, the dark count, or number of electrons per second per pixel generated in the absence of an input light signal. The size of the dark count is reduced by cooling the detector and can often be reduced to a negligible level. For CCDs an additional noise contributor is the readout noise, usually given as some number of equivalent electrons rms. These noise factors are included in the discussion of signal-to-noise ratio and relations for detection limits in later sections.

For specifics on particular detectors the reader should consult data sheets prepared by manufacturers and references devoted to discussions of detectors.

B. MODULATION TRANSFER FUNCTION OF IDEAL PIXEL

The detection characteristics of a pixel array are determined in large part by the pixel size, which, in turn, can be described in terms of a detector MTF. An ideal detector is one for which the counts from a given pixel are completely independent of those from neighboring pixels; hence an ideal pixel can be represented as a rectangular well.

Consider such a pixel of dimensions a and b in the x and y directions, respectively. We derive an MTF by defining this rectangular aperture as a point spread function (PSF) according to

$$i(x, y) = 1, \quad |x| \le a/2 \quad \text{and} \quad |y| \le b/2,$$
$$= 0, \quad |x| \ge a/2 \quad \text{or} \quad |y| \ge b/2, \quad (16.1.1)$$

where the pixel center is at $x = y = 0$. Substituting Eq. (16.1.1) into Eq. (11.1.6) gives the normalized MTF as

$$T_a(\nu_x, \nu_y) = \operatorname{sinc}(\pi \nu_x a) \operatorname{sinc}(\pi \nu_y b), \quad (16.1.2)$$

where $\operatorname{sinc}(z) = (\sin z)/z$. The spatial frequencies in Eq. (16.1.2) are defined following Eq. (11.1.6), where γ is the orientation of the input sine target, discussed in Section 11.I.A, relative to the x, y coordinate system. If the lines of the sine target are parallel to x or y, one sinc function in Eq. (16.1.2) becomes one.

Assuming square pixels of side a and $\nu_x = \nu$, Eq. (16.1.2) becomes

$$T_a(\nu) = \operatorname{sinc}(\pi \nu a), \quad T_a(\nu'_n) = \operatorname{sinc}(\pi \nu'_n), \quad (16.1.3)$$

where we define a normalized detector frequency $\nu'_n = \nu a$. Note that the MTF and normalized detector frequency are independent of wavelength. In the discussion to follow we take Eqs. (16.1.3) as the representation of an ideal pixel array.

The system MTF, including detector, is the product $T(\nu) T_a(\nu)$, where $T(\nu)$ includes the factors in Eq. (11.1.14) for a system with aberrations. Following the procedure in Chapter 11, we rewrite this product in terms of the normalized spatial frequency of the system. Denoting the system MTF by T_s we get

$$T_s(\nu_n) = T(\nu_n) \operatorname{sinc}\left[\pi\left(\frac{a}{\lambda F}\right)\nu_n\right] = T(\nu_n) \operatorname{sinc}\left(\frac{\pi\nu_n}{\Delta}\right), \quad (16.1.4)$$

where $\Delta = \lambda F/a =$ number of pixels per length λF. From Eq. (10.1.11) we see that λF is the approximate radius of the Airy disk.

Figure 16.1 shows pixel MTFs for three different values of Δ. Note that the MTF is negative for some $\nu_n < 1$ when $\Delta < 1$. System MTFs for a perfect system with a circular aperture and $\varepsilon = 0.33$ are shown in Fig. 16.2 for each detector MTF in Fig. 16.1. The effect of the detector MTF is evident by inspection of the curves in Fig. 16.2. For $\Delta = 2$ there is little change of the MTF due to the optics only, with progressively greater change as Δ decreases. Note that high-frequency information is lost or masked for $\Delta = 0.5$. This is not surprising, given that a single pixel spans an Airy disk diameter in this case.

According to the Nyquist criterion for discrete sampling, two samples per resolution element are required for unambiguous resolution of images that are just resolved according to the Rayleigh criterion. With a linear

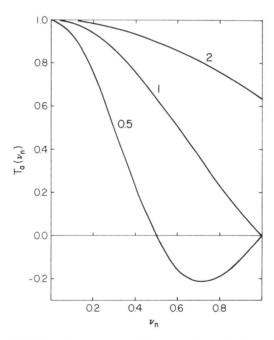

Fig. 16.1. Pixel MTF for different values of Δ based on Eq. (16.1.4). $\Delta = \lambda F/a$, where a is the pixel size.

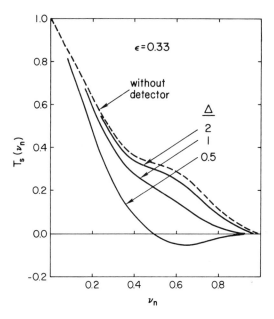

Fig. 16.2. System MTF including detector for circular aperture with central obscuration. Detector MTF is shown in Fig. 16.1.

separation at the Rayleigh limit given by λF, the Nyquist criterion is satisfied with a pixel size $a = \lambda F/2$, hence $\Delta = 2$. Strictly speaking, this criterion applies to sampling in one direction, but it is also appropriate for panoramic sampling.

Given $\Delta = \lambda F/a$, the curves in Fig. 16.1 change in proportion to wavelength with F/a held constant. If, for example, the curve for $\Delta = 2$ is appropriate for wavelength λ, then the $\Delta = 1$ curve is appropriate for $\lambda/2$ with the same pixel size and focal ratio. In this case, therefore, an image that is properly sampled at longer wavelengths may not be adequately sampled at shorter wavelengths.

We now apply these results to the HST cameras assuming (1) a perfect telescope and (2) a pixel MTF of the form given in Eq. (16.1.4). The first assumption is not far from correct for the near ultraviolet and longer wavelengths but is a poor assumption in the far ultraviolet. The second assumption presumes an ideal pixel and may not be strictly valid for a real detector. Nevertheless, we proceed with these assumptions to illustrate the general characteristics of the HST cameras in selected modes.

Listed in Table 16.1 are the focal ratios of the direct imaging modes of the wide-field planetary camera (WFPC) and faint object camera (FOC) and the width of a pixel projected on the sky for each mode. Note that the

I. Detector Characteristics

Table 16.1

Selected Characteristics of Hubble Space Telescope Cameras[a]

WFPC			FOC		
F	a (arc-sec)	F/a[b]	F	a (arc-sec)	F/a
12.9	0.100	0.86	48	0.044	1.92
30	0.043	2.00	96	0.022	3.84
			288	0.007	11.52

[a] WFPC: $a = 15$ μm, FOC: $a = 25$ μm.
[b] The unit of F/a is μm^{-1}.

pixel size of the $f/12.9$ wide-field mode of the WFPC is larger than the FWHM, given in Fig. 11.15, for all wavelengths to which it is sensitive. At the opposite extreme, we see that the pixel size of the $f/288$ mode of the FOC is three or more times smaller than the FWHM for all wavelengths within its range. The 3.8 square arc-sec field covered in this FOC mode is, however, much smaller than the 2.63 square arc-min field covered in the $f/12.9$ wide-field mode and illustrates the inevitable trade-off between angular resolution and field coverage.

Taking the entries for F/a in Table 16.1, we get the lines representing Δ for each of the camera modes shown in Fig. 16.3. The lines for the FOC

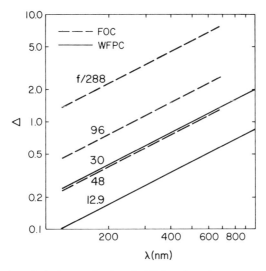

Fig. 16.3. Number of pixels spanning length λF for selected camera modes of the Hubble Space Telescope.

are terminated at a wavelength of 650 nm, the approximate long-wave cutoff of the spectral response of the FOC detectors. For each mode the range of wavelengths for which $\Delta > 2$ is the range in which the image of a point source is oversampled. When $\Delta = 2$ there are approximately 12 pixels covering the Airy disk.

It is important to note that detection of a single point source does not require a large number of pixels covering the Airy disk. For the $f/12.9$ mode of the WFPC, for example, most of the energy of a single star image is recorded on four pixels, which is sufficient for many types of observations.

C. APPROXIMATE PIXEL MODULATION TRANSFER FUNCTION

An approximate relation often used to represent the MTF of a square pixel is a Gaussian profile of the form

$$T_a(\nu) = \exp[-0.282(\pi\nu a)^2], \qquad (16.1.5)$$

where the constant 0.282 is chosen to make $T_a = 0.5$ at $\nu = 1/2a$. Figure 16.4 shows T_a for a sinc function and Gaussian, the former according to Eqs. (16.1.3) and the latter from Eq. (16.1.5). If curves like those in Fig. 16.2 are generated using Eq. (16.1.5) rather than (16.1.3), the resulting system MTFs are little different except that the curve for $\Delta = 0.5$ does not go

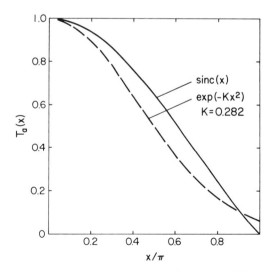

Fig. 16.4. Plot of sinc and Gaussian representations of pixel MTF. K is chosen to make the Gaussian 0.5 at $x = \pi/2$. See Eq. (16.1.5).

negative. The general comments made on Fig. 16.2 are also true for the modified MTFs.

Because the MTF and PSF are a Fourier transform pair, the PSF for Eq. (16.1.5) is also a Gaussian. This form of the PSF is often used to represent the profile of a stellar image degraded by atmospheric turbulence, a topic we discuss in the next section.

II. EFFECTS OF ATMOSPHERIC TURBULENCE

The magnitude limits that can be reached by a ground-based telescope depend on many factors, of which the image size of a stellar source due to atmospheric effects is an important one. In this section we describe some of the effects of the atmosphere on images, including selected relations based on the theory of atmospheric turbulence. The relations given by this theory are not derived; for thorough discussions of the theory of turbulence applied to optical astronomy the reader should consult the references by Roddier and Coulman.

Although the general characteristics of images and the wavefronts that produce them have been known for a long time, it is only recently that technological advances have made it possible to overcome the resolution limits of long-exposure photography. These advances have led to techniques such as speckle imaging and speckle interferometry for achieving high angular resolution with ground-based telescopes. For information on these techniques, the reader should consult the reference by McAlister.

The most notable effect of a turbulent atmosphere is a blurred image in the focal plane of a telescope. For a large telescope the image size, often called the seeing disk, is usually larger than the diffraction disk. The angular radius of the diffraction disk α_d, derived from Eq. (10.1.11), is

$$\alpha_d(\text{arc-sec}) = 0.25\lambda\,(\mu m)/D(m). \qquad (16.2.1)$$

If the radius α of the seeing disk is 1 arc-sec and $\lambda = 0.5$ μm, we get $\alpha = \alpha_d$ for a 12-cm telescope. Thus for visible wavelengths and 1-m or larger telescopes, the image size is determined by atmospheric effects. Because $\alpha_d \propto \lambda/D$ the seeing disk may be comparable to the diffraction disk at infrared wavelengths for large telescopes, especially at the longest wavelengths that reach the ground.

The effect of atmospheric turbulence on stellar images is usually separated into two distinct phenomena. *Seeing* is the term used to describe random changes in the direction of light entering a telescope, while *scintillation* refers to random fluctuations in the intensity. Both of these effects arise

314   16. System Noise and Detection Limits

from variations in the index of refraction, which give rise to a distorted wavefront. A cross section of such a wavefront reaching the ground at a given instant of time is shown schematically in Fig. 16.5.

We first describe the effects of seeing and scintillation on a stellar image as observed with the eye. Scintillation is most evident to the unaided eye as the phenomenon called "twinkling." In a telescope the twinkling is usually not seen, and a photometer is needed to record the fluctuations in intensity. In general, the larger the aperture the smaller are the fractional changes in the intensity.

The effect of seeing is a function of the telescope aperture. In good seeing, with a 10-cm aperture or less, the Airy disk of a star moves randomly about its mean position in the focal plane with excursions of one or two arc-seconds. In a large telescope, 1 m or larger, a blurred image is seen with little or no motion of the image as a whole. If the eye could follow the rapid changes within the image, it would see a changing pattern of speckles, each speckle having a size comparable to an Airy disk. A given speckle pattern is stationary over times on the order of 10–50 msec, with two patterns similar only for point sources within about 10 arc-sec of one another.

From these observations we deduce that the curvature of the wavefront is negligible over distances of the order of 10 cm, with instantaneous slopes of 1 or 2 arc-sec from an undistorted wavefront. The image seen in a large telescope is thus the average over many sections of the wavefront, each with a different instantaneous slope.

The demonstration that wavefront distortions arise from variations in index of refraction was discussed in Chapter 3 from the point of view of Fermat's principle. This approach was adequate for showing the origin of seeing, but a more fruitful approach is one based on the theory of atmospheric turbulence. We now present selected results derived from a statistical approach, with results taken from the reference by Roddier.

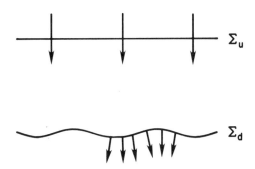

Fig. 16.5. Cross sections of undistorted wavefront Σ_u at top of atmosphere and distorted wavefront Σ_d at ground, after passage through turbulent atmosphere.

II. Effects of Atmospheric Turbulence

We first consider the image of a point source that has been broadened by seeing to a width large compared to the diffraction width. An approximate form for the distribution of energy within such an image is a Gaussian, with the normalized intensity given by

$$i(\alpha) = \exp(-\alpha^2/2\sigma'^2), \qquad (16.2.2)$$

where α is the angular distance from the image peak and σ' is the rms deviation in a given direction from the peak. To find the MTF we substitute Eq. (16.2.2) into Eq. (11.1.8), adjust the normalization factor to give $T(0) = 1$, and get

$$T(\nu) = \exp(-2\pi^2 \sigma'^2 \nu^2). \qquad (16.2.3)$$

If the unit of σ' is arc-seconds, the unit of ν is cycles per arc-second. Note the correspondence between these relations for $i(\alpha)$ and $T(\nu)$ with Eqs. (11.1.17) and (11.1.18). Equation (11.1.18) gives the pointing degradation and its product with the telescope MTF is the system MTF in the presence of pointing error. Equation (16.2.3) can be taken as the system MTF provided the telescope MTF is essentially unity over the range where $T(\nu)$ is effectively nonzero.

As an illustration, we choose σ' to give an image whose FWHM is 1 arc-sec. Setting $i(\alpha) = 0.5$ with $\alpha = 0.5$ arc-sec in Eq. (16.2.2) gives $\sigma' = 0.423$ arc-sec. Substituting this value of σ' in Eq. (16.2.3), we find $T(\nu) = 0.029$ for $\nu = 1$ cycle/arc-sec. This is the effective cutoff frequency and is small compared to the diffraction cutoff frequency D/λ for a large telescope. For a 4-m telescope at $\lambda = 500$ nm we get $\nu_c = D/\lambda = 38.8$ cycles/arc-sec, and in this case we are justified in taking Eq. (16.2.3) as the system MTF.

Although Eq. (16.2.3) is a reasonable approximation to the MTF of a large ground-based telescope, it is an ad hoc relation based solely on a statistical approach. An approach based on the physics of the atmospheric turbulence gives an MTF that leads to a better description of the observed image profile. The resulting degradation function, as given by Roddier, is

$$T(\nu) = \exp[-3.44(\lambda \nu/r_0)^{5/3}], \qquad (16.2.4)$$

where ν is the angular frequency and r_0 is a wavelength-dependent length that is a measure of the seeing quality. As noted by Roddier, this $T(\nu)$ is appropriate for a long-exposure image.

The parameter r_0 is defined such that the angular resolution of a telescope is set by the atmosphere when $D > r_0$ and set by the telescope when $D < r_0$. For a large telescope limited by seeing, the limiting angular resolution is

$$\alpha_0 = 1.27(\lambda/r_0). \qquad (16.2.5)$$

The specific form of r_0 is given by

$$r_0 = 0.185\lambda^{6/5}(\cos\gamma)^{3/5}S^{-3/5}, \qquad (16.2.6)$$

where γ is the zenith angle and S is a function integrated through the atmosphere which is a measure of the turbulence. The reader should consult the reference by Roddier for details on the function S.

It is evident from Eq. (16.2.6) that r_0 increases slowly with increasing wavelength and decreases with increasing zenith angle. At $\gamma = 45°$, r_0 is approximately 20% smaller than at the zenith. In the examples to follow, we assume γ is zero.

Figure 16.6 shows $T(\nu)$ from Eq. (16.2.4) for $\lambda = 0.5$ μm and $r_0 = 12.9$ cm, with the latter value chosen to give $\alpha_0 = 1$ arc-sec. The image profile is the Fourier transform of $T(\nu)$, which, because $T(\nu)$ is nearly (but not exactly) a Gaussian function, is also nearly a Gaussian. Results presented by Roddier show that the image core is close to this form, but the decrease in the wings is slower than that of a Gaussian function.

Table 16.2 gives values of r_0, α_0, and α_d for several wavelengths, assuming the same turbulence for each. The values of α_d are calculated from Eq. (16.2.1) assuming a 4-m telescope. The improvement in resolution with increasing wavelength is evident from these results. Note also that diffraction, negligible in the visible, grows in significance in the infrared and is larger than the turbulence effect at $\lambda = 10$ μm.

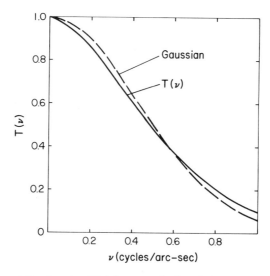

Fig. 16.6. Degradation function $T(\nu)$ for atmospheric turbulence according to Eq. (16.2.4) with $\lambda = 0.5$ μm and $r_0 = 12.9$ cm. Gaussian is a plot of Eq. (16.2.3) with $\sigma' = 0.375$ arc-sec.

III. Signal-to-Noise Ratio

Table 16.2

Angular Resolution for Constant Turbulence

λ (μm)	r_0 (cm)	α_0 (arc-sec)	α_d (arc-sec)
0.5	12.9	1.00	0.03
2.2	76.3	0.74	0.14
10.0	470	0.55	0.63

All of the results above are based on the long-exposure MTF given in Eq. (16.2.4). Roddier gives an expression for a short-exposure MTF, and the reader should consult this reference for details.

This discussion of the effects of atmospheric turbulence is only a brief overview of a large subject area but should suffice to give the reader an idea of the basic concepts.

III. SIGNAL-TO-NOISE RATIO

The performance of a system is most often given in terms of a quantity called the *signal-to-noise ratio* (SNR). The signal is the total number of detected photons on a given pixel, denoted here by n_s. If the signal is recorded a large number of times under identical conditions, the mean signal is $\langle n_s \rangle$ with a statistical fluctuation in the number of detected photons about the mean. This fluctuation is a consequence of the random arrival rate of the photons at the detector and the random selection of those that are detected.

The variance or mean square noise in the signal is $\langle n_s \rangle$, and the standard deviation or noise is $\sqrt{\langle n_s \rangle}$. For an ideal detector, one giving only counts from the incident light, the SNR in the presence of photon noise is

$$\mathrm{SNR} = \langle n_s \rangle / \sqrt{\langle n_s \rangle} = \sqrt{\langle n_s \rangle}. \tag{16.3.1}$$

For a real detector measuring a signal in the presence of a background source, the mean square noise from uncorrelated sources is the sum of the variances of the separate sources, and the SNR is

$$\mathrm{SNR} = \langle n_s \rangle / \sqrt{\langle n_s \rangle + \langle n_b \rangle + \langle n_d \rangle}, \tag{16.3.2}$$

where $\langle n_b \rangle$ is the noise from the background and $\langle n_d \rangle$ the number of extraneous counts from the detector. The extra detector counts are due to dark counts and, in the case of a CCD detector, readout noise. The *fractional accuracy* for a given SNR is defined as 1/SNR; thus an observation with

1% accuracy requires a SNR of 100. It is evident from Eq. (16.3.2) that better accuracy is achieved with smaller $\langle n_d \rangle$.

We now consider two limiting cases of SNR for an ideal detector, one with zero background, the other with background large compared to the signal. Let S and B denote the incident signal and background flux, respectively, in photons per second on an ideal detector. If the exposure time is t and the quantum efficiency is Q, then the SNRs are

$$\text{SNR} = \sqrt{SQt}, \qquad B \ll S, \qquad (16.3.3a)$$

$$\text{SNR} = \sqrt{SQt} \cdot \sqrt{S/B}, \qquad B \gg S. \qquad (16.3.3b)$$

It is instructive to examine these relations for SNR from two different perspectives. Consider first the situation where S and B are constant, and observations are made of the same source with different detectors and/or exposure times. In both cases we see that the $\text{SNR} \propto \sqrt{Qt}$; hence a larger Q with a given exposure time or a longer exposure with the same Q gives a larger SNR. It is also evident that increasing the SNR by a factor of k requires a Qt product that is k^2 times larger.

Alternatively, consider two sources with signal fluxes S_1 and S_2, respectively, observed against the same background B to the same SNR level. From Eqs. (16.3.3a) and (16.3.3b) we get

$$\frac{S_1}{S_2} = \frac{Q_2 t_2}{Q_1 t_1}, \qquad \text{signal-limited}, \qquad (16.3.4a)$$

$$\frac{S_1}{S_2} = \left(\frac{Q_2 t_2}{Q_1 t_1}\right)^{1/2}, \qquad \text{background-limited}. \qquad (16.3.4b)$$

As a numerical example, let $t_1 = t_2$ and $Q_2/Q_1 = 6.3$. In the signal-limited case $S_2 = S_1/6.3$; in the background-limited case we have $S_2 = S_1/2.5$. Thus the more sensitive detector can observe a source that is two stellar magnitudes fainter in the signal-limited case, but only one magnitude fainter in the other case, for the same exposure and SNR.

From Eq. (16.3.3) it is evident that the SNR achieved with a given detector is larger with a longer exposure time. Consider a single exposure of length t compared to k separate exposures, each of length t/k. For each short exposure, the SNR is $\sqrt{k}$ smaller than for the single long exposure for either case in Eq. (16.3.3). By replacing t by $k \cdot t/k$ in Eqs. (16.3.3) we see that

$$\text{SNR}_1 = \sqrt{k \cdot \text{SNR}_k}, \qquad (16.3.5)$$

where the subscripts 1 and k refer to the long and short exposures, respectively. It follows, therefore, that the SNR of k added exposures is the same as that of a single long exposure, where the total exposure time is the same for both.

IV. Detection Limits 319

As a final limiting case, consider the situation where the detector noise is large compared to either the background or signal. Assuming the detector noise is due to both dark counts and readout noise, the SNR for the detector-limited case can be written as

$$\text{SNR} = SQt/\sqrt{Ct + R^2}, \qquad (16.3.6)$$

where C is the dark count per second and R the rms readout noise. If readout noise is negligible compared to dark count, the addition of k separate short exposures leads to the result given in Eq. (16.3.5). If R is dominant, however, then the SNR for k added exposures is smaller than that of a single long exposure by a factor of $\sqrt{k}$.

This treatment is sufficient to illustrate how system performance is specified in terms of SNR. We now use this parameter to give a more detailed analysis for both photometry and spectroscopy.

IV. DETECTION LIMITS

Most telescope/instrument systems are used for observations that are at or near the limits of the system. These limits may be due to source faintness, sky background, limited observing time, detector noise, or any combination of these. It is therefore important to know how each of these affects the magnitude limit that can be reached at a given SNR. Treatments like the one that follows have been given by several authors, including Baum, Code, and Bowen. References are listed at the end of the chapter.

In this section we consider three types of observations and the relation between source brightness, exposure time, and SNR in the presence of the various factors that degrade the SNR. Types of observations discussed include stellar photometry, slit-limited spectroscopy at various resolutions, and slitless spectroscopy. For each observation mode we illustrate the general results with graphs for the HST and large ground-based telescopes of various diameters, using detector characteristics suitable for each. We assume in all cases that the light is collected by a single telescope; situations in which an array of telescopes sends light to one or more instruments are deferred to Chapter 17.

We begin with the expression for the photon flux collected by a telescope of diameter D and transmitted to the detector. For a star of apparent magnitude m, the signal flux is

$$S = N\tau \frac{\pi}{4}(1-\varepsilon^2)D^2\Delta\lambda \cdot 10^{-0.4m},$$
$$= 0.7 N\tau D^2 \Delta\lambda \cdot 10^{-0.4m}, \qquad (16.4.1)$$

where we set $\pi(1-\varepsilon^2)/4 = 0.7$, assuming a typical ε for a Cassegrain telescope. This factor is included in all the relations that follow. The remaining factors in Eq. (16.4.1) are defined as follows: $N = 10^4$ photons/(sec cm^2 nm) for a zero-magnitude A0 star at a wavelength of 550 nm, τ is the system transmittance from the top of the atmosphere to the detector (not including slit losses), and $\Delta\lambda$ is the bandpass of the instrument used. For photometry the bandpass is defined by a filter; for spectroscopy the bandpass is set by the spectrometer.

The photon flux from the sky background is given by

$$B = 0.7 N\tau D^2 \Delta\lambda' \cdot 10^{-0.4m'} \phi\phi', \qquad (16.4.2)$$

where N and τ are defined above, $\Delta\lambda'$ is the bandpass of sky on the detector, m' is sky brightness in magnitudes per arc-second squared, and $\phi\phi'$ is the detector area in arc-seconds squared projected on the sky. For stellar photometry and slit spectroscopy $\Delta\lambda' = \Delta\lambda$; for slitless spectroscopy the two bandpasses are different.

In terms of photon flux, quantum efficiency Q, and exposure time t, we write Eq. (16.3.2) as

$$\text{SNR} = \frac{\kappa S Q t}{\sqrt{(\kappa S + B)Qt + Ct + R^2}} = \frac{\kappa S Q t}{\sqrt{\kappa S Q t + \langle n_u \rangle}}, \qquad (16.4.3)$$

where C and R are the dark counts per second and rms readout noise, respectively, as used in Eq. (16.3.6), and $\langle n_u \rangle$ is the sum of all contributors to the noise.

The factor κ in Eq. (16.4.3) is included to account for factors not included in the transmittance of the system. In some photometric modes, for example, some fraction of the flux in a stellar image may not fall on a given pixel or group of pixels. For the HST, for example, the fraction of the energy on a set of pixels centered on the image depends on the camera mode. The same is true for a ground-based telescope measuring an image with a Gaussian profile. For slit spectroscopy part of the image at the entrance slit may be intercepted by the slit jaws and not reach the detector, or the signal of interest may be the core of an absorption line. The factor κ can account for these factors.

Other useful forms of Eq. (16.4.3) are obtained by solving this relation for either m or t. We choose to solve Eq. (16.4.3) for m, with Eq. (16.4.1) substituted for S. The result is

$$m = -2.5 \log\left[\frac{(\text{SNR})^2}{1.4 N \kappa \tau \Delta\lambda D^2 Qt}\left(1 + \left(1 + \frac{4\langle n_u \rangle}{(\text{SNR})^2}\right)^{1/2}\right)\right]. \qquad (16.4.4)$$

Representative results obtained from Eq. (16.4.4) for various combinations of parameters in different observation modes are given in the sections that follow.

A. STELLAR PHOTOMETRY

Before considering specific telescope and detector combinations, it is instructive to look at two limiting cases for a noise-free detector: signal-limited and background-limited. In the former case we assume $\langle n_u \rangle$ is negligible; in the latter case $\langle n_u \rangle = BQt$ and is large compared to the signal. In the signal-limited case Eq. (16.4.4) becomes

$$m = -2.5 \log\left[\frac{(\text{SNR})^2}{0.7 N \kappa \tau \Delta \lambda D^2 Qt}\right], \qquad (16.4.5)$$

while in the background-limited case

$$m = 0.5 m' - 1.25 \log\left[\frac{(\text{SNR})^2 \phi \phi'}{0.7 N \kappa^2 \tau \Delta \lambda D^2 Qt}\right]. \qquad (16.4.6)$$

We first consider the situation where observations for a fixed bandpass are made with different telescopes and/or detectors to the same SNR level. We also assume $\phi = \phi'$ and constant sky brightness. Starting with Eq. (16.4.5) or (16.4.6), we find the difference of the magnitudes reached as a function of the remaining variables, for the same SNR. For the signal-limited case we get

$$m_2 - m_1 = 2.5 \log\left[\left(\frac{D_2}{D_1}\right)^2 \frac{\tau_2 Q_2 t_2}{\tau_1 Q_1 t_1}\right], \qquad (16.4.7)$$

and for the background-limited case

$$m_2 - m_1 = 1.25 \log\left[\left(\frac{D_2 \phi_1}{D_1 \phi_2}\right)^2 \frac{\tau_2 Q_2 t_2}{\tau_1 Q_1 t_1}\right]. \qquad (16.4.8)$$

From Eq. (16.4.7) we see that doubling the telescope diameter with all other parameters held constant gives $\Delta m = 1.5$; from Eq. (16.4.8) the same conditions give $\Delta m = 0.75$. Thus the faintness of a star observed to the same SNR is proportional to the telescope area in the signal-limited case but only proportional to the telescope diameter in the background-limited case.

We see from Eq. (16.4.8) that the faintness of a star observed to the same SNR is inversely proportional to the image diameter with all other parameters constant. For ground-based telescopes the importance of good seeing in reaching faint magnitudes is therefore evident. In the event that the image diameter is determined by diffraction rather than seeing, Eq. (16.4.8) is modified by replacing ϕ_1/ϕ_2 by D_2/D_1, and the faintness of a star observed to the same SNR is again proportional to the telescope area.

We now consider the situation where the same telescope and detector are used for observations made to different SNR levels. In this case the result for signal-limited observations is

$$m_2 - m_1 = -5 \log(\text{SNR}_2/\text{SNR}_1), \qquad (16.4.9)$$

and for background-limited observations

$$m_2 - m_1 = -2.5 \log(\text{SNR}_2/\text{SNR}_1). \qquad (16.4.10)$$

Therefore the slope in a log(SNR) versus magnitude plot is different by a factor of two in the two regions.

Returning now to Eq. (16.4.4), we give results for the HST and ground-based telescopes of different apertures, each with the same detector and filter characteristics. The parameters used for the calculations are given in Table 16.3, with a CCD as the detector.

We choose $\kappa = 0.8$ for all telescopes and assume this fraction of the transmitted light contributes to the detected signal. For the wide-field mode of the WFPC, the fraction κ of a stellar image falls on four pixels. For a large ground-based telescope, the fraction κ of an image with a Gaussian profile typically covers many pixels. The number of pixels k spanning the image is given, in appropriate units, as

$$k = \frac{\text{width}}{\text{pixel size}} = \frac{5FD(\text{m})\phi(\text{arc-sec})}{\Delta(\mu\text{m})}. \qquad (16.4.11)$$

We assume the total readout noise is proportional to k^2, hence there is no on-chip summing before readout. The dark count is, of course, proportional to k^2.

Results obtained from Eq. (16.4.4) with the parameters in Table 16.3 are shown in Fig. 16.7 for an exposure time of 2400 sec, the approximate time available in the dark part of one HST orbit. From the results in Fig. 16.7 it is evident that the ground-based observations are primarily in the background-limited region, while those with the WFPC are in the transition between the signal-limited and background-limited regimes.

Table 16.3

Detector and Telescope Parameters

Detector	$\Delta = 15 \, \mu\text{m}$,	$R = 10$ counts/pixel,	
	$C = 0.01$ counts/(pixel sec),	$\kappa = 0.8$,	
	$Q = 0.5$		
Telescope	$\tau = (0.9)^2 = 0.81$		
Filter	$\tau = 0.8$,	$\Delta\lambda = 100$ nm (V-band)	
Relay optics	$\tau = 0.5$		
Other	$m'\,[\text{mag}/(\text{arc-sec})^2]$	ϕ (arc-sec)	F
WFPC	23	0.2	12.9
Ground	22	1.0	2.5

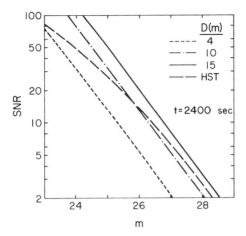

Fig. 16.7. Representative SNR-apparent magnitude diagram for photometry. Parameters are given in Table 16.3.

It is also of interest to note that the noise contribution of the detector is negligible, compared to the sky background, for each of the ground-based telescopes at the chosen focal ratio. For each of these telescopes the sky contribution is approximately 50 times larger than the detector noise. For the HST, on the other hand, both the sky noise and the detector noise are smaller, especially the former. The net result for this camera mode of the HST is detector noise that is about 25% larger than the sky noise. At small SNR the effect of the detector noise is to reduce the limiting magnitude by about 0.4 magnitudes.

The results for the ground-based telescopes in Fig. 16.7 are based on a sky brightness of 22 mag/square arc-sec. If the sky is fainter by 0.5 magnitudes, the curves for these telescopes are shifted to the right by 0.25 magnitudes, as shown by Eq. (16.4.6). This change is not as pronounced as that obtained with better seeing; for an image diameter smaller by a factor of two the curves are shifted to the right by 0.75 magnitudes.

The results in Fig. 16.7 are intended only as an illustration of the relation between SNR and limiting magnitude for one given set of parameters. Because of the many variables that influence this relation, each particular telescope-filter combination requires its own set of calculations.

B. SLIT SPECTROSCOPY—STELLAR SOURCES

The calculation of SNR for slit spectrometers proceeds by using Eqs. (16.4.3) and (16.4.4) modified for different source or observing conditions.

For a stellar source there are two cases: the star image fits entirely within the slit, or part of the image falls on the slit jaws. If the star image is entirely within the slit we set $\kappa = \kappa_0$; if part of the image is intercepted by the slit we set $\kappa = \kappa_0(\phi/\phi')$, where ϕ' is the diameter of the image and ϕ is the slit width projected on the sky. The slit-limited case is the usual one with large telescopes, especially at high resolution. The factor κ_0 is the level of the signal relative to the continuum. For an emission line $\kappa_0 > 1$; in the core of an absorption line $\kappa_0 < 1$.

Incorporating this factor into Eq. (16.4.4) gives

$$m = -2.5 \log\left[\frac{(\text{SNR})^2(\phi'/\phi)}{1.4N\kappa_0\tau\Delta\lambda D^2 Qt}\left(1 + \left(1 + \frac{4\langle n_u\rangle}{(\text{SNR})^2}\right)^{1/2}\right)\right].$$
(16.4.12)

In the case where $\phi = \phi'$, Eq. (16.4.12) is the same as Eq. (16.4.4) at the continuum level and the limiting cases for a noise-free detector are given by Eqs. (16.4.5) and (16.4.6). The only difference for a spectrometer is $\Delta\lambda = P\omega'$, where P is the plate factor and ω' the projected slit width in the spectrometer focal plane.

In the slit-limited case we use Eq. (12.2.1) to express ϕ in terms of the projected slit width, hence $\phi = \omega'/rDF_2$. This is required because ϕ and $\Delta\lambda$ are not independent of one another. With this substitution into Eq. (16.4.12) we see that one factor of D is canceled. The limiting cases for a noise-free detector are

$$m = -2.5 \log\left[\frac{(\text{SNR})^2 rF_2\phi'}{0.7N\kappa_0\tau P\omega'^2 DQt}\right]$$
(16.4.13)

in the signal-limited case and

$$m = 0.5m' - 1.25 \log\left[\frac{(\text{SNR})^2 rF_2\phi'^3}{0.7N\kappa_0^2\tau P\omega'^2 DQt}\right]$$
(16.4.14)

in the background-limited case.

From these relations we see that the faintness of a star observed to the same SNR is proportional to D in the signal-limited case and proportional to $\sqrt{D}$ in the background-limited case. Note that for a diffraction-limited telescope we have $\phi' \propto 1/D$, and the faintness reached at a given SNR is proportional to D^2.

If a given system is used to make observations to different SNR levels, the relations obtained from Eqs. (16.4.13) and (16.4.14) are the same as those given in Eqs. (16.4.9) and (16.4.10), respectively. Hence the slope in a log(SNR) versus magnitude plot is again different by a factor of two in the two regions.

IV. Detection Limits

We now give results derived using Eq. (16.4.12) for ground-based telescopes with representative spectrometers. The parameters of each spectrometer are given in Table 16.4; the detector used is a CCD with parameters given in Table 16.3.

Results obtained from Eq. (16.4.12) with the given parameters are shown in Fig. 16.8 for an exposure time of 2400 sec. The choice of $\kappa_0 = 1$ indicates observations at the continuum level of the stellar spectra. The detector noise is larger than the sky noise in both modes, by a factor of 250 for the echelle and 10 for the grating mode. Entrance slit widths are approximately 1.1 and 0.45 arc-sec for the 4- and 10-m telescopes, respectively.

In both modes the magnitude reached at SNR = 100 is close to that found from Eq. (16.4.13) for the signal-limited case. The curvature evident in Fig. 16.8 as SNR decreases indicates a transition to the detector-limited regime.

Table 16.5 gives parameters for one mode of each of the HST spectrometers. Both the faint object spectrograph (FOS) and high-resolution spectrograph (HRS) use Digicon detectors but with different photocathodes, hence the quantum efficiencies differ. For both spectrometers we choose $\kappa_0 = 0.8$ to account for the fraction of light transmitted by the entrance apertures. We also assume the incident photon flux N for the HRS at $\lambda = 200$ nm is the same as for the FOS at $\lambda = 500$ nm.

Results for these selected FOS and HRS modes are shown in Fig. 16.9 for exposures of 2400 sec. Sky noise is negligible for the HRS mode and is approximately 30% of the detector noise for the FOS mode. The results shown in Fig. 16.9 are only indicative of the performance capabilities of the HST spectrometers and need adjustment for stars of different spectral type, especially in the ultraviolet range of the HRS. For details on these

Table 16.4

Spectrometer Parameters

	Grating 600 grooves/mm, $m = 1$	Echelle $\tan \delta = 2.0$
F_2	1.5	2.0
r	0.9	0.7
P (Å/mm)	55.6	3.68
R (at 500 nm)	3000	45 000
τ(system)	0.15	0.10

Other: $\phi' = 1$ arc-sec, $\kappa_0 = 1$, $d_1 = 200$ mm
 Projected slit ω' spans two pixels (30 μm)
 Sky background = 22 mag/(arc-sec)^2

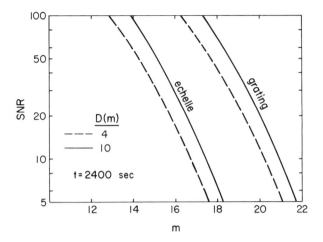

Fig. 16.8. Representative SNR-apparent magnitude diagram for spectrometry. Parameters for the detector are in Table 16.3, for spectrometers in Table 16.4.

and other modes, including sensitivities to stars of different spectral type, the reader should consult the HST Instrument Handbook.

C. SLIT SPECTROSCOPY—EXTENDED SOURCES

The photon flux of an extended source collected by a telescope of diameter D and transmitted to the detector is given by

$$S = 0.7 N\tau D^2 \Delta\lambda \cdot 10^{-0.4m} \phi\phi', \quad (16.4.15)$$

where m is the source brightness in magnitudes per square arc-second and $\phi\phi'$ is again the detector area in square arc-second projected on the sky. For a source whose spectrum is a composite of stellar spectra, the parameter N has the nominal value given following Eq. (16.4.1); for an emission line source the product $N\Delta\lambda$ is adjusted to the proper flux value. The flux from the sky background is given by Eq. (16.4.2).

Table 16.5

Hubble Space Telescope Spectrometer Parameters

FOS	$Q = 0.2$, $\tau(\text{system}) = 0.3$, R (at 500 nm) = 1200
HRS	$Q = 0.15$, $\tau(\text{system}) = 0.1$, R (at 200 nm) = 1.E5
Other	$\phi = \phi' = 0.25$ arc-sec, $\kappa_0 = 0.8$, sky background = 23 mag/(arc-sec)^2, $\Delta = 50$ μm, $C = 0.001$ counts/(diode sec)

IV. Detection Limits

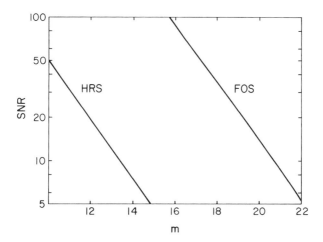

Fig. 16.9. SNR-apparent magnitude diagram for selected HST spectrometer modes. Parameters are given in Table 16.5.

For both source and sky the flux is proportional to $D^2\phi\phi'$, and using Eq. (12.2.1) we get $D^2\phi\phi' = \omega'h'/rF_2^2$. Hence the SNR depends on the camera focal ratio but is independent of the telescope diameter. Substituting this result in Eq. (16.4.3), we find that

$$\text{SNR} \propto (rF_2^2)^{-1/2}. \qquad (16.4.16)$$

It is evident from Eq. (16.4.16) that SNR is larger for smaller F_2. This result is in accord with that given in Eq. (12.2.16) for the illumination of an image. Thus observation of an extended source to a given SNR level, through either a spectrometer or a filter, is done in a shorter time with a faster camera.

D. SLITLESS SPECTROSCOPY

For observations of stellar sources in the slitless mode, the source and background signals are given by Eqs. (16.4.1) and (16.4.2), respectively. In this mode $\Delta\lambda = P\omega'$ for the star signal, while $\Delta\lambda'$ is the bandpass for all wavelengths transmitted to the detector. For the background-limited case, assuming a noise-free detector, the only one considered here, we find

$$m = 0.5m' - 1.25\log\left[\frac{(\text{SNR})^2\phi\phi'}{0.7N\tau\Delta\lambda D^2 Qt}\left(\frac{\Delta\lambda'}{\Delta\lambda}\right)\right], \qquad (16.4.17)$$

where κ is set equal to one. Note that Eq. (16.4.17) is simply Eq. (16.4.6) modified for the case of different spectral and sky bandpasses. The effect of the bandpass ratio is to give a brighter magnitude reached for a given SNR, assuming all other parameters are the same.

To illustrate the effect we assume a nonobjective transmission grating mode with a spectral resolution $R = 100$. If $\Delta\lambda'$ is taken as $\lambda/2$, where λ is the wavelength at the blaze peak, then the bandpass ratio $\Delta\lambda'/\Delta\lambda$ is approximately 50. The difference in magnitude with and without this factor is $\Delta m = -1.25 \log(50)$, or about 2.1 magnitudes. Although the limiting magnitude is brighter for this slitless mode, the gain is 50 spectral elements per source compared to one obtained with a narrow filter in a photometry mode.

V. CONCLUDING COMMENTS

The discussion in this chapter is intended to show the limits that can be reached in detecting sources in different modes, assuming a single telescope of diameter D. In the next chapter we extend these results for spectrometric modes to telescopes with more than one primary mirror, such as the multiple-mirror telescope.

REFERENCES

Baum, W. (1962). "Stars and Stellar Systems II, Astrophysical Techniques," Chap. 1. Univ. of Chicago Press, Chicago, Illinois.
Bowen, I. (1964). *Astron. J.* **69**, 816.
Code, A. (1973). "Annual Review of Astronomy and Astrophysics," Vol. 11, p. 239. Annual Reviews, Palo Alto, California.
Roddier, F. (1981). "Progress in Optics," Vol. 19, Chap. 5. North-Holland, Amsterdam.

BIBLIOGRAPHY

Additional information on atmospheric effects
Coulman, C. (1985). "Annual Review of Astronomy and Astrophysics," Vol. 23, p. 19. Annual Reviews, Palo Alto, California.
Woolf, N. (1982). "Annual Review of Astronomy and Astrophysics," Vol. 20, p. 367. Annual Reviews, Palo Alto, California.
Review of scientific results from speckle interferometry and an extensive bibliography
McAlister, H. (1985). "Annual Review of Astronomy and Astrophysics," Vol. 23, p. 59. Annual Reviews, Palo Alto, California.
See listings in the bibliography in Chapter 11 for characteristics of instruments on the Hubble Space Telescope.

Chapter 17 | Multiple-Aperture Telescopes

The usual configuration of an optical telescope is one with a single circular aperture and an angular resolution in the diffraction limit of order λ/D. This dependence on wavelength and diameter applies to any telescope; thus the largest single-aperture radio telescope has an angular resolution several orders of magnitude poorer than that of any optical telescope of modest size. To overcome this limitation, radio telescope configurations with multiple apertures have long been used to achieve high angular resolution. Given the resolution limit of λ/D, this means a large effective D, hence a long baseline between separate telescopes. For ground-based optical telescopes, on the other hand, angular resolution is usually limited by the atmosphere and the need for separate telescopes is less evident. With the demand for larger optical telescopes, however, increased attention has been given to the possibility of multiple-aperture arrangements to achieve large light-gathering power at less cost.

The history of single-mirror telescopes indicates a cost that scales approximately as the 2.5 power of the diameter. Studies of multiple-aperture telescopes indicated that large effective apertures could be achieved at significantly reduced cost, a conclusion that was first borne out with the construction of the multiple-mirror telescope (MMT) of the Smithsonian Astrophysical Observatory. The MMT has six single 1.8-m telescopes mounted in a common frame with an effective diameter of about 4.5 m. A photograph of the MMT is shown in Fig. 17.1. The National New Technology telescope (NNTT) proposed by Kitt Peak National Observatory is

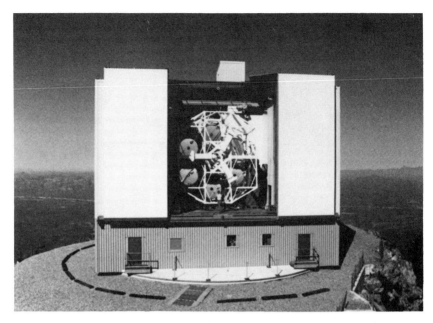

Fig. 17.1. The multiple-mirror telescope located on Mount Hopkins, Arizona. (Photograph courtesy of Multiple-Mirror Telescope Observatory.)

of a similar type with four 7.5-m telescopes mounted in a common frame, as shown in Fig. 17.2.

The MMT type is one approach in the design of large telescopes; a second approach is the so-called segmented-mirror telescope or SMT. In an SMT design the primary mirror consists of many separate segments whose combined surfaces are equivalent to a single mirror. The 10-m Keck telescope now being built by the University of California and California Institute of Technology is an SMT type with hexagonal segments. The SMT design is not a multiple-aperture telescope, as we define it.

In this chapter we discuss some of the optical characteristics of the diffraction images given by multiple-aperture telescopes (MAT) of different configurations. We also consider different ways in which a ground-based MAT can be used with a slit spectrometer, including a comparison with a single-mirror telescope of the same effective aperture.

I. DIFFRACTION IMAGES

In this section we describe the diffraction image given by selected MAT configurations, each of which has some number N of identical telescopes

I. Diffraction Images   331

Fig. 17.2. Late 1985 model of the proposed National New Technology telescope. (Photograph courtesy of National Optical Astronomy Observatories.)

with circular apertures. Arrangements discussed include linear arrays with an even number of equally spaced telescopes, a square array with $N = 4$, and a hexagonal array with $N = 6$. For a discussion of these and other types of arrays, including an atlas of diffraction images, the reader should consult the excellent article by Meinel et al. listed in the references.

The calculation of the diffraction image of an MAT is based on the "array theorem," as it is called by Hecht and Zajac. Consider a set of identical telescopes or, equivalently, exit pupils distributed as shown in Fig. 17.3. For perfect telescopes the wavefront at each exit pupil is part of a spherical surface whose center is at O in Fig. 17.3. The amplitude U at a point P in the image is found by evaluating Eq. (10.1.2), where the integration includes each of the apertures. When this is done for N circular

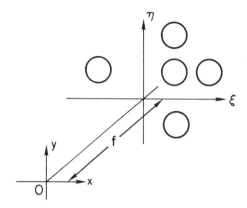

Fig. 17.3. Schematic layout of telescopes, each of focal length f, with pupils in the $\xi\eta$-plane and combined image in the xy plane.

apertures, each of diameter D, the result is

$$U(x, y) = A \frac{2J_1(v)}{v} \sum_{n=1}^{N} \exp[-iK(\xi_n x + \eta_n y)], \qquad (17.1.1)$$

where A is a normalization constant, $v = \pi r/\lambda F = \pi D\alpha/\lambda$ [see Eq. (10.1.12)], $K = 2\pi/\lambda f$ with f the telescope focal length, ξ_n and η_n are coordinates at the center of the nth telescope, and x and y are coordinates at point P on the image.

Note that the factor involving v in Eq. (17.1.1) is the amplitude for a single circular aperture, as given in Eq. (10.1.9). The sum in Eq. (17.1.1) represents the superposition of the amplitudes from the N apertures treated as coherent sources, where $K(\xi_n x + \eta_n y)$ is the phase difference between a wave from the center of the nth telescope and one from the origin of the (ξ, η) coordinate system.

The intensity at point P of the diffraction pattern is given by $I(P) = |U(P)|^2$, with the normalized intensity $i(P) = I(P)/I(0)$. In effect, the intensity of the diffraction pattern is that of a single telescope modulated by an interference pattern produced by the separate telescopes. This is similar to the result given in Eq. (13.3.1) for a diffraction grating, with the intensity the product of a blaze function and an interference factor. We now evaluate Eq. (17.1.1) for the arrays of interest.

A linear array of N equally spaced telescopes, where N is even, is shown in Fig. 17.4. We choose $\xi = 0$ in the center of the array for convenience in evaluating the sum Σ in Eq. (17.1.1). The center-to-center spacing of adjacent telescopes is γD, with $\gamma > 1$ to ensure no overlap. The ξ-coordinates at the centers of the telescopes are $\pm \gamma D/2$, $\pm 3\gamma D/2$, ..., $\pm(2N-1)\gamma D/2$.

I. Diffraction Images 333

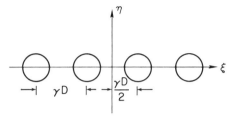

Fig. 17.4. Linear array of telescopes, each of diameter D, with spacing γD between centers.

Substituting into Eq. (17.1.1) we find

$$\Sigma = 2 \sum_{n=1}^{N/2} \cos((2n-1)\gamma v_x) = \frac{\sin(N\gamma v_x)}{\sin(\gamma v_x)}, \quad (17.1.2)$$

where $v_x = \pi Dx/\lambda f$. Therefore the normalized intensity is given by

$$i(P) = \frac{1}{N^2}\left[\frac{2J_1(v)}{v}\right]^2\left[\frac{\sin(N\gamma v_x)}{\sin(\gamma v_x)}\right]^2. \quad (17.1.3)$$

The smallest value of v_x for which $i(P) = 0$ is given by $N\gamma v_x = \pi$, hence $\alpha_x = \lambda/DN\gamma$, where α_x is the angular resolution in the x direction. Note that the total distance spanned by the telescopes of the array is $\gamma(N-1)D + D$, hence the angular resolution in this direction is essentially that of a single telescope of this diameter. The resolution in the y direction is the same as that of a single telescope of diameter D, as is evident by letting $v_x = 0$ in Eq. (17.1.3).

Results obtained from Eq. (17.1.3) are shown in Fig. 17.5 for $N = 2$ and two values of γ, in a slice across the image in the x direction. Note the narrowing of the central peak in Fig. 17.5 as γ increases. In this case we also see that there are more subsidiary peaks under the diffraction envelope when γ is larger, with a larger fraction of the total energy in these peaks. These additional peaks make it more difficult to achieve the resolution given by the central peak.

In Fig. 17.6 we show a slice across the image in the x direction for $N = 4$ and $\gamma = 2$. Note that the principal peaks under the diffraction envelope are "sharper," a consequence of the larger $N\gamma$ product, with weak subsidiary bands between the principal peaks. In general, there are $N - 2$ weak bands between adjacent strong bands.

Following the same procedure for the square array shown in Fig. 17.7, the normalized intensity is given by

$$i(r,\theta) = \frac{1}{16}\left[\frac{2J_1(v)}{v}\right]^2\left[\frac{\sin(2\gamma v\cos\theta)}{\sin(\gamma v\cos\theta)}\frac{\sin(2\gamma v\sin\theta)}{\sin(\gamma v\sin\theta)}\right]^2, \quad (17.1.4)$$

where $v = \pi Dr/\lambda f$, $x = r\cos\theta$, and $y = r\sin\theta$.

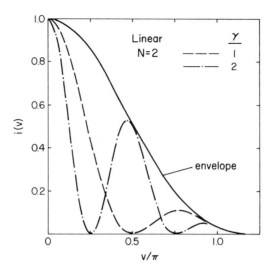

Fig. 17.5. Normalized PSFs for pairs of two telescopes. Angular distance from peak to first minimum is $\lambda/2\gamma D$.

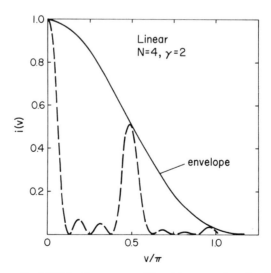

Fig. 17.6. Normalized PSF for linear array of four telescopes. Angular distance from peak to first minimum is $\lambda/N\gamma D$.

I. Diffraction Images

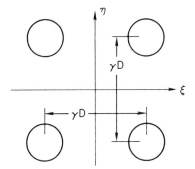

Fig. 17.7. Square array of telescopes of diameter D, spacing γD.

Results from Eq. (17.1.4) are shown in Fig. 17.8 for $\gamma = 2$. Note that the profile along the x and y directions is the same as that given in Fig. 17.5 for $\gamma = 2$. This is expected because the array spans the same distance, $2D$ between centers, in both cases. In general, the FWHM of the central peak is inversely proportional to γ.

The final array we consider is the hexagonal array of six telescopes shown in Fig. 17.9, with the centers of the telescopes on a circle of radius γD and hence a center-to-center distance between adjacent telescopes of γD. The coordinates at the centers are $(\pm \gamma D, 0)$ and $(\pm \gamma D/2, \pm \sqrt{3}\, \gamma D/2)$, and the sum Σ in (17.1.1) is

$$\Sigma = 2 \cos(2\gamma v \cos \theta) + 4 \cos(\gamma v \cos \theta) \cos(\sqrt{3}\, \gamma v \sin \theta).$$

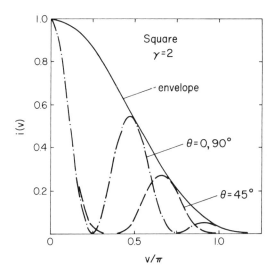

Fig. 17.8. Profiles of normalized PSFs for square array with $\gamma = 2$.

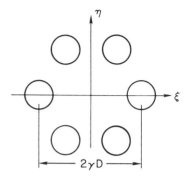

Fig. 17.9. Hexagonal array of telescopes of diameter D, span $2\gamma D$.

The normalized intensity for this array is

$$i(r, \theta) = \frac{1}{36}\left[\frac{2J_1(v)}{v}\right]^2 \Sigma^2. \qquad (17.1.5)$$

For this six-telescope array, the image profile is the same at 60° intervals. Profiles are shown in Fig. 17.10 for $\gamma = 1$, hence touching telescopes, at $\theta = 0$ and 30°. With this choice of γ, opposite telescopes in the array have a center-to-center separation of $2D$, and the span of the array is $3D$. With larger γ the FWHM of the central peak shrinks in direct proportion to the increase in γ. The value of γ for the MMT is approximately 1.4.

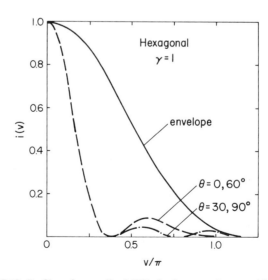

Fig. 17.10. Profiles of normalized PSFs for hexagonal array with $\gamma = 1$.

II. Slit Spectroscopy with Arrays   337

A hexagonal array with six touching telescopes, each of diameter D, is a fair approximation to a single telescope of aperture $3D$ and central obscuration of diameter D. The reader can verify that the average of the $i(v)$ curves in Fig. 17.10 does follow the PSF curve for $\varepsilon = 0.33$ in Fig. 10.2 fairly closely.

These examples are sufficient to illustrate the general features of diffraction images of telescope arrays, with Eq. (17.1.1) as the basis for treating other configurations. For a discussion of arrays utilized as stellar interferometers, the reader should consult the review by Labeyrie. Selected references on the MMT are also given at the end of the chapter.

II. SLIT SPECTROSCOPY WITH ARRAYS

The combination of an MAT and a slit spectrometer has certain differences from a single telescope used with the same instrument. In this section we explore characteristics of different MAT-spectrometer configurations and their effects on limiting magnitude. We consider only the case of seeing-limited stellar spectroscopy, with the diameter of a stellar image large compared to the diffraction image and larger than the entrance slit width.

The configuration for a single telescope feeding a slit spectrometer is shown in Fig. 12.4; the characteristics of this arrangement are discussed in detail in Chapter 12. Important relations given in Chapter 12, and reproduced here for convenient reference, are

$$\omega' = r\phi DF_2, \qquad h' = \phi' DF_2, \tag{17.2.1}$$

$$\Delta\lambda = P\omega' = r\phi D/d_1 A, \tag{17.2.2}$$

where ω' and h' are the projected slit width and height, respectively, and $\Delta\lambda$ is the spectral purity. For the definition of other parameters in these relations, see Section 12.II.

A. IMAGE FORMAT AT ENTRANCE SLIT

A schematic diagram of an MAT directing light to the slit of a spectrometer is shown in Fig. 17.11. Each telescope in the array has diameter D, focal length f, and focal ratio F. The total span of the array is denoted by D_a and the focal ratio of the envelope of the beams is F_a. For the MMT the focal ratios F and F_a are 31.6 and 9, respectively, and $D_a = 6.4$ m.

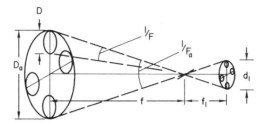

Fig. 17.11. Square array with slit at combined focus. D, Diameter of individual telescope; D_a, diameter of array; d_1, diameter of spectrometer collimator.

One method of directing the beams from the separate telescopes places the images on top of one another at the slit. In this so-called superposed mode the focal ratio of the collimator beam is F_a and the optics of the spectrometer must be large enough to accept this beam divergence. The beam from each telescope falls on a different part of the collimator and following optics in this mode.

A second method of placing the images on the slit is the so-called stacked mode, in which images of the separate telescopes are placed side by side along the slit. This is shown schematically in Fig. 17.12 for four images. Because the beams are spatially separated at the slit, each can be redirected with a small prism and superposed on the collimator. With the beams superposed on the collimator, the focal ratio of this beam is F. The reference by Chaffee and Latham describes the characteristics of the image stacker used at the MMT.

The image size at the entrance slit is the same for both of these modes, and the fraction of the light that passes the slit is the same, assuming the same ϕ and ϕ' in Eq. (17.2.1). However, the projected image is different, with the length of the image N times longer perpendicular to the main dispersion direction for the stacked mode. If the observation is detector-limited due to noise from dark counts, the limiting magnitude, with all other parameters the same, is not as faint when the image stacker is used. If the observation is signal- or sky-limited, however, the two modes are equivalent for a given ϕ and ϕ'.

We now examine the assumption of having the same ϕ for the two modes. From Fig. 17.11 we see that $f/f_1 = D_a/d_1$. For a slit width $\omega = f\phi$

Fig. 17.12. Schematic of image stacker to combine separate telescope beams, each of focal ratio F, at spectrometer collimator.

and a projected slit width $\omega' = \omega r(f_2/f_1)$, we find $\omega' = r\phi D_a F_2$. Hence the projected slit width for the superposed mode is determined in part by the span of the MAT.

For the stacked mode, on the other hand, ω' is given by Eq. (17.2.1) with D the diameter of one of the telescopes in the array. To maintain the same ω' in the two modes requires a smaller ϕF_2 product in the superposed mode by the factor D/D_a.

Let us illustrate the difference between these two modes by taking the MMT as an example. Assume the required ω' is 30 μm for a slit width of $\phi = 2$ arc-sec. For the superposed mode with $D_a = 6.4$ m we find $rF_2 = 0.47$; for the stacked mode and $D = 1.82$ m we get $rF_2 = 1.64$. The small camera focal ratio required for the superposed mode is not a practical one, whereas that for the stacked mode is easily achieved. Assuming the smallest practical $rF_2 = 1$, the only way to maintain the projected slit width for the superposed mode is to reduce ϕ and thus lose a significant fraction of the light at the slit.

For the stacked mode it is important to note that the stacked images can be reimaged onto the actual spectrometer slit at a different focal ratio without changing the relation for ω' given by Eq. (17.2.1). This is shown schematically in Fig. 17.13, where a lens L images the stacked images at IS on the entrance slit ES and changes the focal ratio from F to F_1. The image stacker at the MMT includes a reimaging lens, which converts the $f/31.6$ beam to an $f/9$ beam to the collimator.

B. COMPARISON OF ARRAY WITH SINGLE TELESCOPE

We now compare the performance of an MAT with a single-aperture telescope of the same collecting area. In these comparisons we assume observations of stellar sources of the same duration made to the same SNR level. We also assume similar detectors used on both types of telescopes.

For the same collecting area $D_s^2 = ND^2$, where D_s is the diameter of the single telescope. For this telescope the relations for m given in Eqs. (16.4.12)–(16.4.14) apply without change. For the MAT with N telescopes, the only change required in Eq. (16.4.12) is the replacement of D^2 with ND^2.

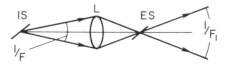

Fig. 17.13. Reimaging system to change focal ratio of stacked beams. IS, Image stacker; L, lens; ES, entrance slit of spectrometer.

We now take Eq. (16.4.12) for each of these telescopes, with κ_0, Q, t, τ, ϕ', and SNR the same for each. For a noise-free detector in the slit-limited case ($\phi < \phi'$), the difference in limiting magnitude is

$$m_a - m_s = -C \log[(\phi\Delta\lambda)_s/(\phi\Delta\lambda)_a], \qquad (17.2.3)$$

where a and s denote the MAT and single telescope, respectively. The constant C is 2.5 for the signal-limited case and 1.25 for the background-limited case.

For the detector-limited case, where readout noise is negligible compared to dark count, the magnitude difference is

$$m_a - m_s = -1.25 \log\left[\frac{(rF_2^2)_a}{(rF_2^2)_s} \frac{(\phi\Delta\lambda)_s}{(\phi\Delta\lambda)_a}\right]. \qquad (17.2.4)$$

For the comparison of the MAT and the single telescope we assume that each is used with the same spectrometer set to give the same spectral purity. In this case $\phi_s/\phi_a = D/D_s = 1/\sqrt{N}$ for the stacked mode and $\phi_s/\phi_a = D_a/D_s$ for the superposed mode.

For the stacked mode $m_a - m_s$ is 1.25 log N for the signal-limited case and 0.625 log N for the other cases. Therefore the gain in faintness of the MAT is proportional to $\sqrt{N}$ in the signal-limited case and proportional to $N^{1/4}$ in the background-limited case. The superposed mode has $D_a > D_s$ and the single-aperture telescope reaches a fainter magnitude than the array in this mode.

It is therefore evident that an MAT with an image stacker has the best performance of these three configurations when the same spectrometer is used on each.

Assuming the same spectral purity, a single-aperture telescope reaches the same magnitude limit as the MAT with image stacker when $\phi_s = \phi_a$. Using Eqs. (17.2.1) and (17.2.2) we find that these conditions are satisfied when P is smaller and ω' is larger, each by a factor $\sqrt{N}$, for the spectrometer on the single telescope. For the same angular dispersion A, this means that d_1 is also $\sqrt{N}$ larger. Therefore the single telescope and the MAT have the same performance when the spectrometer on the former is $\sqrt{N}$ times larger. Note, however, that the projected slit widths are no longer the same.

Although the comparisons made using Eqs. (17.2.3) and (17.2.4) have been phrased in terms of MAT versus single-aperture telescope, the results are the same for an SMT–single telescope comparison. With appropriate tilt of each segment or groups of segments, a stellar image can be stacked along a slit in the same way as with an MMT design. The Keck ten-meter telescope (TMT) has 36 hexagonal segments and up to 36 separate images could be placed side by side along a slit.

III. CONCLUDING COMMENTS

The striving of astronomers for larger light-gathering power, greater observing efficiency, and better angular resolution has been instrumental in the development of multiple-aperture and segmented-mirror telescopes. Although the bulk of observing will continue to be carried out with single-aperture telescopes for the foreseeable future, it is evident that many observations will be possible only with the new generation of telescopes.

REFERENCES

Chaffee, F., and Latham, D. (1982). *Publ. Astron. Soc. Pac.* **94**, 386.
Meinel, A., Meinel, M., and Woolf, N. (1983). "Applied Optics and Optical Engineering," Vol. 9, Chap. 5. Academic Press, New York.

BIBLIOGRAPHY

Beckers, J., Ulich, B., and Williams, J. (1982). *Proc. SPIE Int. Soc. Opt. Eng.* **332**, 2. Bellingham, Washington.
Code, A. (1973). "Annual Review of Astronomy and Astrophysics," Vol. 11, p. 239. Annual Reviews, Palo Alto, California.
Labeyrie, A. (1978). "Annual Review of Astronomy and Astrophysics," Vol. 16, p. 77. Annual Reviews, Palo Alto, California.
Nelson, J. (1982). *Proc. SPIE Int. Soc. Opt. Eng., SPIE* **332**, 109. Bellingham, Washington.
"Optical Telescopes of the Future" (1977). Proceedings of ESO Conference. European Space Organization, Geneva, Switzerland.

Also see first listing in bibliography in Chapter 6.

Index

A

Aberration
  astigmatism, 50, 51, 64-67, 76-78
  character of, 72, 73
  chromatic, 57-59, 125-128
  classical, 190-193
  coma, 50, 51, 69, 70, 78
  distortion, 50, 51, 70, 78
  field curvature, 52, 82-88
  introduction to, 41-52
  multisurface systems, 79, 80
  orthogonal, 192-196
  ray and wavefront, 67-69
  spherical
    fifth-order, 44, 131
    third-order, 43-46, 78
Aberration, angular, see Angular aberration
Aberration coefficients
  decentered pupil
    general surface, 89
    mirror, 90
  displaced stop or pupil
    general surface, 77
    mirror, 77
  grating
    astigmatism, 265
    coma, 266
  stop at surface
    general surface, 71
    mirror, 71
    plane-parallel plate, 130
Aberration coefficients, element
  aspheric Schmidt plate, 123, 129, 130
  concentric meniscus lens, 153
  diffraction grating, 263-266, 297
  grism, 300, 301
  paraboloid, 78, 274
  prism, 296
  prism-grating, 303
  Schmidt-Cassegrain
    primary, 144
    secondary, 145
  spherical mirror, 78
Aberration coefficients, grating mounting
  Czerny-Turner
    monochromator, 277, 278
    spectrograph, 279
  Monk-Gillieson, 282
  nonobjective
    grating, 297
    grism, 300, 301
    prism-grating, 303
Aberration coefficients, system
  Bouwers camera, 153
  prime focus corrector, 166, 168
  Schmidt
    focal reducer, 176
    general, 123
  Schmidt-Cassegrain
    all-spherical, 150
    flat field anastigmat, 146, 147
    general, 144, 145
  three-mirror telescope, 115
    Paul-Baker, 115-117

Aberration coefficients, system *(cont)*
  two-mirror telescope
    corrected Ritchey-Chretien, 171
    general, 81, 99
    hybrid, 106
    with misaligned secondary, 111
Aberration, transverse, *see* Transverse aberration
Achromatic corrector
  Maksutov, 156, 157
  Schmidt, 135-138
Airy disk, 182, 183, 188, 189
Airy pattern
  asympotic form, 184, 185
  encircled energy, 185-187
  general, 182
  peak intensity, 188, 191
  point spread function, 180-185
  radii of dark rings, 183
  Strehl ratio, 191
Alignment errors
  decenter, 88, 89, 110-112
  despace, 113, 114
  Schmidt camera, 90-92
  tilt, 88, 89, 110-112
  two-mirror telescope, 110-114
Anamorphic magnification
  definition, 225
  diffraction grating, 240
  prism, 236
Anastigmatic telescope
  Schmidt-Cassegrain, 146-150
  two-mirror, 108, 109
Angular aberration, 45, 46, 50
  corrected Ritchey-Chretien, 172
  fifth-order spherical
    aspheric plate, 131
    spherical mirror, 131
  paraboloid, 95
  prime focus plus aspheric plate, 166
  relation to transverse, 69
  Schmidt with misaligned plate, 91
  Schmidt-Cassegrain aplanat, 150
  two-mirror telescope,
    aplanatic, 103
    classical, 101
    despaced secondary, 113, 114
    general, 99
    hybrid, 107
    misaligned secondary, 111
Angular dispersion, *see* Dispersion
Angular magnification, *see* Magnification
Angular resolution, 39, 197, 202, 333
  limit due to atmosphere, 313-317
  limit due to diffraction, 183, 313

Aperture stop, 18, 19, *see also* Stop, aperture
Aplanatic condition, 72
Aplanatic telescope, 102-105, 107-109
Array, 329-337
  general, 331, 332
  hexagonal, 335-337
  linear, 332-334
  slit spectroscopy, 337-340
    superposed images, 338
    stacked images, 338, 339
  square, 333, 335
Aspheric corrector
  aberration coefficients, 91, 129
  achromatic, 134-138
  aspheric coefficient, 62, 124, 131
  chromatic aberration, 57-59, 125-128
  fifth-order spherical, 131
  prime focus, 166-169
  radius of curvature, 58, 127
  Ritchey-Chretien focus, 170-173
  size, 128
  surface figure, 56, 58, 125
  surface parameters, 125
  zonal deviations, 126
Astigmatism, *see also* Aberration; Aberration coefficients
  angular, 50, 51, 69
  longitudinal, 64, 77
  sagittal, 64, 76
  tangential, 64, 76
  transverse, 67, 73, 78
Atmosphere
  degradation function, 315
  differential refraction, 26
  long-exposure modulation transfer function (MTF), 315, 316
  refraction, 26
  turbulence, 27, 313-317
    scintillation, 313, 314
    seeing, 27, 313, 314
    speckle, 314
Autocorrelation, 203, 206, 208

B

Back focal distance, 17, 97
Baker-Schmidt telescope, 142, 146-150, 159
  aberration coefficients, 146
  chromatic aberration, 148, 149
  conic constants, 147, 148
  semisolid, 158, 159
Blaze, grating
  angle, 242

Index

function, 244–249
wavelength, 245
Bouwers camera, 152–156
  aberration coefficient, 153
  chromatic aberration, 155
  concentric meniscus corrector, 152–154
  parameters, 155
Brightness, 227, 228
  extended source, 229
  stellar source, 229

C

Camera
  Bouwers, 152–156
  Maksutov, 156–158
  Schmidt, 55–59, 123–141
  Schmidt-Cassegrain, 142–152
Cassegrain telescope, 15, 16, 53–55, 81, 97–107
  aberration coefficients, 99
  angular aberrations, 99, 101–103
  diameter of secondary, 21
  exit pupil, 18, 19
  power, 17
  type,
    classical, 53, 100, 101
    Couder, 108
    Dall-Kirkham, 55, 107
    hybrid, 106, 107
    Ritchey-Chretien, 102–105
Catadioptric telescope
  Bouwers, 152–156
  Maksutov, 156–158
  Schmidt, solid, semisolid, 138–141
  Schmidt-Cassegrain, semisolid, 158, 159
Chief ray, 18
Chromatic aberration
  meniscus corrector, 155
  Schmidt camera, 57–59, 125–128
  solid Schmidt, 139
Classical Cassegrain, *see* Cassegrain telescope
Collimator
  paraboloid
    off-axis, 273–275
    on-axis, 273, 274
  spherical mirror, 276
Coma, *see also* Aberration; Aberration coefficients

angular, 50, 51, 69
sagittal, 72, 73, 78
spectral, 283, 296, 297
tangential, 72, 73
transverse, 70, 73, 78
Conic constant
  definition, 37, 38
  relation to eccentricity, 38
Conic section, 37, 38
  focal length of reflecting, 42, 43
Conjugate points, 9
Contrast, 200
Coudé focus, *see* Focus
Couder telescope, *see* Cassegrain telescope
Cross dispersion, 285–288
  modes, 288–290
Curvature
  in inhomogeneous medium, 25
  of field, *see* Field curvature
  spectrum line, 261
Cutoff frequency, 201, 202, *see also* Modulation transfer function
Czerny-Turner mounting, 276–281
  monochromator, 277, 278
  spectrograph, 279–281

D

Dall-Kirkham telescope, *see* Cassegrain telescope
Decenter, *see* Alignment errors
Defocus, 192, 194, 195
Degradation function, 206, 208, 209
  atmosphere, 315
  microripple, 209
  midfrequency ripple, 208
  root-mean-square pointing error, 209
Despace, *see* Alignment errors
Detection limits, 319–328
  stellar photometry, 321–323
  stellar spectroscopy, 323–328
    slit-limited, 324–327
    slitless, 327, 328
Detector
  charge coupled device (CCD), 307, 317, 322
  digicon, 325, 326
  modulation transfer function (MTF), 308–313
  quantum efficiency, 309
  signal-to-noise ratio, 317–319

Differential refraction, see Atmosphere
Diffraction
　focus, 194, 195
　　shift from paraxial focus, 194
　image
　　annular aperture, 182-184
　　array, 331-337
　　circular aperture, 182-184
　　rectangular aperture, 232
　　slit, 232
　integral
　　array, 332
　　in presence of aberrations, 189, 190
　　perfect, 180-182
　limit, 39, 183, 191, 197
　variables, 181, 184
Diffraction grating
　anamorphic magnification, 240
　angles, 238, 239
　angular dispersion, 239
　blaze
　　angle, 242
　　function, 244-249
　　wavelength, 245
　constant, 238
　efficiency, 243-249
　equation, 238, 260
　free spectral range, 241
　holographic, 251
　Littrow mode, 242
　$L \cdot R$ product, 249-251
　polarization, 249
　reflection, 238, 239, 260
　resolution, 240, 241
　sign convention, 238, 239, 260
　transmission, 238, 239, 260
Diffraction grating mounting, see also Aberration coefficients, element; Aberration coefficients, grating mounting
　concave
　　cross-disperser, 289, 290
　　inverse Wadsworth, 269, 270
　　Rowland, 267, 268
　　Wadsworth, 268, 269
　plane
　　cross-disperser, 289
　　Czerny-Turner, 276-281
　　Ebert-Fastie, 281
　　Monk-Gillieson, 282-284
　　with Schmidt camera, 275, 276
　transmission
　　cross-disperser, 289
　　nonobjective, 297, 303

Dispersion
　angular, 222
　prism, 32, 236
　grating, 239
　curves, glass, 237
　linear, 222, 223
Distortion, see also Aberration; Aberration coefficients
　angular, 50, 51

E

Ebert-Fastie mounting, see Diffraction grating mounting
Eccentricity, 37, 38
　relation to conic constant, 38
Echelle, 239, 241-243, 284-294
　blaze
　　angle, 242
　　function, 247, 249
　　wavelength, 245
　blaze peak
　　efficiency, 248, 249
　　equations, 287
　cross-dispersion modes, 288-290
　design example, 293
　effective groove width, 248
　free spectral range, 285
　order
　　length, 285
　　separation, 287
　polarization, 249
　spectrometer configuration
　　modified Czerny-Turner, 291, 292
　　with Schmidt camera, 292
　spectrum format, 285-288
Echellette, 294
Ellipsoid, 35, 36, 38
Encircled energy (EE), 185-187, 204
　annular aperture, 185, 186
　asymptotic approximation, 187
　circular aperture, 185, 186
　definition, 185
　in presence of
　　figure error, 196
　　random wavefront error, 211
　　rms pointing error, 212
　relation to modulation transfer function (MTF), 204
Entrance pupil, 18

Étendue
　definition, 227, 228
　diffraction limit, 234
　Fabry-Perot, 253
　Fourier transform spectrometer, 255
　spectrometer, 228
Exit pupil
　definition, 18
　two-mirror telescope, 18, 19

F

Fabry-Perot interferometer, 251-254
　comparison with echelle, 253, 254
　étendue, 253
　free spectral range, 252
　$L \cdot R$ product, 253
　spectral purity, 252
　spectral resolution, 252
Fermat's principle
　aberration compensation example
　　Cassegrain telescope, 53-55
　　Schmidt camera, 55-59
　application to
　　atmosphere, 26, 27
　　conic mirrors, 34-37
　　diffraction grating, 257-260
　　general surface, 61-65
　　prism, 31, 32
　　spherical surface, 29
　　thin lens, 30, 31
　general statement, 22, 23
　laws of refraction and reflection, 28, 29
　physical interpretation, 32, 33
Field curvature, 82-88
　introduction to, 52
　median, 86
　Petzval, 83, 84
　sagittal, 85, 86
　tangential, 85, 86
Field curvature, element
　paraboloid, 97
　spherical mirror, 87
Field curvature, system
　grating, mountings, 266
　　Czerny-Turner, 281
　　Monk-Gillieson, 266
　　nonobjective, 297
　　Rowland, 266, 267
　　Wadsworth, 266
　prime focus with corrector, 167
　Schmidt-Cassegrain aplanat, 150

two-mirror telescope
　aplanatic, 103
　classical, 101
　general, 100
Field flattener lens
　aberrations, 162
　Ritchey-Chretien telescope, 162-164
　Schmidt camera, 87, 88, 164, 165
Field lens, 162, 165
Field stop, 18
Flux, 188, 189, 228, 229, 319, 326
Focal length
　mirror, 12, 42, 43
　thick lens, 13
　thin lens, 14, 30
　two-mirror telescope, 17
Focal ratio, 17
Focal reducer, 173-178
　general configuration, 173-175
　Schmidt camera example, 175-178
　types, 175
Focus
　Cassegrain, 16
　coude, 106, 107, 304
　diffraction, 194
　Gregorian, 16
　Nasmythe, 106, 107, 304
Folded Schmidt camera, 275, 276
Fourier transform spectrometer, 224, 254-256
Fraunhofer diffraction
　annular aperture, 182
　array, 333-337
　circular aperture, 182
　diffraction grating, 244
　rectangular aperture, 232
　slit, 232
Free spectral range
　diffraction grating, 241
　Fabry-Perot, 252
Frequency, see Spatial frequency

G

Gaussian equation, see Paraxial equation
Gaussian profile
　image motion, 209
　intensity, 315
　modulation transfer function (MTF), 312, 315

Glass
  dispersion, 236, 237
  index of refraction, 32
Grating, see Diffraction grating
Gregorian telescope, 15-17, 20, 21, 97-105
  aberration coefficients, 99
  angular aberrations, 99
  aplanatic, 102-105
  classical, 100, 101
  exit pupil, 18, 19
Grism, 299-302

H

Hubble Space Telescope, 102, 213-219, 234, 322-327
  astigmatism, 219
  faint object camera (FOC), 310-312
  faint object spectrograph (FOS), 325-327
  high-resolution spectrograph (HRS), 325-327
  high-speed photometer (HSP), 213
  image characteristics, predicted, 214-218
  instrument complement, 213
  intensity, 188, 189
  parameters, 213
  wide-field planetary camera (WFPC), 310-312, 322, 323
Hybrid telescope, 106, 107
Hyperboloid, 36-38

I

Illumination, 229, 230
  extended source, 230
  stellar source, 230
Image
  brightness, see Illumination
  diffraction, see Diffraction image
  format at spectrometer slit, 337
  slicer, 231
  stacker, 338, 339
Index of refraction, 8, 32
Intensity, 180-185, 187-189
  asymptotic average, 184, 185
  average over Airy disk, 188
  multiple-aperture telescope
    hexagonal, 336, 337
    linear, 333, 334
    square, 333, 335
  normalized, 182
  peak, 188, 190, 191
  radial dependence, 182, 184, 185
  Strehl, 191
Interferometer
  Fabry-Perot, 251-254
  Michelson, 254
Inverse Wadsworth, see Diffraction grating mounting

L

Lagrange invariant, 11, 228
Lateral magnification, see Magnification
Limiting magnitude
  array versus single telescope, 339, 340
  general form, 320, 324
  stellar photometry
    background-limited, 321-323
    signal-limited, 321-323
  slit-limited spectroscopy
    extended source, 326, 327
    stellar source, 323-326
  slitless spectroscopy, 327, 328
Luminosity, 227, 228
Luminosity-resolution product
  Fabry-Perot, 253
  general, 229
  grating, 249
  Fourier transform spectrometer, 255

M

Magnification
  anamorphic, see Anamorphic magnification
  angular, 10
  lateral or transverse
    mirror, 12
    refracting surface, 10
    thin lens, 14
Maksutov camera, 156-158
Meniscus corrector
  achromatic, 156
  concentric, 152-155
Michelson interferometer, 224, 254
Modulation transfer function (MTF), 199-212, 308-313
  annular aperture, 203-205
  atmosphere, 315
  cutoff frequency, 201, 202
  definition, 200
  and degradation functions, 206-212

Index   349

in presence of aberrations, 207
pixel, 308–313
relation to
  encircled energy, 204
  point spread function, 204
Monk–Gillieson mounting, 282–284
Monochromator
  Czerny–Turner, 276–278
  Ebert–Fastie, 281
  Monk–Gillieson, 282–284
Multiple aperture telescope, *see* Array
Multiple-mirror telescope (MMT), 329, 330

N

National New Technology Telescope (NNTT), 329–331
Noise, *see also* Signal-to-noise ratio
  dark count, 307, 317–319
  photon, 317, 318
  readout, 307, 317–319
Nonobjective
  grating, 297–299
  grism, 299–302
  prism, 295–297
  prism-grating, 303, 304
Normalized intensity, *see* Intensity
Normalized parameters, two-mirror telescopes
  definitions, 16, 17
  table, 97, 98
Nyquist criterion, 309, 310

O

Objective mode, 231, 232, 237
Objective prism, 237
Obscuration, 21, 105
Off-axis paraboloid, *see* Paraboloid
Optical path difference (OPD)
  diffraction grating, 262
  multisurface system, 75, 79
  relation to transverse aberration, 68, 80
  single surface, 46, 64, 67, 68
Optical path length (OPL)
  general, 23
  grating, 259, 260
  refracting surface, 32, 33, 62, 63
Optical transfer function (OTF), *see* Transfer function

P

Paraboloid, 34
  angular aberrations, 95
  image surface curvature, 97
  off-axis, 273–275
Paraxial equation
  mirror, 12
  refracting surface, 9, 29
  thick lens, 13
  thin lens, 14
Paul–Baker telescope, 115–117
Petzval curvature
  general, 83, 84
  spherical mirror, 87
    plus thin lens, 87
  two-mirror telescope, 88, 100
Phase transfer function, *see* Transfer function
Photometry, 321–323
Pixel, 307
Pixel modulation transfer function
  approximation, 312, 313
  ideal square well, 308–312
Plane-parallel plate
  aberration coefficients, 130
  image displacement, 14, 15
Plate factor, 223
Point spread function (PSF), 180–185
  annular aperture, 182
  array, *see* Intensity
  asymptotic approximation, 184, 185
  Gaussian profile, 315
  in presence of aberrations, 195, 196
  relation to modulation transfer function, 203, 204
  with random,
    rms pointing error, 211
    wavefront error, 210
Power,
  mirror, 12
  refracting surface, 9
  separated thin lens doublet, 14
  thick lens, 13
  thin lens, 14
  two-mirror telescope, 17
Prime focus telescope, 95–97
Prime focus corrector, 165–170
  aspheric plate, 166–168
  multiple aspheric plates, 168, 169
  Wynne triplet, 169, 170

Prism,
  aberration coefficients, 295, 296
  angular dispersion, 31, 32, 236
  deviation, 56, 57
Pupil
  entrance, 18
  exit, 18
  decentered, 88–90
  displaced, 74
  two-mirror telescope, 18, 19

Q

Quantum efficiency, 307

R

Radius of curvature
  image surfaces, 82–88
  sign convention, 8
  spectral image, 261
Ray coordinate system, 7, 258
Rayleigh criterion, 197, 201, 233
Reflection, law of 11, 28
Refraction
  atmospheric, 26
  differential atmospheric, 26
  law of, 8, 28, 64
Resolution
  angular, 39, 197, 202, 333
  atmospheric limit, typical, 95, 313–317
  diffraction limit, 183, 191, 197, 202, 233
  spectral, 226, 237, 240, 241
Ritchey-Chretien telescope, 102–105, 162–164, 170–173
  angular aberrations, 103
  conic constants, 102
  corrected at Cassegrain focus, 170–172
  field curvature, 103
  field-flattened, 162–164
  modified flat-field, 172, 173
Root-mean-square wavefront error, *see* Wavefront
Rowland mounting, 264–268

S

Sagittal
  astigmatic image location, 64, 76
  astigmatism, 67, 78

coma, 72, 73, 78
image surface curvature, 85, 86
Scale, telescope, 19
Schmidt camera, 55–59, 90–92, 122–141
  aberration coefficients, 123
  with misaligned corrector, 90–92
  achromatic, 134–138
  all-reflecting, 159
  aspheric coefficient, 62, 124, 131
  chromatic aberration, 57–59, 125–128
  field-flattened, 164, 165
  fifth-order spherical aberration, 131
  focal reducer, 175–178
  folded, 178, 275, 276
  solid, semisolid, 138–142
    chromatic blur, 139
    effective focal length, 139
Schmidt telescope, *see* Schmidt camera
Schmidt-Cassegrain camera, 142–152, 158, 159
  aberration coefficients
    flat-field, 146
    general, 143–145
  all-spherical mirrors, 150, 151
  anastigmatic flat-field, 146–149
  aplanat, 150, 151
  chromatic aberration, 148, 149
  semisolid, 158, 159
Schwarzschild telescope, *see* Telescope type
Secondary mirror
  alignment errors and aberrations, 110–114
  diameter
    Schmidt-Cassegrain, 146
    two-mirror, 21
  distance from focus, 20
  neutral point, 112
Seeing, atmospheric, 27, 313–317
Segmented mirror telescope, 330, 340
Semisolid Schmidt camera, *see* Schmidt camera
Shift, focal surface, 19, 20
Sign convention, 7, 8, 259
  angles, 8
  distances, 8
  indices of refraction, 12
  surface radii, 8, 82
Signal-to-noise ratio (SNR), 317–319
  background-limited case, 318
  definition, 317
  detector-limited case, 319
  fractional accuracy, 317

ideal case, 317
signal-limited case, 318
Slitless spectroscopy
 limiting magnitude, 327, 328
 spectral purity, 231
Slit spectroscopy, limiting magnitude
 extended source, 327, 328
 stellar source, 323-326, 340
Snell's law
 reflection, 11, 28
 refraction, 8, 28, 64
Solid Schmidt camera, see Schmidt camera
Space telescope, see Hubble Space Telescope
Spatial frequency, 200
 cutoff, 201
 normalized, 202
Speckle, 313, 314
Spectral purity
 definition, 225, 226
 diffraction limit, 234
 slitless mode, 231
Spectral resolution
 definition, 226
 diffraction limit, 234
 Fabry-Perot, 252
 grating, 240, 241
 prism, 237
Spectrometer parameters
 anamorphic magnification, 225, 236, 240
 dispersion
  angular, 222, 236, 239
  linear, 222, 223
 étendue, 228, 234
 flux, 228, 229
 flux-resolution product, 229, 234
 illumination, 229, 230
 luminosity, 227, 228
 luminosity-resolution product, 229, 234
 plate factor, 223
 projected slit, 224, 225
 spectral purity, 226, 234
 spectral resolution, 226, 234, 237, 240, 241
 speed, 230
Spectrometer type
 Czerny-Turner, 276-281
 Ebert-Fastie, 281
 Fabry-Perot, 251-254
 Fourier transform, 254-256
 inverse Wadsworth, 269, 270

Monk-Gillieson, 226, 282-284
Rowland, 267, 268
Wadsworth, 268, 269
Spectrum line
 curvature, 261
 tilt, 261
Speed, see Spectrometer parameters
Spherical aberration, see also Aberration; Aberration coefficients
 angular, 45
 circle of least confusion, 47, 48
 definition, 43
 fifth-order, in collimated light
  aspheric plate, 131
  conic mirror, 44
 spherical mirror, 44, 131
 transverse, 43, 44, 73, 78
Stop, aperture
 definition, 18
 displaced from surface, 74
 Schmidt camera, 55, 124
 Schmidt-Cassegrain camera, 143
Stop, field, 18
Stop-shift statements, 75
Strehl ratio, 191, 209, 214, 217
Super-Schmidt camera, see Telescope type
Surface curvature, image, 82-86
 median, 86
 Petzval, 84
 sagittal, 85, 86
 tangential, 85, 86
Surface equation
 conic mirror, 37, 38, 43
 general, 62

T

Tangential
 astigmatic image location
  general, 64, 76
  grating, 264
 astigmatism, 67, 78, 265
 coma, 72, 73, 78
 image surface curvature
  general, 85, 86
  grating, 266
Telescope type
 all-reflecting
  Schmidt, 159
  Schmidt-Cassegrain, 159
 aplanatic

Schmidt-Cassegrain, 150, 151
two-mirror, 102, 103, 108, 109
Array, 331-337
Baker-Schmidt, 146-149
Bouwers meniscus, 152-156
Cassegrain, 15, 16, 53-55, 97-105
classical, 53-55, 100, 101
Couder, 108
Dall-Kirkham, 55, 107
Gregorian, 15, 16, 97-105
hybrid, 106, 107
inverse Cassegrain, 108, 109
Maksutov, 156-158
multiple-mirror, 329-331
paraboloid, 95-97
Paul-Baker, 115-117
prime focus corrected, 165-170
Ritchey-Chretien, 102-105
corrected, 170-172
field-flattened, 162-164
Schmidt, 55-59, 122-141
achromatic, 134-138
field-flattened, 164, 165
solid, 138-141
super, 157
Schmidt-Cassegrain, 142-152
solid, 158, 159
Schwarzschild, 109
segmented mirror, 330, 340
three-mirror, 114-117
two-mirror, 15-17, 53-55, 97-109
Thick lens, 13
Thick plate, 14, 15
aberration coefficients, 130
image displacement, 15
Thin lens
focal length, 14, 30, 31
paraxial equation, 14
power, 14
Three-mirror telescope, 114-117
Paul-Baker, 115-117
Transfer function, *see also* Modulation transfer function
modulation, 200, 203
relation to encircled energy, 204
relation to point spread function, 204
optical, 202
phase, 202
Transverse aberration
chromatic
Schmidt camera, 59, 125-128
Schmidt-Cassegrain camera, 148, 149

definitions, 46, 67-69, 73, 78
diffraction grating, 263-265
grating mounting
Czerny-Turner
monochromator, 278
spectrograph, 280
inverse Wadsworth, 269
Monk-Gillieson, 265, 266, 283
Rowland, 265, 266
Wadsworth, 265, 266
in limit of small object distance, 162
multisurface system, 80
nonobjective mode
grating, 297, 299
grism, 301
prism, 296
relation to angular, 69
single surface, stop at surface
general, 71
mirror, 72
Transverse magnification, *see* Magnification
Turbulence, *see* Atmosphere
Two-mirror telescope, 15-17, 53-55, 97-109
aberration coefficients, 99
alignment errors, 110-114
angular aberrations, 99
comparison between types, 103-106
conic constants, 98, 99, 102
image surface curvatures, 100
normalized parameters, 16, 17, 97
parameter combinations, 98
Petzval curvature, 88, 100
power, 17

V

Vignetting, 250, 275

W

Wadsworth mounting, 268, 269
Wavefront
aberration, 67-69, 189, 190
definition, 39
root-mean-square error, 191-193, 195, 196
Wynne triplet, 169, 170

Z

Zernike polynominals, 194